CAMPBELL

essential biology

CAMPBELL

essential biology

7e

Eric J. Simon • **Jean L. Dickey** • **Jane B. Reece**

New England College   Clemson, South Carolina   Berkeley, California

with contributions from
Rebecca S. Burton
Alverno College

Pearson

330 Hudson Street, NY NY 10013

Courseware Portfolio Management, Director: *Beth Wilbur*
Courseware Portfolio Management, Specialist:
  Alison Rodal
Courseware Director, Content Development:
  Ginnie Simione Jutson
Courseware Sr. Analyst: *John Burner*
Developmental Editor: *Susan Teahan*
Courseware Editorial Assistant: *Alison Candlin*
Managing Producer: *Mike Early*
Content Producer: *Lori Newman*
Senior Content Developer: *Sarah Jensen*
Rich Media Content Producers: *Tod Regan, Ziki Dekel*
Full-Service Vendor: *Integra Software Services, Inc.*

Copyeditor: *Joanna Dinsmore*
Compositor: *Integra Software Services, Inc.*
Design Manager: *Mark Ong*
Cover and Interior Design: *TT Eye*
Illustrators: *Lachina*
Rights & Permissions Project Manager: *Ben Ferrini*
Rights & Permissions Management: *Cenveo*
Photo Researcher: *Kristin Piljay*
Manufacturing Buyer: *Stacey Weinberger*
Product Marketing Manager: *Christa Pelaez*
Field Marketing Manager: *Kelly Galli*
Cover Photo Credit: *Pascal Pittorino/naturepl.com/
  Getty Images*

Library of Congress Cataloging-in-Publication Data

Names: Simon, Eric J. (Eric Jeffrey), 1967- author.
Title: Campbell essential biology / Eric J. Simon, Jean L. Dickey, Jane B.
  Reece, Rebecca S. Burton.
Other titles: Essential biology
Description: 7[th edition]. | Hoboken : Pearson Education, Inc., [2019] |
  Revision of: Campbell essential biology / Eric J. Simon, New England
  College, Jean L. Dickey, Clemson, South Carolina, Kelly A. Hogan,
  University of North Carolina, Chapel Hill, Jane B. Reece, Berkeley,
  California. 2016. 6th edition. | Includes index.
Identifiers: LCCN 2017047482 | ISBN 9780134765037
Subjects: LCSH: Biology--Textbooks.
Classification: LCC QH308.2 .C343 2019 | DDC 570—dc23
LC record available at https://lccn.loc.gov/2017047482

1 18

ISBN 10: 0-134-76503-6; ISBN 13: 978-0-134-76503-7 (Student edition)

ISBN 10: 0-134-81413-4; ISBN 13: 978-0-134-81413-1 (Books a la Carte)

www.pearson.com

About the Authors

ERIC J. SIMON

is a professor in the Department of Biology and Health Science at New England College (Henniker, New Hampshire). He teaches introductory biology to science majors and nonscience majors, as well as upper-level courses in tropical marine biology and careers in science. Dr. Simon received a B.A. in biology and computer science, an M.A. in biology from Wesleyan University, and a Ph.D. in biochemistry from Harvard University. His research focuses on innovative ways to use technology to increase active learning in the science classroom, particularly for nonscience majors. Dr. Simon is also the author of the introductory biology textbook *Biology: The Core*, 2nd Edition, and a coauthor of *Campbell Biology: Concepts & Connections*, 9th Edition.

To my lifelong friends BZ, SR, and SR, who have taught me the value of loyalty and trust during decades of unwavering friendship

JEAN L. DICKEY

is Professor Emerita of Biological Sciences at Clemson University (Clemson, South Carolina). After receiving her B.S. in biology from Kent State University, she went on to earn a Ph.D. in ecology and evolution from Purdue University. In 1984, Dr. Dickey joined the faculty at Clemson, where she devoted her career to teaching biology to nonscience majors in a variety of courses. In addition to creating content-based instructional materials, she developed many activities to engage lecture and laboratory students in discussion, critical thinking, and writing, and implemented an investigative laboratory curriculum in general biology. Dr. Dickey is the author of *Laboratory Investigations for Biology*, 2nd Edition, and is a coauthor of *Campbell Biology: Concepts & Connections*, 9th Edition.

To my mother, who taught me to love learning, and to my daughters, Katherine and Jessie, the twin delights of my life

JANE B. REECE

was Neil Campbell's longtime collaborator and a founding author of *Campbell Essential Biology* and *Campbell Essential Biology with Physiology*. Her education includes an A.B. in biology from Harvard University (where she was initially a philosophy major), an M.S. in microbiology from Rutgers University, and a Ph.D. in bacteriology from the University of California, Berkeley. At UC Berkeley, and later as a postdoctoral fellow in genetics at Stanford University, her research focused on genetic recombination in bacteria. Dr. Reece taught biology at Middlesex County College (New Jersey) and Queensborough Community College (New York). Dr. Reece's publishing career began in 1978 when she joined the editorial staff of Benjamin Cummings, and since then, she played a major role in a number of successful textbooks. She was the lead author of *Campbell Biology* Editions 8–10 and a founding author of *Campbell Biology: Concepts & Connections*.

To my wonderful coauthors, who have made working on our books a pleasure

NEIL A. CAMPBELL

(1946–2004) combined the inquiring nature of a research scientist with the soul of a caring teacher. Over his 30 years of teaching introductory biology to both science majors and nonscience majors, many thousands of students had the opportunity to learn from him and be stimulated by his enthusiasm for the study of life. He is greatly missed by his many friends in the biology community. His coauthors remain inspired by his visionary dedication to education and are committed to searching for ever-better ways to engage students in the wonders of biology.

Preface

Biology education has been transformed in the last decade. The non-majors introductory biology course was (in most cases) originally conceived as a slightly less deep and broad version of the general biology course. But a growing recognition of the importance of this course—one that is often the most widely enrolled within the department, and one that serves as the sole source of science education for many students—has prompted a reevaluation of priorities and a reformulation of pedagogy. Many instructors have narrowed the focus of the course from a detailed compendium of facts to an exploration of broader themes within the discipline—themes such as the central role of evolution and an understanding of the process of science. For many educators, the goals have shifted from communicating a great number of bits of information toward providing a deep understanding of fewer, but broader, principles. Luckily for anyone teaching or learning biology, opportunities to marvel at the natural world and the life within it abound. Furthermore, nearly everyone realizes that the subject of biology has a significant impact on his or her own life through its connections to medicine, biotechnology, agriculture, environmental issues, forensics, and many other areas. Our primary goal in writing *Campbell Essential Biology* is to help teachers motivate and educate the next generation of citizens by communicating the broad themes that course through our innate curiosity about life.

Goals of the Book

Although our world is rich with "teachable moments" and learning opportunities, an explosion of knowledge threatens to bury a curious person under an avalanche of information. "So much biology, so little time" is the universal lament of biology educators. Neil Campbell conceived of *Campbell Essential Biology* as a tool to help teachers and students focus on the most important areas of biology. To that end, the book is organized into four core areas: cells, genes, evolution, and ecology. Dr. Campbell's vision, which we carry on and extend in this edition, has enabled us to keep *Campbell Essential Biology* manageable in size and thoughtful in the development of the concepts that are most fundamental to understanding life. We've aligned this new edition with today's "less is more" approach in biology education for nonscience majors—where the emphasis is on fewer topics but broader themes—while never allowing the important content to be diluted.

We formulated our approach after countless conversations with teachers and students in which we noticed some important trends in how biology is taught. In particular, many instructors identify three goals: (1) to engage students by relating biology content to their lives and the greater society; (2) to help students understand the process of science by teaching critical thinking skills that can be used in everyday life; and (3) to demonstrate how biology's broader themes—such as evolution and the relationship of structure to function—serve to unify the entire subject. To help achieve these goals, every chapter of this book includes several important features. First, a chapter-opening essay called Biology and Society highlights a connection between the chapter's core content and students' lives. Second, an essay called The Process of Science (in the body of the chapter) describes how the scientific process has illuminated the topic at hand, using a classic or modern experiment as an example. Third, a chapter-closing Evolution Connection essay relates the chapter to biology's unifying theme of evolution. Fourth, the broad themes that unify all subjects within biology are explicitly called out (in blue) multiple times within each chapter. Finally, to maintain a cohesive narrative throughout each chapter, the content is tied together with a unifying chapter thread, a relevant high-interest topic that is touched on several times in the chapter and woven throughout the three feature essays. Thus, this unifying chapter thread ties together the pedagogical goals of the course, using a topic that is compelling and relevant to students.

New to This Edition

This latest edition of *Campbell Essential Biology* goes even further than previous editions to help students relate the material to their lives, understand the process of science, and appreciate how broad themes unify all aspects of biology. To this end, we've added significant new features and content to this edition:

- **A new approach to teaching the process of science.** Conveying the process of science to nonscience-major undergraduate students is one of the most important goals of this course. Traditionally, we taught the scientific method as a predefined series of steps to be followed in an exact order (observation, hypothesis, experiment, and so forth). Many instructors have shifted away from such a specific flow chart to a more nuanced approach that involves multiple pathways, frequent restarts, and other features that more accurately reflect how science is actually undertaken. Accordingly, we have revised the way that the process of science is discussed within our text, both in Chapter 1 (where the process is discussed in detail) and in The Process of Science essay in every chapter of the textbook. Rather than using specific terms in a specific order to describe the process, we now divide it into three broad interrelated areas: background, method, and results. We believe that this new approach better conveys how science actually proceeds and demystifies the topic for non-scientists. Chapter 1 also contains important information that promotes critical thinking, such as discussion of control groups, pseudoscience, and recognizing reliable sources of information. We believe that providing students with such critical-thinking tools is one of the most important outcomes of the nonscience-major introductory course.
- **Major themes in biology incorporated throughout the book.** In 2009, the American Association for the Advancement of Science published a document that served as a call to action in undergraduate biology education. The principles of this document, which

is titled "Vision and Change," are becoming widely accepted throughout the biology education community. "Vision and Change" presents five core concepts that serve as the foundation of undergraduate biology. In this edition of *Campbell Essential Biology,* we repeatedly and explicitly link book content to themes multiple times in each chapter, calling out such instances with boldfaced blue text. For example, in Chapter 4 (A Tour of the Cell), the interrelationships of cellular structures are used to illustrate the theme of interactions within biological systems. The plasma membrane is presented as an example of the relationship between structure and function. The cellular structures in the pathway from DNA to protein are used to illustrate the importance of information flow. The chloroplasts and mitochondria serve as an example of the transformations of energy and matter. The DNA within these structures is also used to illustrate biology's overarching theme of evolution. Students will find three to five examples of themes called out in each chapter, which will help them see the connections between these major themes and the course content. To reinforce these connections, this edition of *Campbell Essential Biology* includes new end-of-chapter questions and Mastering Biology activities that promote critical thinking relating to these themes. Additionally, PowerPoint© lecture slides have been updated to incorporate chapter examples and offer guidance to faculty on how to include in these themes within classroom lectures.

■ **Updated connections to students' lives.** In every edition of *Campbell Essential Biology*, we seek to improve and extend the ways that we connect the course content to students' lives. Accordingly, every chapter begins with an improved feature called Why It Matters showing the relevance of the chapter content from the very start. Additionally, with every edition, we introduce some new unifying chapter threads intended to improve student relevance. For example, this edition includes new threads that discuss evolution in a human-dominated world (Chapter 14) and the importance of biodiversity to human affairs (Chapter 20). As always, we include some updated Biology and Society chapter-opening essays (such as "A Solar Revolution" in Chapter 7), The Process of Science sections (such as a recent experiment investigating the efficacy of radiation therapy to treat prostate cancer, in Chapter 2), and Evolution Connection chapter-closing essays (such as an updated discussion of biodiversity hot spots in Chapter 20). As we always do, this edition includes many content updates that connect to students' lives, such as information on cutting-edge cancer therapies (Chapter 8) and recent examples of DNA profiling (Chapter 12).

■ **Developing data literacy through infographics.** Many nonscience-major students express anxiety when faced with numerical data, yet the ability to interpret data can help with many important decisions we all face. Increasingly, the general public encounters information in the form of infographics, visual images used to represent data. Consistent with our goal of preparing students to approach important issues critically, this edition includes a series of new infographics, or Visualizing the Data figures. Examples include the elemental composition of the human body (Chapter 2), a comparison of calories burned through exercise versus calories consumed in common foods (Chapter 5), and ecological footprints (Chapter 19). In addition to the printed form, these infographics are available as an interactive feature in the eText and as assignable tutorial questions within Mastering Biology.

■ **Helping students to understand key figures.** For this new edition, a key figure in each chapter is supplemented by a short video explaining the concept to the student. These Figure Walkthrough videos will be embedded in the eText and will be assignable in Mastering Biology. The animations are written and narrated by authors Eric Simon and Jean Dickey, as well as teacher and contributor Rebecca Burton.

Attitudes about science and scientists are often shaped by a single, required science class—*this* class. We hope to nurture an appreciation of nature into a genuine love of biology. In this spirit, we hope that this textbook and its supplements will encourage all readers to make biological perspectives a part of their personal worldviews. Please let us know how we are doing and how we can improve the next edition of *Campbell Essential Biology.*

ERIC SIMON
Department of Biology and
Health Science
New England College
Henniker, NH 03242
SimonBiology@gmail.com

JEAN DICKEY
Clemson, SC
dickeyj@clemson.edu

JANE B. REECE
Berkeley, California

The following Visual Walkthrough highlights key features of *Campbell Essential Biology 7e.*

Develop and practice science literacy skills

Learn how to view your world using scientific reasoning with *Campbell Essential Biology*. See how concepts from class and an understanding of how science works can apply to your everyday life. Engage with the concepts and practice science literacy skills with Mastering Biology and Pearson eText.

NEW! New and updated Process of Science essays present scientific discovery as a flexible and non-linear process.

Each essay summarizes the **background, method,** and **results** from a scientific study.

New Thinking Like a Scientist questions appear at the end of each Process of Science essay and involve applying a scientific reasoning skill.

Examples of new Process of Science topics include:

- Chapter 4: How Was the First 21st-Century Antibiotic Discovered? p. 61
- Chapter 9: What Is the Genetic Basis of Short Legs in Dogs? p.156
- Chapter 11: Can Avatars Improve Cancer Treatment? p.210
- Chapter 16: What Killed the Pines? p.330
- Chapter 20: Does Biodiversity Protect Human Health? p.446

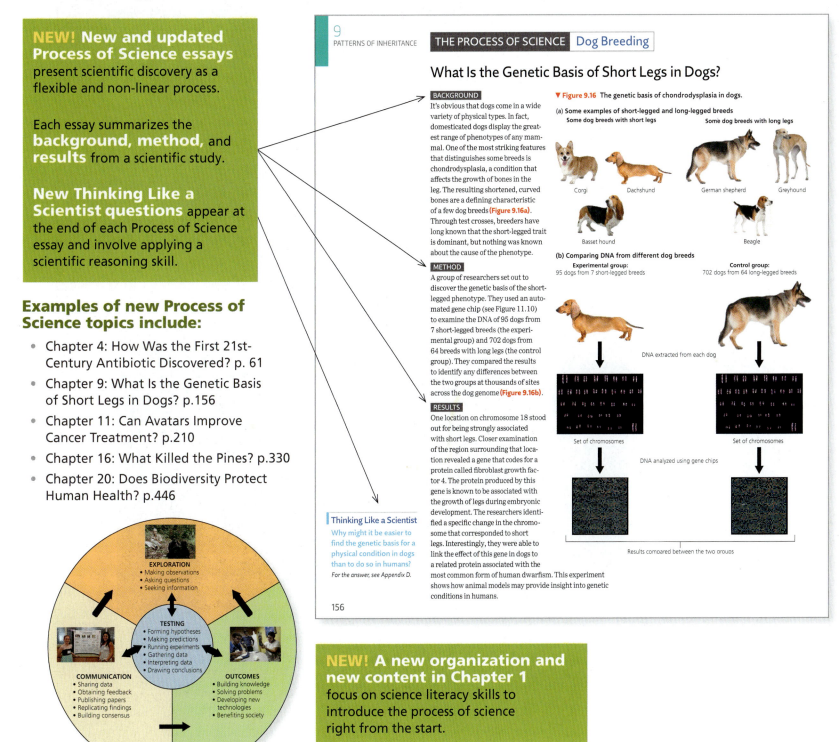

THE PROCESS OF SCIENCE   Dog Breeding

What Is the Genetic Basis of Short Legs in Dogs?

BACKGROUND
It's obvious that dogs come in a wide variety of physical types. In fact, domesticated dogs display the greatest range of phenotypes of any mammal. One of the most striking features that distinguishes some breeds is chondrodysplasia, a condition that affects the growth of bones in the leg. The resulting shortened, curved bones are a defining characteristic of a few dog breeds (Figure 9.16a). Through test crosses, breeders have long known that the short-legged trait is dominant, but nothing was known about the cause of the phenotype.

METHOD
A group of researchers set out to discover the genetic basis of the short-legged phenotype. They used an automated gene chip (see Figure 11.10) to examine the DNA of 95 dogs from 7 short-legged breeds (the experimental group) and 702 dogs from 64 breeds with long legs (the control group). They compared the results to identify any differences between the two groups at thousands of sites across the dog genome (Figure 9.16b).

RESULTS
One location on chromosome 18 stood out for being strongly associated with short legs. Closer examination of the region surrounding that location revealed a gene that codes for a protein called fibroblast growth factor 4. The protein produced by this gene is known to be associated with the growth of legs during embryonic development. The researchers identified a specific change in the chromosome that corresponded to short legs. Interestingly, they were able to link the effect of this gene in dogs to a related protein associated with the most common form of human dwarfism. This experiment shows how animal models may provide insight into genetic conditions in humans.

Thinking Like a Scientist
Why might it be easier to find the genetic basis for a physical condition in dogs than to do so in humans?
For the answer, see Appendix D.

▼ Figure 9.16 The genetic basis of chondrodysplasia in dogs.

(a) Some examples of short-legged and long-legged breeds
Some dog breeds with short legs — Corgi, Dachshund, Basset hound
Some dog breeds with long legs — German shepherd, Greyhound, Beagle

(b) Comparing DNA from different dog breeds
Experimental group: 95 dogs from 7 short-legged breeds
Control group: 702 dogs from 64 long-legged breeds
DNA extracted from each dog
Set of chromosomes
DNA analyzed using gene chips
Results compared between the two groups

156

EXPLORATION
- Making observations
- Asking questions
- Seeking information

TESTING
- Forming hypotheses
- Making predictions
- Running experiments
- Gathering data
- Interpreting data
- Drawing conclusions

COMMUNICATION
- Sharing data
- Obtaining feedback
- Publishing papers
- Replicating findings
- Building consensus

OUTCOMES
- Building knowledge
- Solving problems
- Developing new technologies
- Benefiting society

NEW! A new organization and new content in Chapter 1 focus on science literacy skills to introduce the process of science right from the start.

Explore biology with . . .

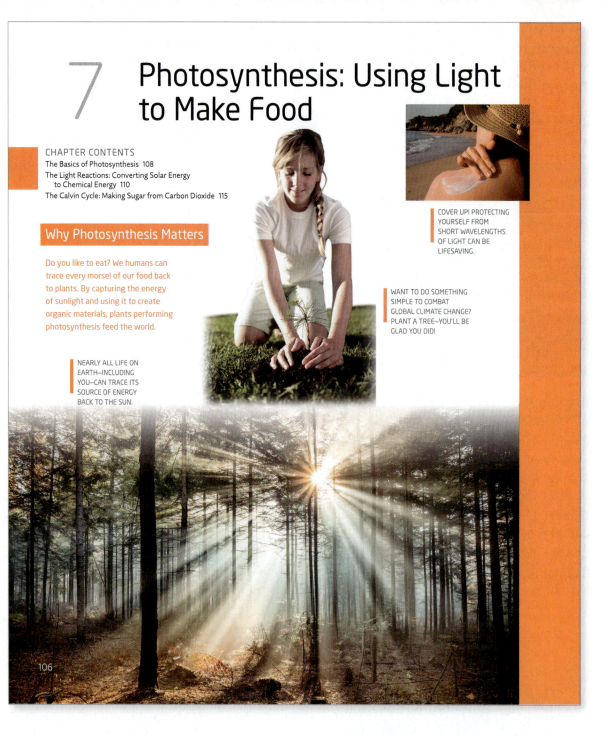

Photosynthesis: Using Light to Make Food

CHAPTER CONTENTS

The Basics of Photosynthesis 108
The Light Reactions: Converting Solar Energy
  to Chemical Energy 110
The Calvin Cycle: Making Sugar from Carbon Dioxide 115

Why Photosynthesis Matters

Do you like to eat? We humans can trace every morsel of our food back to plants. By capturing the energy of sunlight and using it to create organic materials, plants performing photosynthesis feed the world.

COVER UP! PROTECTING YOURSELF FROM SHORT WAVELENGTHS OF LIGHT CAN BE LIFESAVING.

WANT TO DO SOMETHING SIMPLE TO COMBAT GLOBAL CLIMATE CHANGE? PLANT A TREE—YOU'LL BE GLAD YOU DID!

NEARLY ALL LIFE ON EARTH—INCLUDING YOU—CAN TRACE ITS SOURCE OF ENERGY BACK TO THE SUN.

106

Why It Matters Photo Collages have been updated to give real-world examples to convey why abstract concepts like cellular respiration or photosynthesis matter.

... the most relevant, real-world examples

New and Updated Chapter Threads weave a compelling topic throughout each chapter, highlighted in the Biology and Society, The Process of Science, and Evolution Connection essays.

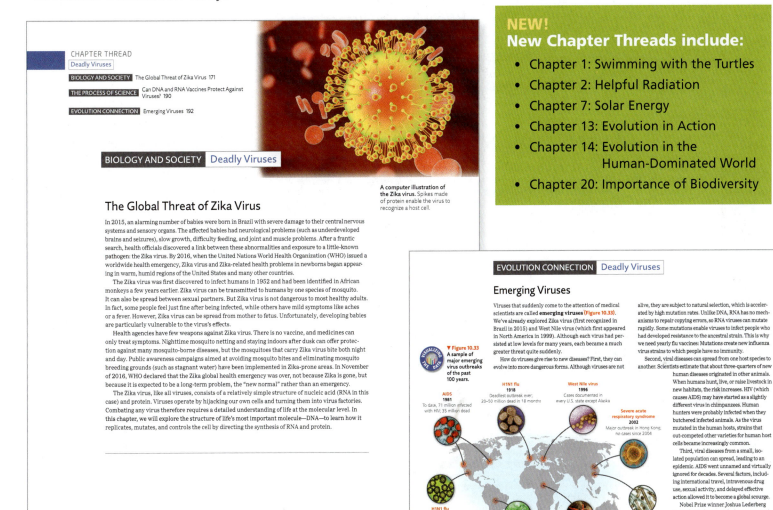

CHAPTER THREAD
Deadly Viruses
BIOLOGY AND SOCIETY The Global Threat of Zika Virus 171
THE PROCESS OF SCIENCE Can DNA and RNA Vaccines Protect Against Viruses? 190
EVOLUTION CONNECTION Emerging Viruses 192

BIOLOGY AND SOCIETY Deadly Viruses

A computer illustration of the Zika virus. Spikes made of protein enable the virus to recognize a host cell.

The Global Threat of Zika Virus

In 2015, an alarming number of babies were born in Brazil with severe damage to their central nervous systems and sensory organs. The affected babies had neurological problems (such as underdeveloped brains and seizures), slow growth, difficulty feeding, and joint and muscle problems. After a frantic search, health officials discovered a link between these abnormalities and exposure to a little-known pathogen: the Zika virus. By 2016, when the United Nations World Health Organization (WHO) issued a worldwide health emergency, Zika virus and Zika-related health problems in newborns began appearing in warm, humid regions of the United States and many other countries.

The Zika virus was first discovered to infect humans in 1952 and had been identified in African monkeys a few years earlier. Zika virus can be transmitted to humans by one species of mosquito. It can also be spread between sexual partners. But Zika virus is not dangerous to most healthy adults. In fact, some people feel just fine after being infected, while others have mild symptoms like aches or a fever. However, Zika virus can be spread from mother to fetus. Unfortunately, developing babies are particularly vulnerable to the virus's effects.

Health agencies have few weapons against Zika virus. There is no vaccine, and medicines can only treat symptoms. Nighttime mosquito netting and staying indoors after dusk can offer protection against many mosquito-borne diseases, but the mosquitoes that carry Zika virus bite both night and day. Public awareness campaigns aimed at avoiding mosquito bites and eliminating mosquito breeding grounds (such as stagnant water) have been implemented in Zika-prone areas. In November of 2016, WHO declared that the Zika global health emergency was over, not because Zika is gone, but because it is expected to be a long-term problem, the "new normal" rather than an emergency.

The Zika virus, like all viruses, consists of a relatively simple structure of nucleic acid (RNA in this case) and protein. Viruses operate by hijacking our own cells and turning them into virus factories. Combating any virus therefore requires a detailed understanding of life at the molecular level. In this chapter, we will explore the structure of life's most important molecule—DNA—to learn how it replicates, mutates, and controls the cell by directing the synthesis of RNA and protein.

NEW!
New Chapter Threads include:

- Chapter 1: Swimming with the Turtles
- Chapter 2: Helpful Radiation
- Chapter 7: Solar Energy
- Chapter 13: Evolution in Action
- Chapter 14: Evolution in the Human-Dominated World
- Chapter 20: Importance of Biodiversity

EVOLUTION CONNECTION Deadly Viruses

Emerging Viruses

Viruses that suddenly come to the attention of medical scientists are called **emerging viruses** (Figure 10.33). We've already explored Zika virus (first recognized in Brazil in 2015) and West Nile virus (which first appeared in North America in 1999). Although each virus had persisted at low levels for many years, each became a much greater threat quite suddenly.

How do viruses give rise to new diseases? First, they can evolve into more dangerous forms. Although viruses are not alive, they are subject to natural selection, which is accelerated by high mutation rates. Unlike DNA, RNA has no mechanisms to repair copying errors, so RNA viruses can mutate rapidly. Some mutations enable viruses to infect people who had developed resistance to the ancestral strain. This is why we need yearly flu vaccines: Mutations create new influenza virus strains to which people have no immunity.

Second, viral diseases can spread from one host species to another. Scientists estimate that about three-quarters of new human diseases originated in other animals. When humans hunt, live, or raise livestock in new habitats, the risk increases. HIV (which causes AIDS) may have started as a slightly different virus in chimpanzees. Human hunters were probably infected when they butchered infected animals. As the virus mutated in the human hosts, strains that out-competed other varieties for human host cells became increasingly common.

Third, viral diseases from a small, isolated population can spread, leading to an epidemic. AIDS went unnamed and virtually ignored for decades. Several factors, including international travel, intravenous drug use, sexual activity, and delayed effective action allowed it to become a global scourge.

Nobel Prize winner Joshua Lederberg warned: "We live in evolutionary competition with microbes. There is no guarantee that we will be the survivors." If we are to be victorious in the fight against emerging viruses, we must understand molecular biology and evolutionary processes.

▼ Figure 10.33 A sample of major emerging virus outbreaks of the past 100 years.

VISUALIZING THE DATA

AIDS 1981 To date, 71 million infected with HIV; 35 million dead

H1N1 flu 1918 Deadliest outbreak ever; 20–50 million dead in 18 months

West Nile virus 1996 Cases documented in every U.S. state except Alaska

Severe acute respiratory syndrome 2002 Major outbreak in Hong Kong; no cases since 2004

H1N1 flu 2009 A combination of bird, swine, and human viruses

Zika fever 2015 Transmitted by mosquitoes, spread by sexual contact

Ebola 1976 Biggest outbreak from 2014 to 2016 in West Africa

Avian flu 1997 Rarely occurs in North America

192

Biology and Society essays

relating biology to everyday life are either new or updated. Some new topics:

- Chapter 7: A Solar Revolution p. 107
- Chapter 10: The Global Threat of Zika Virus p. 171
- Chapter 14: Humanity's Footprint p. 269
- Chapter 17: Evolving Adaptability p. 337

Evolution Connection essays

demonstrate the importance of evolution as a theme throughout biology, by appearing in every chapter. Some new topics:

- Chapter 1 Turtles in the Tree of Life p. 18
- Chapter 10 Emerging Viruses   p. 192
- Chapter 20 Saving the Hot Spots p. 449

Complex biological processes are explained . . .

Mastering™ Biology is an online homework, tutorial, and assessment platform that improves results by helping students quickly master concepts.

A wide range of interactive, engaging, and assignable activities, many of them contributed by *Campbell Essential Biology* authors, encourage active learning and help with understanding tough course concepts.

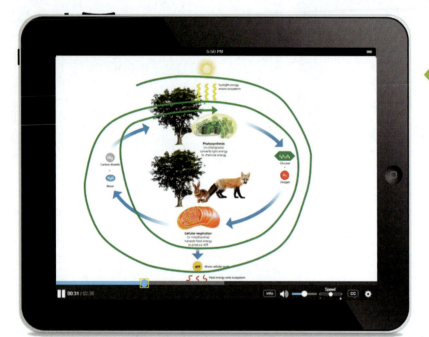

NEW! 20 Figure Walkthrough Videos, created and narrated by the authors, give clear, concise explanations of key figures in each chapter. The videos are embedded in the Pearson eText, accessible through QR codes in the print text, and assignable in Mastering Biology.

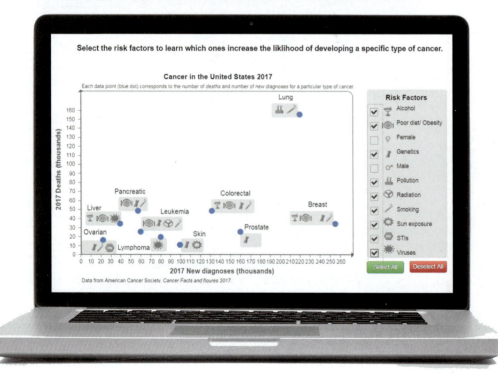

NEW! Visualizing the Data coaching activities bring the infographic figures in the text to life and are embedded in the eText and assignable in Mastering Biology.

. . . with engaging visuals and narrated examples in Mastering Biology

12 Topic Overview videos, created by the authors, introduce key concepts and vocabulary. These brief, engaging videos introduce topics that will be explored in greater depth in class.

Topics include:

- Macromolecules
- Ecological Organization
- Mechanisms of Evolution
- An Introduction to Structure and Function
- Interactions Between the Respiratory and Circulatory Systems
- DNA Structure and Function
- . . . And more!

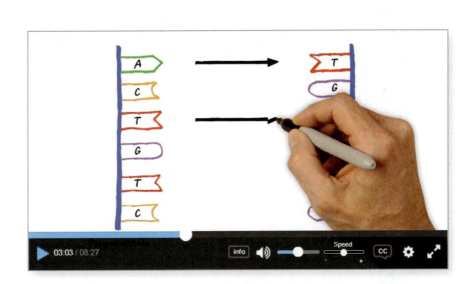

Part A

Can you match the terms to their definitions?

Drag the terms on the left to the appropriate blanks on the right to complete the sentences.

Reset | Help

| RNA |
| replication |
| base |
| translation |
| DNA |
| transcription |

[ ] serves as the molecular basis for life.

DNA copies itself via the process of [ ].

RNA is produced from DNA via the process of [ ].

Proteins are produced from RNA via the process of [ ].

There are five examples of a [ ] : A, G, C, T, and U.

One way that [ ] is different from DNA is that it contains Us instead of Ts.

BioInteractive Short Films from HHMI, Everyday Biology Videos, Video Tutors, BioFlix® 3D animations, and MP3 Audio Tutors support key concept areas covered in the text and provide coaching by using personalized feedback on common wrong answers.

New approaches to teaching and learning . . .

Ready-to-Go Teaching Modules make use of teaching tools for before, during, and after class, including new ideas for in-class activities. These modules incorporate the best that the text, Mastering Biology, and Learning Catalytics have to offer and can be accessed through the Instructor Resources area of Mastering Biology.

Learning Catalytics™ helps generate class discussion, customize lectures, and promote peer-to-peer learning with real-time analytics. Learning Catalytics acts as a student response tool that uses students' smartphones, tablets, or laptops to engage them in more interactive tasks and thinking.

- Help your students develop critical thinking skills
- Monitor responses to find out where your students are struggling
- Rely on real-time data to adjust your teaching strategy

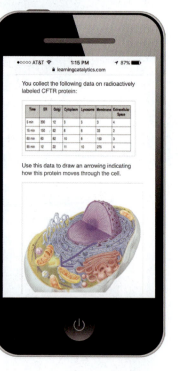

. . . and the resources to accomplish them

Extensive resources save instructors valuable time both in course preparation and during class. Instructor materials can be accessed and downloaded from the Instructor Resources area of Mastering Biology.
www.pearson.com/mastering/biology

New! Identifying Major Themes end-of-chapter questions in the text and coaching activities in Mastering Biology give instructors resources to integrate Vision and Change biological themes into their course.

Revised Guided Reading Activities in the Mastering Biology Study Area and Instructor Resources offer a simple resource that encourages students to get the most out of each text chapter. These worksheets accompany each chapter of the text and are downloadable from Mastering Biology.

Identifying Major Themes--Chapter 18

Part A

Can you identify the major theme illustrated by each of the following examples? If necessary, you can review the themes in Chapter 1 of your book.

Match the themes on the left with the examples on the right. Not all themes will be used.

Reset | Help

Information flow	Solar energy from sunlight, captured by chlorophyll during the process of photosynthesis, powers most ecosystems. Pathways that transform energy and matter
	After a period of lower-than-average rainfall, drought-resistant individuals may be more prevalent in a plant population. Evolution
	Reptilian scales and the waxy coating on many leaves reduce water loss. Relationship of structure to f
	Other organisms may compete its physical and chemical enviro

Submit | My Answers | Give Up

Correct

IDENTIFYING MAJOR THEMES

For each statement, identify which major theme is evident (the relationship of structure to function, information flow, pathways that transform energy and matter, interactions within biological systems, or evolution) and explain how the statement relates to the theme. If necessary, review the themes (see Chapter 1) and review the examples highlighted in blue in this chapter.

11. The highly folded membranes of the mitochondria make these organelles well suited to carry out the huge number of chemical reactions required for cellular respiration to proceed.

12. Cellular respiration and photosynthesis are linked, with each process using inputs created by the other.

13. Your body uses many different intersecting chemical pathways that, all together, constitute your metabolism.

For answers to Identifying Major Themes, see Appendix D.

Complete the following questions as you read the chapter content—Cellular Respiration: Aerobic Harvest of Food Energy:

1. The majority of a cell's ATP is produced within which of the following organelles?

    a. mitochondria

    b. nucleus

    c. ribosome

    d. Golgi apparatus

2. Students frequently have the misconception that plant cells don't perform cellular respiration. Briefly explain the basis of this misconception.

3. Briefly explain why the overall equation for cellular respiration has multiple arrows. Use the following figure, which illustrates the equation for cellular respiration, to help you answer.

$C_6H_{12}O_6$ + 6 O_2 → → → 6 CO_2 + 6 H_2O + approx. 32 ATP

The **Instructor Exchange** in the Instructor Resources area of Mastering Biology provides successful, class-tested active learning techniques and analogies from biology instructors around the nation, offering a springboard for quick ideas to create more compelling lectures. Contributor Kelly Hogan moderates contributions to the exchange.

Engage with biology concepts anytime, anywhere with Pearson eText

New to *Campbell Essential Biology* 7th edition/*Campbell Essential Biology with Physiology* 6th edition, the Pearson eText includes videos, interactives, animations, and audio tutors that bring the text to life and help you understand key concepts. Get all the help you need in one integrated digital experience.

NEW! Over 100 rich media resources, many of them created by the author team, are included in the Pearson eText and accessible on smartphones, tablets, and computers. Examples of the rich media include: Figure Walkthrough videos, Topic Overview videos, MP3 Audio Tutors, Video Tutors, and BioFlix Tutorials.

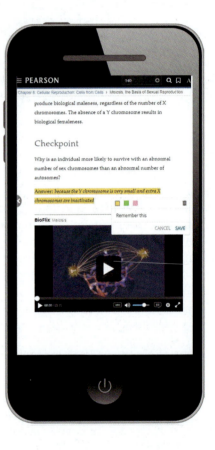

Pearson eText Mobile App offers offline access and can be downloaded for most iOS and Android phones/tablets from the Apple App Store or Google Play.

- Seamlessly integrated videos and other rich media
- Accessible (screen-reader ready)
- Configurable reading settings, including resizable type and night reading mode
- Instructor and student note-taking, highlighting, bookmarking, and search

Acknowledgments

Throughout the process of planning and writing *Campbell Essential Biology*, the author team has had the great fortune of collaborating with an extremely talented group of publishing professionals and educators. We are all truly humbled to be part of one of the most experienced and successful publishing teams in biology education. Although the responsibility for any shortcomings lies solely with the authors, the merits of the book and its supplements reflect the contributions of a great many dedicated colleagues.

First and foremost, we must acknowledge our huge debt to Neil Campbell, the founding author of this book and a source of ongoing inspiration for each of us. Although this edition has been carefully and thoroughly revised—to update its science, its connections to students' lives, its pedagogy, and its currency—it remains infused with Neil's original vision and his commitment to share biology with introductory students.

This edition benefited significantly from the efforts of contributor Rebecca S. Burton from Alverno College. Using her years of teaching expertise, Becky made substantial improvements to two chapters, contributed to the development of new and revised Chapter Thread essays, and helped shape the emphasis on the unifying themes throughout the text and in Mastering Biology. We thank Becky for bringing her considerable talents to bear on this edition!

This book could not have been completed without the efforts of the *Campbell Essential Biology* team at Pearson Education. Leading the team is courseware portfolio management specialist Alison Rodal, who is tireless in her pursuit of educational excellence and who inspires all of us to constantly seek better ways to help teachers and students. Alison stands at the interface between the book development team and the educational community of professors and students. Her insights and contributions are invaluable. We also thank the Pearson Science executive team for their supportive leadership, in particular, senior vice president of portfolio management Adam Jaworski, director of portfolio management Beth Wilbur, and directors of courseware content development Barbara Yien and Ginnie Simione Jutson.

It is no exaggeration to say that the talents of the best editorial team in the industry are evident on every page of this book. The authors were continuously guided with great patience and skill by courseware senior analyst John Burner and senior developmental editor Susan Teahan. We owe this editorial team—which also includes the wonderfully capable and friendly editorial assistant Alison Candlin—a deep debt of gratitude for their talents and hard work.

Once we formulated our words and images, the production and manufacturing teams transformed them into the final book. Senior content producer Lori Newman oversaw the production process and kept everyone and everything on track. We also thank the managing content producer Mike Early for his careful oversight. Every edition of *Campbell Essential Biology* is distinguished by continuously updated and beautiful photography. For that we thank photo researcher Kristin Piljay, who constantly dazzles us with her keen ability to locate memorable images.

For the production and composition of the book, we thank senior project editor Margaret McConnell of Integra Software Services, whose professionalism and commitment to the quality of the finished product is visible throughout. The authors owe much to copyeditor Joanna Dinsmore and proofreader Pete Shanks for their keen eyes and attention to detail. We thank design manager Mark Ong and designer tani hasegawa of TT Eye for the beautiful interior and cover designs; and we are grateful to Rebecca Marshall and Courtney Coffman and the artists at Lachina for rendering clear and compelling illustrations. We also thank rights and permissions project manager Matt Perry at Cenveo and the manager of rights and permissions Ben Ferrini. In the final stages of production, the talents of manufacturing buyer Stacy Weinberger shone.

Most instructors view the textbook as just one piece of the learning puzzle, with the book's supplements and media completing the picture. We are lucky to have a *Campbell Essential Biology* supplements team that is fully committed to the core goals of accuracy and readability. Content producer Lori Newman expertly coordinated the supplements, a difficult task given their number and variety. We also thank media project manager Ziki Dekel for his work on the excellent Instructor Resources and eText that accompanies the text. We owe particular gratitude to the supplements authors, especially the indefatigable and eagle-eyed Ed Zalisko of Blackburn College, who wrote the Instructor Guide and the PowerPoint© Lectures; the highly skilled and multitalented Doug Darnowski of Indiana University Southeast, who revised the Quiz Shows and Clicker Questions; and Jean DeSaix of the University of North Carolina at Chapel Hill, Justin Shaffer of the University of California, Irvine, Kristen Miller of the University of Georgia, and Suann Yang of SUNY Geneseo, our collaborative team of Test Bank authors for ensuring excellence in our assessment program. In addition, the authors thank Reading Quiz authors Amaya Garcia Costas of Montana State University and Cindy Klevickis of James Madison University; Reading Quiz accuracy reviewer Veronica Menendez; Practice Test author Chris Romero of Front Range Community College; and Practice Test accuracy reviewer Justin Walgaurnery of the University of Hawaii.

We wish to thank the talented group of publishing professionals who worked on the comprehensive media program that accompanies *Campbell Essential Biology*. The team members dedicated to Mastering Biology are true "game changers" in the field of biology education. We thank rich media content producers Ziki Dekel and Tod Regan for coordinating our multimedia plan. Vital contributions were also made by associate Mastering media producer Kaitlin Smith and web developer Barry Offringa. We also thank Sarah Jensen, senior content developer, for her efforts to make our media products the best in the industry.

As educators and writers, we are very lucky to have a crack marketing team. Product marketing manager Christa Pelaez and field marketing manager Kelli Galli seemed to be everywhere at once as they helped us achieve

10 The Structure and Function of DNA · 170

CHAPTER THREAD
Deadly Viruses · 171

BIOLOGY AND SOCIETY   The Global Threat of Zika Virus · 171

DNA: Structure and Replication · 172
DNA and RNA Structure · 172
Watson and Crick's Discovery of the Double Helix · 173
DNA Replication · 175

From DNA to RNA to Protein · 176
How an Organism's Genotype Determines Its Phenotype · 176
From Nucleotides to Amino Acids: An Overview · 177
The Genetic Code · 178
Transcription: From DNA to RNA · 179
The Processing of Eukaryotic RNA · 180
Translation: The Players · 180
Translation: The Process · 182
Review: DNA → RNA → Protein · 183
Mutations · 184

Viruses and Other Noncellular Infectious Agents · 186
Bacteriophages · 186
Plant Viruses · 188
Animal Viruses · 188

THE PROCESS OF SCIENCE   Can DNA and RNA Vaccines Protect Against Viruses? · 190

HIV, the AIDS Virus · 190
Prions · 192

EVOLUTION CONNECTION   Emerging Viruses · 192

11 How Genes Are Controlled · 196

CHAPTER THREAD
Cancer · 197

BIOLOGY AND SOCIETY   Breast Cancer and Chemotherapy · 197

How and Why Genes Are Regulated · 198
Gene Regulation in Bacteria · 198
Gene Regulation in Eukaryotic Cells · 200
Cell Signaling · 203
Homeotic Genes · 204
Visualizing Gene Expression · 204

Cloning Plants and Animals · 205
The Genetic Potential of Cells · 205
Reproductive Cloning of Animals · 206
Therapeutic Cloning and Stem Cells · 208

The Genetic Basis of Cancer · 209
Genes That Cause Cancer · 209

THE PROCESS OF SCIENCE   Can Avatars Improve Cancer Treatment? · 210

Cancer Risk and Prevention · 212

EVOLUTION CONNECTION   The Evolution of Cancer in the Body · 213

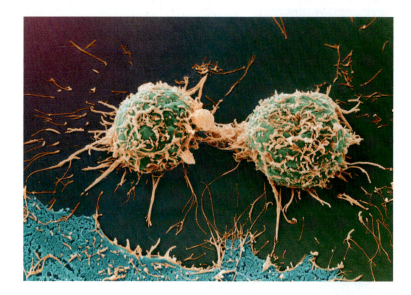

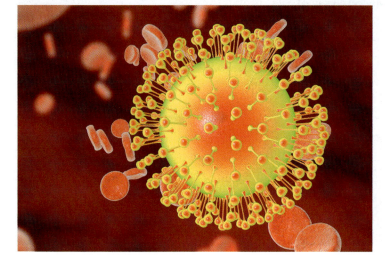

12 DNA Technology 216

CHAPTER THREAD
DNA Profiling 217

BIOLOGY AND SOCIETY Using DNA to Establish Guilt and Innocence 217

Genetic Engineering 218
Recombinant DNA Techniques 218
Gene Editing 220
Medical Applications 221
Genetically Modified Organisms in Agriculture 222
Human Gene Therapy 224

DNA Profiling and Forensic Science 225
DNA Profiling Techniques 225
Investigating Murder, Paternity, and Ancient DNA 228

Bioinformatics 229
DNA Sequencing 229
Genomics 230
Genome-Mapping Techniques 231
The Human Genome 231

THE PROCESS OF SCIENCE Did Nic Have a Deadly Gene? 233
Applied Genomics 233
Systems Biology 234

Safety and Ethical Issues 235
The Controversy over Genetically Modified Foods 235
Ethical Questions Raised by Human DNA Technologies 236

EVOLUTION CONNECTION The Y Chromosome as a Window on History 237

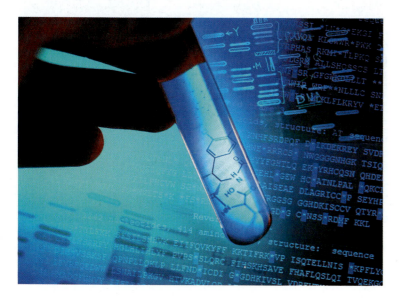

Unit 3 Evolution and Diversity 241

13 How Populations Evolve 242

CHAPTER THREAD
Evolution in Action 243

BIOLOGY AND SOCIETY Mosquitoes and Evolution 243

The Diversity of Life 244
Naming and Classifying the Diversity of Life 244
Explaining the Diversity of Life 245

Charles Darwin and *The Origin of Species* 246
Darwin's Journey 246
Darwin's Theory 248

Evidence of Evolution 248
Evidence from Fossils 248
Evidence from Homologies 250
Evolutionary Trees 251

Natural Selection as the Mechanism for Evolution 252
Natural Selection in Action 253
Key Points about Natural Selection 254

The Evolution of Populations 254
Sources of Genetic Variation 254
Populations as the Units of Evolution 255
Analyzing Gene Pools 256
Population Genetics and Health Science 257
Microevolution as Change in a Gene Pool 257

Mechanisms of Evolution 258
Natural Selection 258
Genetic Drift 258
Gene Flow 260
Natural Selection: A Closer Look 261

THE PROCESS OF SCIENCE Did Natural Selection Shape the Beaks of Darwin's Finches? 262

EVOLUTION CONNECTION The Rising Threat of Antibiotic Resistance 265

14 How Biological Diversity Evolves 268

| CHAPTER THREAD | |
Evolution in the Human-Dominated World 269

BIOLOGY AND SOCIETY Humanity's Footprint 269

The Origin of Species 270
What Is a Species? 271
Reproductive Barriers between Species 272
Mechanisms of Speciation 274

THE PROCESS OF SCIENCE Do Human Activities Facilitate Speciation? 276

Earth History and Macroevolution 279
The Fossil Record 279
Plate Tectonics and Biogeography 281
Mass Extinctions and Explosive Diversifications of Life 283

Mechanisms of Macroevolution 283
Large Effects from Small Genetic Changes 283
The Evolution of Biological Novelty 284

Classifying the Diversity of Life 286
Classification and Phylogeny 286
Classification: A Work in Progress 288

EVOLUTION CONNECTION Evolution in the Anthropocene 289

15 The Evolution of Microbial Life 292

CHAPTER THREAD
Human Microbiota 293

BIOLOGY AND SOCIETY Our Invisible Inhabitants 293

Major Episodes in the History of Life 294

The Origin of Life 296
A Four-Stage Hypothesis for the Origin of Life 296
From Chemical Evolution to Darwinian Evolution 298

Prokaryotes 299
They're Everywhere! 299
The Structure and Function of Prokaryotes 300
The Ecological Impact of Prokaryotes 303
The Two Main Branches of Prokaryotic Evolution: Bacteria and Archaea 304

THE PROCESS OF SCIENCE Are Intestinal Microbiota to Blame for Obesity? 306

Protists 307
Protozoans 308
Slime Molds 309
Unicellular and Colonial Algae 310
Seaweeds 310

EVOLUTION CONNECTION The Sweet Life of Streptococcus mutans 311

16 The Evolution of Plants and Fungi 314

CHAPTER THREAD
Plant-Fungus Interactions 315

BIOLOGY AND SOCIETY The Diamond of the Kitchen 315

Colonizing Land 316
Terrestrial Adaptations of Plants 316
The Origin of Plants from Green Algae 318

Plant Diversity 318
Highlights of Plant Evolution 318
Bryophytes 319
Ferns 321
Gymnosperms 322
Angiosperms 324
Plant Diversity as a Nonrenewable Resource 327

Fungi 328
Characteristics of Fungi 329

THE PROCESS OF SCIENCE What Killed the Pines? 330
The Ecological Impact of Fungi 331
Commercial Uses of Fungi 331

EVOLUTION CONNECTION A Pioneering Partnership 332

17 The Evolution of Animals 336

CHAPTER THREAD
Human Evolution 337

BIOLOGY AND SOCIETY Evolving Adaptability 337

The Origins of Animal Diversity 338
What Is an Animal? 338
Early Animals and the Cambrian Explosion 339
Animal Phylogeny 340

Major Invertebrate Phyla 341
Sponges 341
Cnidarians 342
Molluscs 343
Flatworms 344
Annelids 345
Roundworms 346
Arthropods 347
Echinoderms 353

Vertebrate Evolution and Diversity 354
Characteristics of Chordates 354
Fishes 356
Amphibians 357
Reptiles 358
Mammals 360

The Human Ancestry 361
The Evolution of Primates 361
The Emergence of Humankind 363

THE PROCESS OF SCIENCE What Can Lice Tell Us About Ancient Humans? 366

EVOLUTION CONNECTION Are We Still Evolving? 367

Unit 4 Ecology   371

18 An Introduction to Ecology and the Biosphere   372

CHAPTER THREAD
Climate Change   373

BIOLOGY AND SOCIETY   Penguins, Polar Bears, and People in Peril   373

An Overview of Ecology   374
Ecology and Environmentalism   374
A Hierarchy of Interactions   375

Living in Earth's Diverse Environments   376
Abiotic Factors of the Biosphere   376
The Evolutionary Adaptations of Organisms   378
Adjusting to Environmental Variability   378

Biomes   380
Freshwater Biomes   380
Marine Biomes   382
How Climate Affects Terrestrial Biome Distribution   384
Terrestrial Biomes   385
The Water Cycle   391
Human Impact on Biomes   392

Climate Change   394
The Greenhouse Effect and Global Warming   394
The Accumulation of Greenhouse Gases   395
Effects of Climate Change on Ecosystems   396

THE PROCESS OF SCIENCE   How Does Climate Change Affect Species Distribution?   397

Looking to Our Future   398

EVOLUTION CONNECTION   Climate Change as an Agent of Natural Selection   399

19 Population Ecology   402

CHAPTER THREAD
Biological Invasions   403

BIOLOGY AND SOCIETY   Invasion of the Lionfish   403

An Overview of Population Ecology   404
Population Density   405
Population Age Structure   405
Life Tables and Survivorship Curves   406
Life History Traits as Adaptations   406

Population Growth Models   408
The Exponential Population Growth Model: The Ideal of an Unlimited Environment   408
The Logistic Population Growth Model: The Reality of a Limited Environment   409
Regulation of Population Growth   410

Applications of Population Ecology   412
Conservation of Endangered Species   412
Sustainable Resource Management   412
Invasive Species   413
Biological Control of Pests   414

THE PROCESS OF SCIENCE   Can Fences Stop Cane Toads?   415

Integrated Pest Management   416

Human Population Growth   417
The History of Human Population Growth   417
Age Structures   418
Our Ecological Footprint   419

EVOLUTION CONNECTION   Humans as an Invasive Species   421

20 Communities and Ecosystems 424

CHAPTER THREAD
Importance of Biodiversity 425

BIOLOGY AND SOCIETY Why Biodiversity Matters 425

Biodiversity 426
Genetic Diversity 426
Species Diversity 426
Ecosystem Diversity 427
Causes of Declining Biodiversity 427

Community Ecology 428
Interspecific Interactions 428
Trophic Structure 432
Species Diversity in Communities 435
Disturbances and Succession in Communities 436
Ecological Succession 436

Ecosystem Ecology 437
Energy Flow in Ecosystems 438
Chemical Cycling in Ecosystems 440

Conservation and Restoration Biology 444
Biodiversity "Hot Spots" 444
Conservation at the Ecosystem Level 445

THE PROCESS OF SCIENCE Does Biodiversity Protect Human Health? 446

Restoring Ecosystems 447
The Goal of Sustainable Development 448

EVOLUTION CONNECTION Saving the Hot Spots 449

Appendices
A Metric Conversion Table A-1
B The Periodic Table A-3
C Credits A-5
D Selected Answers A-11

Glossary G-1

Index I-1

CAMPBELL

essential
biology

1 Learning About Life

CHAPTER CONTENTS
The Scientific Study of Life 4
The Properties of Life 10
Major Themes in Biology 11

Why Biology Matters

Nearly everyone has an inborn curiosity about the natural world. Whatever your connection to nature—perhaps you have pets; enjoy visiting parks, zoos, or aquariums; or watch TV shows about interesting creatures—this book will help demonstrate how the study of biology connects to your life.

YOU'RE A SCIENTIST! ALTHOUGH YOU MAY NOT REALIZE IT, YOU USE THE PROCESS OF SCIENCE EVERY DAY.

WHAT THE HECK IS THAT? IF YOU'VE WONDERED WHAT AN UNUSUAL OR ESPECIALLY BEAUTIFUL ANIMAL IS CALLED, YOU'RE CURIOUS ABOUT BIOLOGY.

IS THERE LIFE ON MARS? ONE OF THE MISSIONS OF THE MARS ROVER IS TO SEARCH FOR SIGNS OF LIFE.

CHAPTER THREAD

Swimming with the Turtles

BIOLOGY AND SOCIETY   A Passion for Life 3
THE PROCESS OF SCIENCE   Do Baby Turtles Swim? 8
EVOLUTION CONNECTION   Turtles in the Tree of Life 18

BIOLOGY AND SOCIETY   Swimming with the Turtles

An inborn urge to learn about life. A college student swims with a green sea turtle off the coast of Belize, Central America.

A Passion for Life

Imagine yourself floating gently in a warm, calm ocean. Through the blue expanse, you spy a green sea turtle gliding toward you. You watch intently as it grazes on seagrass. It's easy to be captivated by this serene sea creature, with its paddle-shaped flippers and large eyes. As you follow it, you can't help but wonder about its life—how old it is, where it is traveling, whether it has a mate.

It's very human to be curious about the world around us. Nearly all of us have an inherent interest in life, an inborn fascination with the natural world. Do you have a pet? Are you concerned with fitness or healthy eating? Have you ever visited a zoo or an aquarium for fun, taken a nature hike through a forest, grown some plants, or gathered shells on the beach? Would you like to swim with a turtle? If you answered yes to any of these questions, then you share an interest in biology.

We wrote *Essential Biology* to help you harness your innate enthusiasm for life, no matter how much experience you've had with college-level science (even if it's none!). We'll use this passion to help you develop an understanding of the subject of biology, an understanding that you can apply to your own life and to the society in which you live. Whatever your reasons for taking this course—even if only to fulfill your school's science requirement—you'll soon discover that exploring life is relevant and important to you.

To reinforce the fact that biology does indeed affect you personally, every chapter of *Essential Biology* opens with an essay—called Biology and Society—where you will see the relevance of that chapter's material. Topics as varied as green energy (Chapter 7), pet genetics (Chapter 9), and the importance of biodiversity (Chapter 20) help to illustrate biology's scope and show how the subject of biology is woven into the fabric of society. Throughout *Essential Biology*, we'll continuously emphasize these connections, pointing out many examples of how each topic can be applied to your life and the lives of those you care about.

The Scientific Study of Life

Now that we've established our goal—to examine how biology affects your life—a good place to start is with a basic definition: **Biology** is the scientific study of life. But have you ever looked up a word in the dictionary, only to find that you need to look up some of the words within that definition to make sense of the original word? The definition of *biology*, although seemingly simple, raises questions such as "What is a scientific study?" and "What does it mean to be alive?"

IF YOU'VE WONDERED WHAT AN UNUSUAL OR ESPECIALLY BEAUTIFUL ANIMAL IS CALLED, YOU'RE CURIOUS ABOUT BIOLOGY.

To start your investigation, this first chapter of *Essential Biology* will explain important concepts within the definition of biology. First, we'll place the study of life in the broader context of science. Next, we'll investigate the nature of life. Finally, we'll introduce a series of broad themes that serve as organizing principles for the information you will learn. Most importantly, throughout this chapter (and all of *Essential Biology*), you'll see examples of how biology affects *your* life, highlighting the relevance of this subject to society and everyone in it. ✅

✅ **CHECKPOINT**

Define biology.

■ *Answer: Biology is the scientific study of life.*

▼ **Figure 1.1 Scientific exploration.** Dr. Jane Goodall spent decades recording her observations of chimpanzee behavior during field research in the jungles of Tanzania.

An Overview of the Process of Science

The definition of *biology* as the scientific study of life leads to an obvious first question: What does it mean to study something scientifically? Notice that biology is not defined simply as "the study of life" because there

are many nonscientific ways that life can be studied. For example, meditation is a valid way of contemplating life, but it is not a *scientific* means of studying life, and therefore it does not fall within the scope of biology. How, then, do we tell the difference between science and other ways of trying to make sense of the world?

Science is an approach to understanding the natural world that is based on inquiry—a search for information, evidence, explanations, and answers to specific questions. Scientists seek natural causes for natural phenomena. Therefore, they focus solely on the study of structures and processes that can be verifiably observed and measured.

Exploration

If you wanted to understand something—say, the behavior of a sea turtle—how would you start? You'd probably begin by looking at it. Biology, like other sciences, begins with exploration **(Figure 1.1)**. During this initial phase of inquiry, you may simply watch the subject and record your observations. A more intense exploration may involve extending your senses using tools such as microscopes or precision instruments to allow for careful measurement. Whatever the source, recorded observations are called **data**—the evidence on which scientific inquiry is based. In addition to gathering your own data, you may read books or articles on the subject to learn about previously collected data.

As you proceed with your exploration, your curiosity will lead to questions, such as "Why is it this way?" "How does it work?" "Can I change it?" Such questions are the launching point for the next step in the process of science: testing.

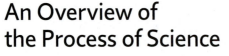

EXPLORATION
• Making observations
• Asking questions
• Seeking information

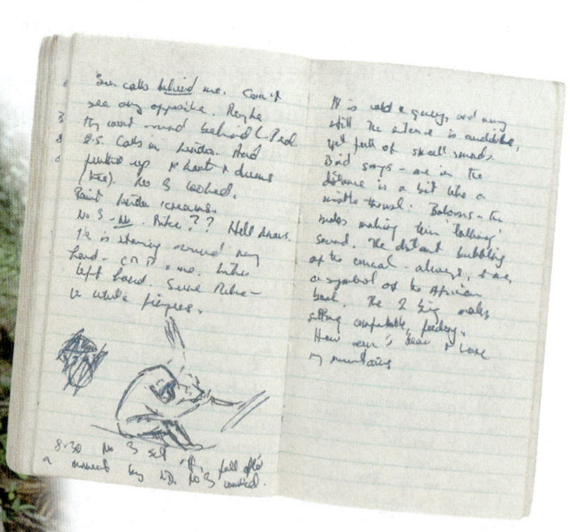

Testing

After making observations and asking questions, you may wish to conduct tests. But where do you start? You could probably think of many possible ways to investigate your subject. But you can't possibly test them all at once. To organize your thinking, you will likely begin by selecting one possible explanation and testing it. In other words, you would make a hypothesis. A **hypothesis** is a proposed explanation for a set of observations. A valid hypothesis must be testable and falsifiable—that is, it must be capable of being demonstrated to be false. A good hypothesis thus immediately leads to predictions that can be tested. Some hypotheses (such as ones involving conditions that can be easily controlled) lend themselves to **experiments**, or scientific tests. Other hypotheses (such as ones involving aspects of the world that cannot be controlled, such as ecological issues) can be tested by making further observations. The results of an experiment will either support or not support the hypothesis.

ALTHOUGH YOU MAY NOT REALIZE IT, YOU USE THE PROCESS OF SCIENCE EVERY DAY.

We all use hypotheses in solving everyday problems, although we don't think of it in those terms. Imagine that you press the power button on your TV remote, but the TV fails to turn on. That the TV does not turn on is an observation. The question that arises is obvious: Why didn't the remote turn on the TV? You probably would not just throw your hands up in the air and say "There's just no way to figure this out!" Instead, you might imagine several possible explanations, but you couldn't investigate them all simultaneously. Instead, you would focus on just one explanation, perhaps the most likely one based on past experience, and test it. That initial explanation is your hypothesis. For example, in this case, a reasonable hypothesis is that the batteries in the remote are dead.

After you've formed a hypothesis, you would make further observations or conduct experiments to investigate this initial idea. In this case, you can predict that if you replace the batteries, the TV will work. Let's say that you conduct this experiment, and the remote still doesn't work. You conclude that this observation does not support your hypothesis. You would then formulate a second hypothesis and test it. Perhaps you hypothesize that the TV is unplugged. You could continue to conduct additional experiments and formulate additional hypotheses until you reach a satisfactory answer to your initial question. As you do this, you are following a series of steps that provide a loose guideline for scientific investigations. These steps are shown in **Figure 1.2** and are sometimes called "the scientific method." They are a rough "recipe" for discovering new explanations, a set of procedures that, if followed, may provide insight into the subject at hand.

The steps are simply a way of formalizing how we usually try to solve problems. If you pay attention, you'll find that you often formulate hypotheses, test them, and draw conclusions. In other words, the process of science is probably your "go-to" method for solving problems. Although the process of science is often presented as a series of linear steps (such as those in Figure 1.2), in reality investigations are almost never this rigid. Different questions will require different paths through the steps. There is no single formula for successfully discovering something new; instead, the process of science suggests a broad outline for how an investigation might proceed. ✓

Communication and Outcomes

The process of science is typically repetitive and nonlinear. For example, scientists often work through several rounds of making observations and asking questions, with each round informing the next, before settling on hypotheses that they wish to test. In fine-tuning their questions, they rely heavily on scientific literature, the published contributions of fellow scientists. By reading about and understanding past studies, they can build on the foundation of existing knowledge.

☑ **CHECKPOINT**

Do all scientific investigations follow the steps in Figure 1.2 in that precise order? Explain.

■ *Answer: No. Different scientific investigations may proceed through the process of science in different ways.*

▶ **Figure 1.2** Testing a common problem using the process of science.

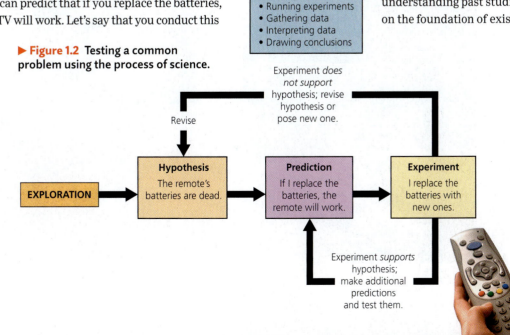

TESTING
• Forming hypotheses
• Making predictions
• Running experiments
• Gathering data
• Interpreting data
• Drawing conclusions

Experiment *does not support* hypothesis; revise hypothesis or pose new one.

Revise

EXPLORATION

Hypothesis
The remote's batteries are dead.

Prediction
If I replace the batteries, the remote will work.

Experiment
I replace the batteries with new ones.

Experiment *supports* hypothesis; make additional predictions and test them.

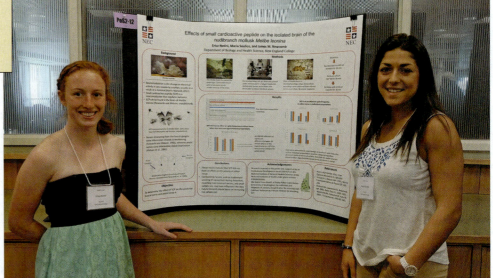

COMMUNICATION
- Sharing data
- Obtaining feedback
- Publishing papers
- Replicating findings
- Building consensus

▶ **Figure 1.3 Scientific communication.** Like these college students, scientists often communicate results to colleagues at meetings.

Additionally, scientists communicate with each other through seminars, meetings, personal communication, and scientific publications **(Figure 1.3)**. Before experimental results are published in a scientific journal, the research is evaluated by qualified, impartial, often anonymous experts who were not involved in the study. This process, intended to provide quality control, is called **peer review**. Reviewers often require authors to revise their paper or perform additional experiments in order to provide more lines of evidence. It is not uncommon for a scientific journal to reject a paper entirely if it doesn't meet the rigorous standards set by fellow scientists. After a study is published, scientists often check each other's claims by attempting to confirm observations or repeat experiments.

Science does not exist just for its own sake. In fact, it is interwoven with the fabric of society **(Figure 1.4)**. Much of scientific research is focused on solving problems that influence our quality of life, such as the push to cure cancer or to understand and slow the process of climate change. Societal needs often determine which research projects are funded. Scientific studies may involve basic research (largely concerned with building knowledge) or they may be more applied (largely concerned with developing new technologies). The ultimate aim of most scientific investigations is to benefit society. This focus on outcomes highlights the connections between biology, your own life, and our larger society.

OUTCOMES
- Building knowledge
- Solving problems
- Developing new technologies
- Benefiting society

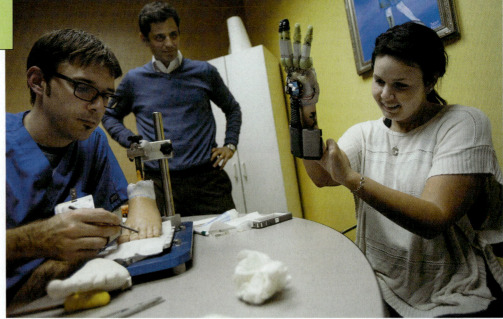

▶ **Figure 1.4 Scientific outcomes.** A 22-year-old woman tries on her new prosthetic hand with individually movable bionic fingers.

Putting all these steps together, **Figure 1.5** presents a more comprehensive model of the process of science. You can see that forming and testing hypotheses (represented in blue) are at the center of science. This core set of activities is the reason that science explains natural phenomena so well. These activities, however, are shaped by exploration (orange) and influenced by communication with other scientists (yellow) and by outcomes (green). Notice that many of these activities connect to others, illustrating that the components of the process of science interact. As in all quests, science includes elements of challenge, adventure, and luck, along with careful planning, reasoning, creativity, patience, and persistence in overcoming setbacks. Such diverse elements of inquiry allow the process of science to be flexible, molded by the needs of each particular challenge.

In every chapter of *Essential Biology*, we include examples of how the process of science was used to study the content presented in that chapter. Some of the questions that will be addressed are Do baby turtles swim (this chapter)? Can avatars improve cancer treatment (Chapter 11)? What can lice teach us about ancient humans (Chapter 17)?

As you become increasingly scientifically literate, you will arm yourself with the tools you need to evaluate claims that you hear. We are all bombarded by information every day—through commercials, social media, websites, magazine articles, and so on—and it can be hard to filter out the bogus from the truly worthwhile. Having a firm grasp of science as a process of inquiry can therefore help you in many ways outside the classroom. ✅

Hypotheses, Theories, and Facts

Since scientists focus on natural phenomena that can be reliably observed and measured, let's explore how the terms *hypothesis, theory,* and *fact* are related.

As previously noted, a hypothesis is a proposed explanation for an observation. In contrast, a scientific **theory** is much broader in scope than a hypothesis. A theory is a comprehensive and well-substantiated explanation. Theories only become widely accepted by scientists if they are supported by a large, varied, and growing body of evidence. A theory can be used to explain many observations. Indeed, theories can be used to devise many new and testable hypotheses. It is important to note that scientists use the word *theory* differently than many people tend to use it in everyday speech, which implies untested speculation ("It's just a theory!"). In fact, the word *theory* is commonly used in everyday speech in the way a scientist uses the word *hypothesis*. It is therefore improper to say that a scientific theory, such as the theory of

▼ **Figure 1.5 An overview of the process of science.** Notice that performing scientific tests lies at the heart of the entire process.

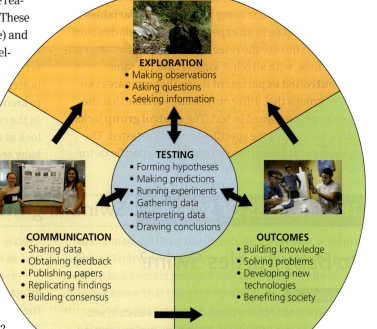

EXPLORATION
- Making observations
- Asking questions
- Seeking information

TESTING
- Forming hypotheses
- Making predictions
- Running experiments
- Gathering data
- Interpreting data
- Drawing conclusions

COMMUNICATION
- Sharing data
- Obtaining feedback
- Publishing papers
- Replicating findings
- Building consensus

OUTCOMES
- Building knowledge
- Solving problems
- Developing new technologies
- Benefiting society

Figure Walkthrough

Mastering Biology
goo.gl/6bRdg9

evolution, is "just" a theory to imply that it is untested or lacking in evidence. In reality, every scientific theory is backed up by a wealth of supporting evidence, or else it wouldn't be referred to as a theory. However, a theory, like any scientific idea, must be refined or even abandoned if new, contradictory evidence is discovered.

A **fact** is a piece of information considered to be objectively true based on all current evidence. A fact can be verified and is therefore distinct from opinions (beliefs that can vary from person to person), matters of taste, speculation, or inference. However, science is self-correcting: New evidence may lead to reconsideration of information previously regarded as a fact.

Many people associate facts with science, but accumulating facts is not the primary goal of science. A dictionary is an impressive catalog of facts, but it has little to do with science. It is true that facts, in the form of verifiable observations and repeatable experimental results, are the prerequisites of science. What advances science, however, are new theories that tie together a number of observations that previously seemed unrelated. The cornerstones of science are the explanations that apply to the greatest variety of phenomena. People like Isaac Newton, Charles Darwin, and Albert Einstein stand out in the history of science not because they discovered a great many facts but because their theories had such broad explanatory power. ✅

✅ **CHECKPOINT**

Why does peer review improve the reliability of a scientific paper?

Answer: A peer-reviewed paper carries a "seal of approval" from impartial experts on the subject.

✅ **CHECKPOINT**

You arrange to meet a friend for dinner at 6 P.M., but when the appointed hour comes, she is not there. You wonder why. Another friend says, "My theory is that she forgot." If your friend were speaking like a scientist, what would she have said?

Answer: "My hypothesis is that she forgot."

7

FEATURES OF SCIENCE	FEATURES OF PSEUDOSCIENCE
Adheres to an established and well-recognized scientific method	Does not adhere to generally accepted processes of science
Repeatable results	Results that cannot be duplicated by others; results that rely on a single person or are solely opinion
Testable claims that can be disproven	Unprovable or untestable claims; reliance on assumptions or beliefs that are not testable
Open to outside review	Rejection of external review or refusal to accept contradictory evidence
Multiple lines of evidence	Overreliance on a small amount of data; underlying causes are not investigated

A field biologist collecting data

A pyramid that is claimed to channel energy

▶ **Figure 1.9** Features of science versus pseudoscience.

be confusing, there are several indicators that you can use to recognize pseudoscience **(Figure 1.9)**. For example, a pseudoscientific study may be based soley or largely on **anecdotal evidence**, an assertion based on a single or a few examples that do not support a generalized conclusion—for example, "Today was unusually cold, so global warming must be a hoax!" A proper scientific investigation is open to outside review (the communication step in Figure 1.5), while pseudoscientific claims often reject external review or refuse to accept contradictory evidence. Often, pseudoscientific claims are based on results that cannot be duplicated by others because they rely on a single person or are solely opinion. A proper scientific study, on the other hand, has repeatable results that stand up to external scrutiny.

One of the best ways to evaluate scientific claims is to consider the source of the information **(Figure 1.10)**. Science depends upon peer review, the evaluation of work by impartial, qualified, often anonymous experts who are not involved in that work. Publishing a study in a peer-reviewed journal is often the best way to ensure that

▼ **Figure 1.10** **Recognizing a reliable source.** The more criteria a given source meets, the more reliable it is.

Souce reliability checklist

- ☐ Is the information current?
- ☐ Is the source primary (and not secondary)?
- ☐ Is/are the author(s) indentifiable and well qualified?
- ☐ Does the author lack potential conflicts of interest?
- ☐ Are references cited?
- ☐ Are any experiments described in enough detail that they could be reproduced?
- ☐ Was the information peer reviewed?
- ☐ Is the information unbiased?
- ☐ Is the intent of the source known and valid?

it will be considered scientifically valid. No matter the source, reliable scientific information can be recognized by being up to date, drawing from known sources of information, having been authored by a reputable expert, and being free of bias.

Now that we have explored the process of science, keep in mind that it has proven to be the most effective method for investigating the natural world. In fact, nearly everything we know about nature was learned through the process of science. ☑

☑ CHECKPOINT

If someone says, "It rained yesterday, so I don't believe that there is a drought," this is an example of what kind of improper thinking?

■ *Answer: a conclusion based on anecdotal evidence*

The Properties of Life

Recall once again the definition at the heart of this chapter: Biology is the scientific study of life. Now that we understand what constitutes a scientific study, we can turn to the next question raised by this definition: What is life? Or, to put it another way, what distinguishes living things from nonliving things? The phenomenon of life seems to defy a simple, one-sentence definition. Yet even a small child instinctively knows that a bug or a plant is alive but a rock is not.

If someone placed an object in front of you and asked whether it was alive, what would you do? Would you poke it to see if it reacts? Would you watch it closely to see if it moves or breathes? Would you dissect it to look at its parts? Each of these ideas is closely related to how biologists actually define

life: We recognize life mainly by what living things do. Using a green sea turtle as an example, **Figure 1.11** highlights the major properties we associate with life: order, cells, growth and development, energy processing, regulation, response to the environment, reproduction, and evolution.

An object is generally considered to be alive if it displays all of these characteristics simultaneously. On the other hand, a nonliving object may display some of these properties, but not all of them. For example, a virus has an ordered structure, but it cannot process energy, nor is it composed of cells. Viruses, therefore, are generally not considered to be living organisms (see Chapter 10 for more information on viruses).

▼ **Figure 1.11 A green sea turtle displays the properties of life.**
An object is considered alive only if it displays all of these properties simultaneously.

Order is apparent in many of the turtle's structures, such as the regular arrangement of plates in the turtle shell.

Like any large organism, the body of a sea turtle is made of trillions of **cells**.

Growth and development into a mature adult sea turtle takes decades.

Scientists who study turtle **evolution** believe that they first appeared nearly 250 million years ago, making their lineage older than the dinosaurs.

As part of the turtle **reproduction** cycle, a female will lay 100-200 eggs into a hole dug in a sandy beach.

The sex of sea turtle hatchlings varies in **response to the environment:** Warmer temperatures favor the development of females, while cooler temperatures favor the development of males.

Energy processing in adult sea turtles depends upon a diet of algae and sea grass.

Although surrounded by salt water, sea turtles carry out **regulation** of the salt level in their body by excreting excess salt through their eyes.

▼ **Figure 1.12 A sample of the diversity of life in a national park in Namibia.**

Even as life on Earth shares recognizable properties, it also exists in tremendous diversity (**Figure 1.12**; see also Chapters 13–17). But must we limit our discussion to life on this planet? Although we have no proof that life has ever existed anywhere other than Earth, biologists speculate that extraterrestrial life, if it exists, could be recognized by the same properties described in Figure 1.11. The Mars rover *Curiosity*, which has been exploring the surface of the red planet since 2012, contains several instruments designed to identify substances that provide evidence of past or present life. For example, *Curiosity* is using a suite of onboard instruments to detect chemicals that could provide evidence of energy processing by microscopic organisms. In 2017, NASA announced that one of Saturn's moons, Enceladus, is the most likely place in our solar system to find extraterrestrial life, due to its abundant water and geothermal activity. NASA hopes to launch a probe there soon. The search continues. ✅

ONE OF THE MISSIONS OF THE MARS ROVER IS TO SEARCH FOR SIGNS OF LIFE.

✅ **CHECKPOINT**

Which properties of life apply to a car? Which do not?

■ Answer: A car demonstrates order, regulation, energy processing, and response to the environment. But a car does not grow, reproduce, or evolve, and it is not composed of cells.

Major Themes in Biology

As new discoveries unfold, biology grows in breadth and depth. However, major themes continue to run throughout the subject. These overarching principles unify all aspects of biology, from the microscopic world of cells to the global environment. Focusing on a few big-picture ideas that cut across many topics within biology can help organize and make sense of all the information you will learn.

This section describes five unifying themes that recur throughout our investigation of biology: the relationship of structure to function, information flow, pathways that transform energy and matter, interactions within biological systems, and evolution. You'll encounter these themes throughout subsequent chapters, with some key examples **highlighted in blue in the text**.

▶ **Figure 1.17**
Transformations of energy and matter in an ecosystem. Nutrients are recycled within an ecosystem, whereas energy flows into and out of an ecosystem.

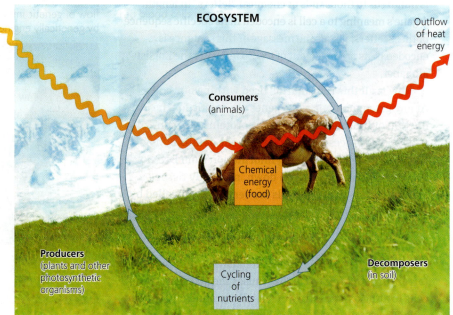

energy. This flow of energy is represented by wavy lines in **Figure 1.17**.

Every object in the universe, both living and nonliving, is composed of matter. In contrast to energy flowing through an ecosystem, matter is recycled within an ecosystem. This recycling is represented by the blue circle in Figure 1.17. For example, minerals that plants absorb from the soil can eventually be recycled back into the soil when plants are decomposed by microorganisms. Decomposers, such as fungi and many bacteria, break down waste products and the remains of dead organisms, changing complex molecules into simple nutrients. The action of decomposers makes nutrients available to be taken up from the soil by plants again, thereby completing the cycle.

Within all living cells, a vast network of interconnected chemical reactions (collectively referred to as metabolism) continually converts energy from one form to another as matter is recycled. For example, as food molecules are broken down into simpler molecules, energy stored in the chemical bonds is released. This energy can be captured and used by the body (to power muscle contractions, for example). The atoms that made up the food can then be recycled (to build new muscle tissue, for example). Within all living organisms, there is a never-ending "chemical square dance" in which molecules swap chemical partners as they receive, convert, and release matter and energy. Within the ocean, for example, sunlight filtering through shallow water enables seagrass to grow. When a sea turtle grazes on seagrass, it obtains energy that it uses to swim and uses the molecules of matter as building blocks for cells within its own body.

Energy transformations can be disrupted, often with dire consequences. Consider what happens if you consume cyanide, one of the deadliest known poisons. Ingesting just 200 milligrams (about half the size of an aspirin tablet) causes death in humans. Cyanide is so toxic because it blocks an essential step within the metabolic pathway that harvests energy from glucose. When even a single protein within this pathway becomes inhibited, cells lose the ability to extract the energy stored in the chemical bonds of glucose. The rapid death that follows is a gruesome illustration of the importance of energy and matter transformations to life. Throughout your study of biology, you

will see more examples of how living organisms regulate the transformation of energy and matter, from microscopic cellular processes such as photosynthesis (Chapter 7) and cellular respiration (Chapter 8), to ecosystem-wide cycles of carbon and other nutrients (Chapter 20), to global cycles of water across the planet (Chapter 18). ✅

Interactions within Biological Systems

The study of life extends from the microscopic level of the molecules and cells that make up organisms to the global level of the entire living planet. We can divide this enormous range into different levels of biological organization, each of which can be viewed from a system perspective. There are many interactions within and between these levels of biological systems.

Imagine zooming in from space to take a closer and closer look at life on Earth. **Figure 1.18** takes you on a tour that spans the levels of biological organization. The top of the figure shows the global level of the entire **biosphere**, which consists of all the environments on Earth that support life—including soil; oceans, lakes, and other bodies of water; and the lower atmosphere. At the other extreme of biological size and complexity are microscopic molecules such as DNA, the chemical responsible for inheritance. Zooming outward from the bottom to the top in the figure, you can see that it takes many molecules to build a cell, many cells to make a tissue, multiple tissues to make an organ, and so on. At each new level, novel properties emerge that are absent from the preceding one. These emergent properties are due to the specific arrangement and interactions of many parts into

▼ **Figure 1.18** Zooming in on life.

2 **Ecosystems**

An ecosystem consists of all living organisms in a particular area and all the nonliving components of the environment with which life interacts, such as soil, water, and light.

3 **Communities**

All organisms in an ecosystem (such as the iguanas, crabs, seaweed, and even bacteria in this ecosystem) are collectively called a community.

1 **Biosphere**

Earth's biosphere includes all life and all the places where life exists.

4 **Populations**

Within communities are various populations, groups of interacting individuals of one species, such as a group of iguanas.

5 **Organisms**

An organism is an individual living thing, like this iguana.

6 **Organ Systems and Organs**

An organism's body consists of several organ systems, each of which contains two or more organs. For example, the iguana's circulatory system includes its heart and blood vessels.

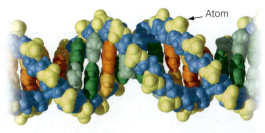

10 **Molecules and Atoms**

Finally, we reach molecules, the chemical level in the hierarchy. Molecules are clusters of even smaller chemical units called atoms. Each cell consists of an enormous number of chemicals that function together to give the cell the properties we recognize as life. DNA, the molecule of inheritance and the substance of genes, is shown here as a computer graphic. Each sphere in the DNA model represents a single atom.

9 **Organelles**

Organelles are functional components of cells, such as the nucleus that houses the DNA.

Nucleus

← Atom

8 **Cells**

The cell is the smallest unit that can display all the characteristics of life.

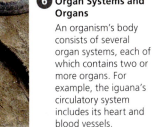

Colorized LM 60x

7 **Tissues**

Each organ is made up of several different tissues, such as the heart muscle tissue shown here. A tissue consists of a group of similar cells performing a specific function.

an increasingly complex system. Such properties are called emergent because they emerge as complexity increases. For example, life emerges at the level of the cell; a test tube full of molecules is not alive. The saying "the whole is greater than the sum of its parts" captures this idea. Emergent properties are not unique to life. A box of camera parts won't do anything, but if the parts are arranged and interact in a

certain way, you can capture photographs. Add structures from a phone, and your camera and phone can interact to gain the ability to send photos to friends. New properties emerge as the complexity increases. Compared with such nonliving examples, however, the unrivaled complexity of biological systems makes the emergent properties of life especially fascinating to study.

✓ **CHECKPOINT**

By definition, all atoms of carbon have exactly 6 _____, but the number of _____ varies from one isotope to another.

■ *Answer: protons; neutrons*

isotopes, meaning that their nuclei remain intact. The isotope carbon-14, on the other hand, is radioactive. A **radioactive isotope** is one in which the nucleus decays spontaneously, shedding particles and energy (radiation).

Uncontrolled exposure to high levels of radiation can be lethal because the particles and energy thrown off by radioactive atoms can damage molecules, especially DNA. Therefore, radiation from decaying isotopes can pose serious health risks. In 1986, the explosion of a nuclear reactor at Chernobyl, Ukraine, released large amounts of radioactive isotopes, killing 30 people within a few weeks. Millions of people in the surrounding areas were exposed, causing

an estimated 6,000 cases of thyroid cancer. The 2011 post-tsunami Fukushima nuclear disaster in Japan did not result in any immediate deaths due to radiation exposure, but scientists are carefully monitoring the people who live in the area for any long-term health consequences.

Although radioactive isotopes can cause harm when uncontrolled, they have many uses in biological research and medicine. The Biology and Society section discussed in general how radiation can be used to diagnose and treat diseases. The Process of Science section will explore one specific medical application of radiation: the treatment of prostate cancer. ✓

THE PROCESS OF SCIENCE | **Helpful Radiation**

How Effective Is Radiation in Treating Prostate Cancer?

BACKGROUND

The prostate is a gland within the male reproductive system that adds fluid to sperm during ejaculation. Among American men, prostate cancer is the most frequently diagnosed cancer (with over 160,000 new cases in 2017) and the second leading cause of cancer deaths (27,000 in 2017), after lung cancer.

METHOD

The most effective treatment for prostate cancer is surgical removal of the prostate. However, removal causes serious side effects. Another treatment option is radioactive seed implantation **(Figure 2.5a)**. In this treatment, several dozen small bits of radioactive metal (most commonly made from iodine-125), each the size of a grain of rice, are placed within the prostate using a needle **(Figure 2.5b)**. The isotopes used will decay into safe, nonradioactive metals in a few months or years. By carefully regulating the dosage, doctors attempt to abolish or slow the tumor without harming surrounding healthy tissues.

RESULTS

Is radioactive seed implantation therapy effective? A study published in 2014 involved over 1,000 British men with prostate cancer who were treated with iodine-125 seed

implants. Seeds were placed inside and around each tumor site. Five years later, 94% of the patients were free of cancer symptoms. While this number is impressive, how do we know if a similar number of men would have recovered even without the implants? Medical research trials include control groups whenever possible. However, there has never been a study involving patients who received radioactive seed implantation and those who received a placebo consisting of nonradioactive seeds because it would be unethical to withhold treatment from cancer patients. Nevertheless, comparisons with other studies that used different methods, such as surgery and external radiation, can be made **(Figure 2.5c)**. Such comparisons are problematic, though, because the groups of patients are not controlled for all characteristics. Therefore, studying the effects of radioactive seed implantation shows how researchers are sometimes unable to run properly controlled experiments when humans are involved.

Thinking Like a Scientist

How might a study on radioactive seed implantation in rats differ from a study on humans?

For the answer, see Appendix D.

▼ **Figure 2.5 Treatment of prostate cancer with radiation.**

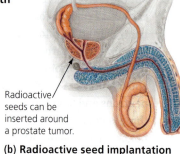

(a) **Radioactive seeds**

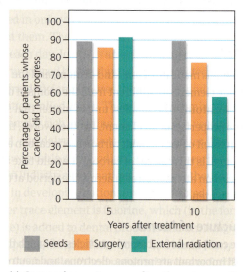

Radioactive seeds can be inserted around a prostate tumor.

(b) **Radioactive seed implantation**

(c) **Comparing outcomes after three types of prostate treatment**

Chemical Bonding and Molecules

Of the three subatomic particles we've discussed—protons, neutrons, and electrons—only electrons are directly involved in chemical reactions. The number of electrons in an atom therefore determines the chemical properties of that atom. Chemical reactions enable atoms to transfer or share electrons. These interactions usually result in atoms staying close together, held by attractions called **chemical bonds**. In this section, we will explore three types of chemical bonds: ionic, covalent, and hydrogen bonds.

Ionic Bonds

Table salt is an example of how the transfer of electrons can bond atoms together. As discussed earlier, the two ingredients of table salt are the elements sodium (Na) and chlorine (Cl). When a chlorine atom and a sodium atom are near each other, the chlorine atom strips an electron from the sodium atom **(Figure 2.6)**. Before this electron transfer, both the sodium and chlorine atoms are electrically neutral. Because electrons are negatively charged, the electron transfer moves one unit of negative charge from sodium to chlorine. This action makes both atoms **ions**, atoms (or molecules) that are electrically charged as a result of gaining or losing electrons. In this case, the loss of an electron results in the sodium ion having a charge of +1, whereas chlorine's gain of an electron results in it having a charge of −1. Negatively charged ions often have names ending in "-ide," like "chloride" or "fluoride." The sodium ion (Na^+) and chloride ion (Cl^-) are held together by an

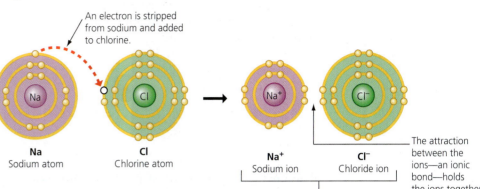

▼ **Figure 2.6 Electron transfer and ionic bonding.** When a sodium atom and a chlorine atom meet, the electron transfer between the two atoms results in two ions with opposite charges.

An electron is stripped from sodium and added to chlorine.

Na Sodium atom

Cl Chlorine atom

Na^+ Sodium ion

Cl^- Chloride ion

Sodium chloride (NaCl)

The attraction between the ions—an ionic bond—holds the ions together.

ionic bond, the attraction between oppositely charged ions. Compounds that are held together by ionic bonds are called ionic compounds, also known as salts. ☑

Covalent Bonds

In contrast to the complete *transfer* of electrons in ionic bonds, a **covalent bond** forms when two atoms *share* one or more pairs of electrons. Of the bonds we've discussed, covalent bonds are the strongest; these are the bonds that hold atoms together in a **molecule**. For example, in **Figure 2.7**, you can see that each of the two hydrogen atoms in a molecule of formaldehyde (CH_2O, a common disinfectant and preservative) shares one pair of electrons with the carbon atom. The oxygen atom shares two pairs of electrons with the carbon, forming a double bond. Notice that each atom of hydrogen (H) can form one covalent bond; oxygen (O) can form two; and carbon (C) can form four.

☑ CHECKPOINT

When a lithium ion (Li^+) joins a bromide ion (Br^-) to form lithium bromide, the resulting bond is a(n) _____ bond.

■ Answer: ionic

▼ **Figure 2.7 Alternative ways to represent a molecule.** A molecular formula, such as CH_2O, tells you the number of each kind of atom in a molecule but not how they are attached together. This figure shows four common ways of representing the arrangement of atoms in molecules.

Name (molecular formula)	Electron configuration Shows how outermost electrons participate in chemical bonding.	Structural formula Represents each covalent bond (a pair of shared electrons) with a line	Space-filling model Shows the shape of a molecule by symbolizing atoms with color-coded balls	Ball-and-stick model Represents atoms with "balls" and bonds with "sticks"

Formaldehyde (CH_2O)

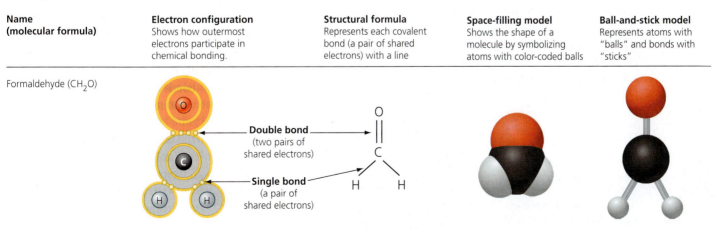

Double bond (two pairs of shared electrons)

Single bond (a pair of shared electrons)

Other liquids have much weaker surface tension; an insect, for example, could not walk on the surface of a cup of gasoline (which is why gardeners sometimes use gasoline to drown bugs removed from bushes).

▼ **Figure 2.12 Sweating as a mechanism of evaporative cooling.**

How Water Moderates Temperature

If you've ever burned your finger on a metal pot while waiting for the water in it to boil, you know that water heats up much more slowly than metal. In fact, because of hydrogen bonding, water has a stronger resistance to temperature change than most other substances.

When water is heated, the heat energy first disrupts hydrogen bonds and then makes water molecules jostle around faster. The temperature of the water doesn't go up until the water molecules start to speed up. Because heat is first used to break hydrogen bonds rather than raise the temperature, water absorbs and stores a large amount of heat while warming up only a few degrees. Conversely, when water cools, hydrogen bonds form in a process that releases heat. Thus, water can release a relatively large amount of heat to the surroundings while the water temperature drops only slightly.

Earth's giant water supply—the oceans, seas, lakes, rivers, and subsurface water—enables temperatures on the planet to stay within limits that permit life by storing a huge amount of heat from the sun during warm periods and giving off heat that warms the air during cold periods. That's why coastal areas generally have milder climates than inland regions. Water's resistance to temperature change also stabilizes ocean temperatures, creating a favorable environment for marine life. You may have noticed that the temperature of the ocean fluctuates much less than the air temperature.

Another way that water moderates temperature is by **evaporative cooling**. When a substance evaporates (changes from a liquid to a gas), the surface of the liquid that remains cools down. This cooling occurs because the molecules with the greatest energy (the "hottest" ones) tend to vaporize first. Think of it like this: If the five fastest runners on your track team quit school, it would lower the average speed of the remaining team. Evaporative cooling helps prevent some land-dwelling creatures from overheating; it's why sweating helps you dissipate excess body heat (**Figure 2.12**). And the expression "It's not the heat; it's the humidity" has its basis in the fact that sweat evaporates more slowly when the air is already saturated with water vapor, so the cooling effect is inhibited.

The Biological Significance of Ice Floating

When most liquids get cold, their molecules move closer together. If the temperature is cold enough, the liquid freezes and becomes a solid. Water, however, behaves differently. When water molecules get cold enough, they move apart, with each molecule staying at "arm's length" from its neighbors, forming ice. A chunk of ice floats because it is less dense than the liquid water in which it is floating. Floating ice is a consequence of hydrogen bonding. In contrast to the short-lived and constantly changing hydrogen bonds in liquid water, the hydrogen bonds that form in solid ice last longer, with each molecule bonded to four neighbors. As a result, ice is a spacious crystal (**Figure 2.13**).

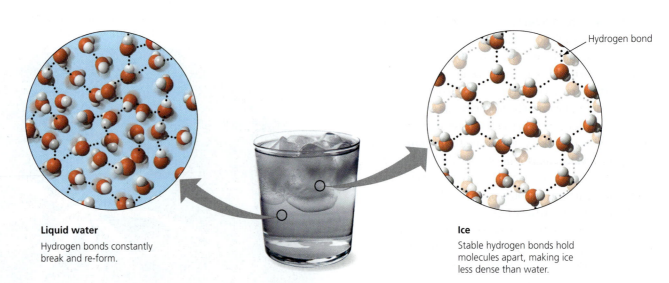

▶ **Figure 2.13 Why ice floats.** Compare the tightly packed molecules in liquid water with the spaciously arranged molecules in the ice crystal. The less dense ice floats atop the denser water.

Liquid water
Hydrogen bonds constantly break and re-form.

Hydrogen bond

Ice
Stable hydrogen bonds hold molecules apart, making ice less dense than water.

How does floating ice help support life on Earth? When a deep body of water cools and a layer of ice forms on top, the floating ice acts as an insulating "blanket" over the liquid water, allowing life to persist under the frozen surface. But imagine what would happen if ice were denser than water: Ice would sink during winter. All ponds, lakes, and even the oceans would eventually freeze solid without the insulating protection of the top layer of ice. Then, during summer, only the upper few inches of the oceans would thaw. It's hard to imagine life surviving under such conditions.

Water as the Solvent of Life

If you've ever stirred sugar into coffee or added salt to soup, you know that you can dissolve sugar or salt in water. This results in a mixture known as a **solution**, a liquid consisting of a homogeneous mixture of two or more substances. The dissolving agent is called the **solvent**, and any substance that is dissolved is called a **solute**. When water is the solvent, the solution is called an **aqueous solution**. The fluids of organisms are aqueous solutions. For example, tree sap is an aqueous solution consisting of sugar and mineral solutes dissolved in water solvent.

Water can dissolve an enormous variety of solutes necessary for life, providing a medium for chemical reactions.

For example, water can dissolve salt ions, as shown in **Figure 2.14**. Each ion becomes surrounded by oppositely charged regions of water molecules. Solutes that are polar molecules, such as sugars, dissolve by orienting locally charged regions of their molecules toward water molecules in a similar way.

We have discussed four special properties of water, each a consequence of water's unique chemical structure. Next, we'll look at aqueous solutions in more detail. ✅

Acids, Bases, and pH

In aqueous solutions, most of the water molecules are intact. However, a very small percentage of the water molecules break apart into hydrogen ions (H^+) and hydroxide ions (OH^-). A balance of these two highly reactive ions is critical for the proper functioning of chemical processes within organisms.

A chemical compound that releases H^+ to a solution is called an **acid**. One example of a strong acid is hydrochloric acid (HCl), the acid in your stomach that aids in digestion of food. In solution, HCl breaks apart into the ions H^+ and Cl^-. A **base** (or alkali) is a compound that accepts H^+ ions and removes them from a solution. Some bases, such as sodium hydroxide (NaOH), do this by releasing OH^-, which combines with H^+ to form H_2O.

To describe the acidity of a solution, chemists use the **pH scale**, a measure of the hydrogen ion H^+ concentration in a solution. The scale ranges from 0 (most acidic) to 14 (most basic). Each pH unit represents a tenfold change in the concentration of H^+. For example, lemon juice at pH 2 has 100 times more H^+ than an equal amount of tomato juice at pH 4. Aqueous solutions that are neither acidic nor basic (such as pure water) are said to be neutral; they have a pH of 7. They do contain some H^+ and OH^-, but the concentrations of the two ions are

☑ **CHECKPOINT**

1. Explain why, if you pour very carefully, you can actually "stack" water slightly above the rim of a cup.
2. Explain why ice floats.

■ *Answers: 1. Surface tension due to water's cohesion keeps the water from spilling over. 2. Ice is less dense than liquid water because the more stable hydrogen bonds lock the molecules into a spacious crystal.*

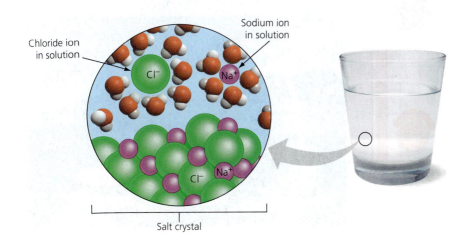

◀ **Figure 2.14 A crystal of table salt (NaCl) dissolving in water.** As a result of electrical charge attractions, H_2O molecules surround the sodium and chloride ions, dissolving the crystal in the process.

Lysosomes

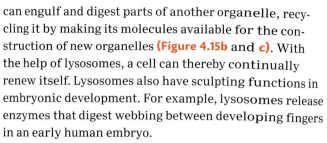

A **lysosome** is a membrane-enclosed sac of digestive enzymes. The enzymes and membranes of lysosomes are made by rough ER and processed in the Golgi apparatus. The lysosome provides a compartment where digestive enzymes can safely break down large molecules without unleashing the digestive enzymes on the cell itself.

Lysosomes perform several digestive functions. Many single-celled protists engulf nutrients into tiny cytoplasmic sacs called food vacuoles. Lysosomes fuse with the food vacuoles, exposing the food to digestive enzymes **(Figure 4.15a)**. Small molecules that result from this digestion, such as amino acids, leave the lysosome and nourish the cell. Lysosomes also help destroy harmful bacteria. For example, your white blood cells ingest bacteria into vacuoles, and lysosomal enzymes that are emptied into these vacuoles rupture the bacterial cell walls. In addition, without harming the cell, a lysosome

can engulf and digest parts of another organelle, recycling it by making its molecules available for the construction of new organelles **(Figure 4.15b** and **c)**. With the help of lysosomes, a cell can thereby continually renew itself. Lysosomes also have sculpting functions in embryonic development. For example, lysosomes release enzymes that digest webbing between developing fingers in an early human embryo.

The importance of lysosomes to cell function and human health is made clear by hereditary disorders called lysosomal storage diseases. A person with such a disease is missing one or more of the digestive enzymes normally found within lysosomes. The abnormal lysosomes become filled with indigestible substances, and this eventually interferes with other cellular functions. Most of these diseases are fatal in early childhood. In Tay-Sachs disease, for example, lysosomes lack a lipid-digesting enzyme. As a result, nerve cells die as they accumulate excess lipids, ravaging the nervous system. Fortunately, lysosomal storage diseases are rare. ✅

▼ **Figure 4.15** Two functions of lysosomes.

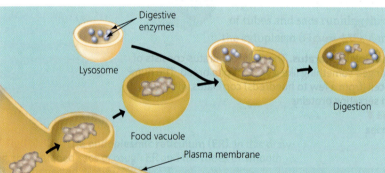

(a) A lysosome digesting food

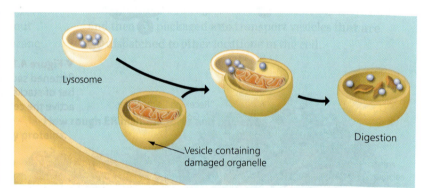

(b) A lysosome breaking down the molecules of damaged organelles

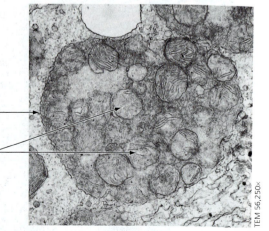

(c) Electron micrograph of lysosome recycling damaged organelles.

Vacuoles

Vacuoles are large vesicles with a variety of functions. For example, Figure 4.15a shows a food vacuole budding from the plasma membrane. Certain freshwater protists have contractile vacuoles that pump out excess water that flows into the cell **(Figure 4.16a)**.

Another type of vacuole is a **central vacuole**, a versatile compartment that can account for more than half the volume of a mature plant cell **(Figure 4.16b)**. A central vacuole stores organic nutrients, such as proteins stockpiled in the vacuoles of seed cells. It also contributes to plant growth by absorbing water and causing cells to expand. In the cells of flower petals, central vacuoles

THE CAFFEINE THAT GIVES COFFEE A KICK PROTECTS COFFEE PLANTS FROM HERBIVORES.

may contain pigments that attract pollinating insects. Central vacuoles may also contain poisons that protect against plant-eating animals. Some important crop plants produce and store large amounts of toxic chemicals—harmful to animals that might graze on the plant but useful to us—such as tobacco plants (which store nicotine) and coffee and tea plants (which store caffeine).

Now that we've explored the organelles of the endomembrane system, **Figure 4.17** reviews how they are related. A product made in one part of the endomembrane system may exit the cell or become part of another organelle without crossing a membrane. Also, membrane made by the ER can become part of the plasma membrane through the fusion of a transport vesicle. So even the plasma membrane is related to the endomembrane system. ☑

☑ CHECKPOINT

Place the following cellular structures in the order they would be used in the production and secretion of a protein: Golgi apparatus, nucleus, plasma membrane, ribosome, transport vesicle.

■ *Answer: nucleus, ribosome, transport vesicle, Golgi apparatus, plasma membrane*

▼ **Figure 4.16 Two types of vacuoles.**

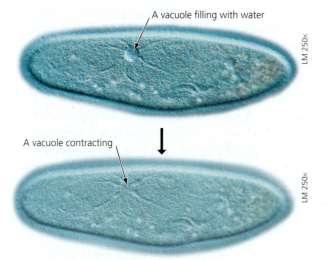

A vacuole filling with water

LM 250x

A vacuole contracting

LM 250x

(a) Contractile vacuole in *Paramecium*. A contractile vacuole fills with water and then contracts to pump the water out of the cell.

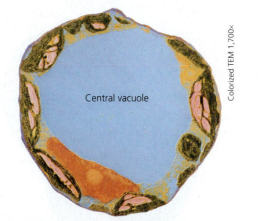

Central vacuole

Colorized TEM 1,700x

(b) Central vacuole in a plant cell. The central vacuole (colorized blue in this micrograph) is often the largest organelle in a mature plant cell.

▼ **Figure 4.17 Review of the endomembrane system.** The dashed arrows show some of the pathways of cell product distribution and membrane migration through transport vesicles.

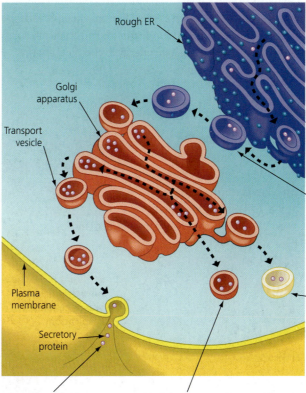

Rough ER

Golgi apparatus

Transport vesicle

Plasma membrane

Secretory protein

Some products are secreted from the cell.

Vacuoles store some cell products.

Transport vesicles carry enzymes and other proteins from the rough ER to the Golgi for processing.

Lysosomes carrying digestive enzymes can fuse with other vesicles.

Chloroplasts and Mitochondria: Providing Cellular Energy

One of the central themes of biology is the transformation of energy: how it enters living systems, is converted from one form to another, and is eventually given off as heat. To follow this flow of energy, we must consider the cellular power stations: chloroplasts and mitochondria.

Chloroplasts

Most of the living world runs on energy from photosynthesis, the conversion of light energy from the sun to the chemical energy of sugar molecules.

Chloroplasts are the photosynthetic organelles of plants and algae.

A chloroplast is divided into compartments by membranes, (**Figure 4.18**). The innermost compartment holds a thick fluid called stroma, which contains DNA, ribosomes, and enzymes. A network of sacs called thylakoids is suspended in the stroma. The sacs are often stacked like poker chips; each stack is called a granum (plural, *grana*). The grana are solar power packs, converting light energy to chemical energy (as detailed in Chapter 7).

Mitochondria

Mitochondria (singular, *mitochondrion*) are the organelles in which cellular respiration takes place; during cellular respiration, energy is harvested from sugars and

transformed into another form of chemical energy called ATP (adenosine triphosphate). Cells use molecules of ATP as a direct energy source. Mitochondria are found in almost all eukaryotic cells, including those of plants and animals.

A mitochondrion is enclosed by two membranes, separated by a narrow space. The inner membrane encloses a thick fluid called the mitochondrial matrix (**Figure 4.19**). The inner membrane has many infoldings called cristae. The cristae create a large surface area in which many molecules that function in cellular respiration are embedded, maximizing ATP output. (You'll learn more about how mitochondria work in Chapter 6.)

Besides providing cellular energy, mitochondria and chloroplasts share another feature: They contain their own DNA that encodes some proteins made by their own ribosomes. Each chloroplast and mitochondrion contains a single circular DNA chromosome that resembles a prokaryotic chromosome. In fact, mitochondria and chloroplasts can grow and pinch in two, reproducing themselves as many prokaryotes do. This and other evidence indicate that they evolved from ancient free-living prokaryotes that established residence within other, larger host prokaryotes. This phenomenon, where one species lives inside a host species, is a special type of symbiosis (see Chapter 16). Over time, mitochondria and chloroplasts became increasingly interdependent with the host prokaryote, eventually evolving into a single organism with inseparable parts. The DNA found within mitochondria and chloroplasts likely includes remnants of this ancient evolutionary event. ✅

✅ CHECKPOINT

1. What does photosynthesis accomplish?
2. What is cellular respiration?
3. Explain what is wrong with the following statement: "Plant cells have chloroplasts, and animal cells have mitochondria."

Answers: 1. the conversion of light energy to chemical energy stored in food molecules 2. a process that converts the chemical energy of sugars and other food molecules to chemical energy in the form of ATP 3. It implies that plant cells do not have mitochondria, when in fact they do.

▼ **Figure 4.18** The chloroplast: site of photosynthesis.

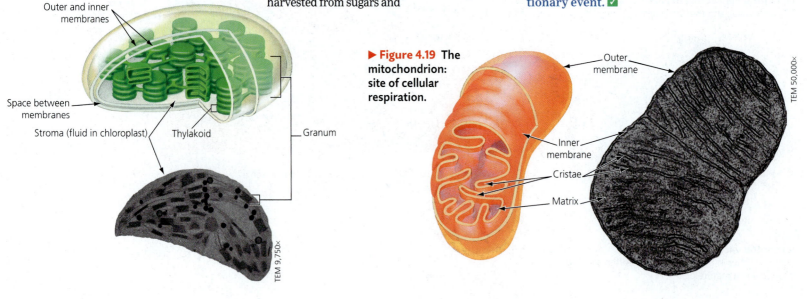

Outer and inner membranes

Space between membranes

Stroma (fluid in chloroplast)

Thylakoid

Granum

TEM 9,750×

▶ **Figure 4.19** The mitochondrion: site of cellular respiration.

Outer membrane

Inner membrane

Cristae

Matrix

TEM 50,000×

The Cytoskeleton: Cell Shape and Movement

If someone asked you to describe a house, you would most likely mention the various rooms and their locations. You probably would not think to mention the foundation and beams that support the house. Yet these structures perform an extremely important function. Similarly, cells have an infrastructure called the **cytoskeleton**, a network of protein fibers extending throughout the cytoplasm. The cytoskeleton serves as both skeleton and "muscles" for the cell, functioning in support and movement.

WITHOUT THE CYTOSKELETON, YOUR CELLS WOULD COLLAPSE IN ON THEMSELVES.

Maintaining Cell Shape

One function of the cytoskeleton is to give mechanical support to the cell and maintain its shape. This is especially important for animal cells, which lack rigid cell walls. The cytoskeleton contains several types of fibers made from different types of protein. One important type of fiber forms **microtubules**, hollow tubes of protein **(Figure 4.20a)**. The other kinds of cytoskeletal fibers, called intermediate filaments and microfilaments, are thinner and solid.

Just as the bony skeleton of your body helps fix the positions of your organs, the cytoskeleton provides anchorage and reinforcement for many organelles in a cell. For instance, the nucleus is held in place by a "cage" of cytoskeletal filaments. Other organelles use the cytoskeleton for movement. For example, a lysosome might reach a food vacuole by gliding along a microtubule track. Microtubules growing from a region called a centrosome guide the movement of chromosomes when cells divide (by means of the mitotic spindle—see Chapter 8). Microfilaments are also involved in cell movements. Different filaments composed of different kinds of proteins contract, resulting in the crawling movement of the protist *Amoeba* **(Figure 4.20b)** and movement of some of our white blood cells. ✅

✅ CHECKPOINT

From which important class of biological molecules are the microtubules of the cytoskeleton made?

Answer: protein

▼ **Figure 4.20** **The cytoskeleton.**

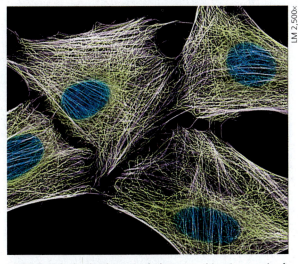

LM 2,500×

(a) Microtubules in the cytoskeleton. In this micrograph of animal cells, the cytoskeleton microtubules are labeled with a fluorescent yellow dye.

(b) Microtubules and movement. The crawling movement of an *Amoeba* is due to the rapid degradation and rebuilding of microtubules.

LM 250×

Flagella and Cilia

In some eukaryotic cells, microtubules are arranged into structures called flagella and cilia, extensions from a cell that aid in movement. Eukaryotic **flagella** (singular, *flagellum*) propel cells with an undulating, whiplike motion. They often occur singly, such as in human sperm cells **(Figure 4.21a)**, but may also appear in groups on the outer surface of protists. **Cilia** (singular, *cilium*) are generally shorter and more numerous than flagella and move in a coordinated back-and-forth motion, like the rhythmic oars of a rowing team. Both cilia and flagella propel various protists through water. For example, *Paramecium* cilia are visible in the SEM in Figure 4.1. Though different in length, number per cell, and beating pattern, cilia and flagella have the same basic architecture.

Some cilia extend from nonmoving cells that are part of a tissue layer, moving fluid over the tissue's surface. For example, cilia lining your windpipe clean your respiratory system by sweeping mucus with trapped debris out of your lungs **(Figure 4.21b)**. Tobacco smoke can paralyze these cilia, interfering with the normal cleansing mechanisms and allowing more toxin-laden smoke particles to reach the lungs. Frequent coughing—common in heavy smokers—then becomes the body's attempt to cleanse the respiratory system.

Because human sperm rely on flagella for movement, it's easy to understand why problems with flagella can lead to male infertility. Interestingly, some men with a type of hereditary sterility also suffer from respiratory problems. Because of a defect in the structure of their flagella and cilia, the sperm of men afflicted with this disorder cannot swim normally within the female reproductive tract to fertilize an egg (causing sterility), and their cilia do not sweep mucus out of their lungs (causing recurrent respiratory infections). ✓

✅ CHECKPOINT

Compare and contrast cilia and flagella.

■ Answer: Cilia and flagella have the same basic structure, are made from microtubules, and aid in movement. Cilia are short and numerous and move back and forth. Flagella are longer, often occurring singly, and they undulate.

▼ **Figure 4.21** **Examples of flagella and cilia.**

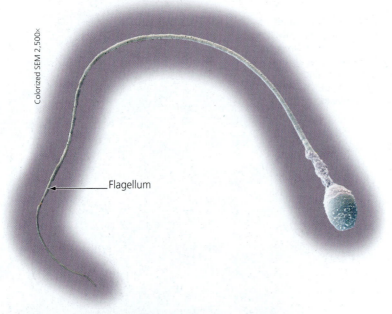

Colorized SEM 2,500×

Flagellum

(a) Flagellum of a human sperm cell. A eukaryotic flagellum undulates in a whiplike motion, driving a cell such as this sperm cell through its fluid environment.

Colorized SEM 3,000×

Cilia

(b) Cilia lining the respiratory tract. The cilia lining your respiratory tract sweep mucus with trapped debris out of your lungs. This helps keep your airway clear and prevents infections.

The Evolution of Bacterial Resistance in Humans

Individuals with variations that make them better suited for the local environment will survive and reproduce more often (on average) than those who lack such variations. When the advantageous variations have a genetic basis, the offspring of individuals with the variations will more often also have the favorable adaptions, giving them a survival and reproductive advantage. In this way, repeated over many generations, natural selection promotes evolution of the population.

Within a human population, the presence of a disease can provide a new basis for measuring those people who are best suited for survival in the local environment. For example, a recent evolutionary study examined people living in Bangladesh. This population has been exposed to the disease cholera—caused by an infectious bacterium—for millennia (Figure 4.22). After cholera bacteria enter a victim's digestive tract (usually through contaminated drinking water), the bacteria produce a toxin that binds to intestinal cells. There, the toxin alters proteins in the plasma membrane, causing the cells to excrete fluid. The resulting diarrhea, which spreads the bacteria by shedding it back into the environment, can cause severe dehydration and death if untreated.

Because Bangladeshis have lived for so long in an environment that teems with cholera bacteria, one might expect that natural selection would favor those individuals who have some resistance to the bacteria. Indeed, recent studies of people from Bangladesh revealed mutations in several genes that appear to confer an increased resistance to cholera. Researchers discovered one mutation in a gene that encodes for the plasma membrane proteins that are the targets of the cholera bacteria. Although the mechanism is not yet understood, these genes appear to offer a survival advantage by making the proteins more resistant to attack by the cholera toxin. Because such genes offer a survival advantage within this population, they have slowly spread through the Bangladeshi population over the past 30,000 years. In other words, the Bangladeshi population is evolving increased resistance to cholera.

In addition to providing insight into the recent evolutionary past, data from this study reveal potential ways that humans might thwart the cholera bacterium. Perhaps pharmaceutical companies can exploit the proteins produced by the identified mutations to create a new generation of antibiotics. If so, this will represent another way that biologists have applied lessons learned from our understanding of evolution to improve human health. It also reminds us that we humans, like all life on Earth, are shaped by evolution due to changes in our environment, including the presence of infectious microorganisms that live around us.

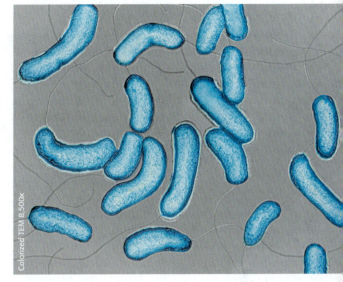

▼ Figure 4.22 *Vibrio cholerae,* the cause of the deadly disease cholera. Some people living in Bangladesh have evolved resistance to this bacterium.

Colorized TEM 8,500×

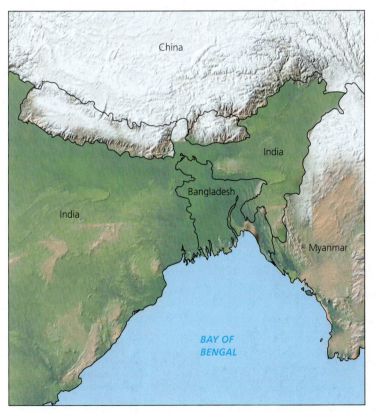

China

India

Bangladesh

India

Myanmar

BAY OF BENGAL

Chapter Review

SUMMARY OF KEY CONCEPTS

The Microscopic World of Cells

The Two Major Categories of Cells

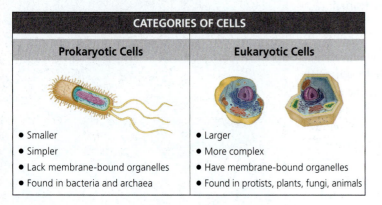

CATEGORIES OF CELLS	
Prokaryotic Cells	**Eukaryotic Cells**
• Smaller	• Larger
• Simpler	• More complex
• Lack membrane-bound organelles	• Have membrane-bound organelles
• Found in bacteria and archaea	• Found in protists, plants, fungi, animals

An Overview of Eukaryotic Cells

Membranes partition eukaryotic cells into functional compartments. The largest organelle is usually the nucleus. Other organelles are located in the cytoplasm, the region outside the nucleus and within the plasma membrane.

Membrane Structure

The Plasma Membrane

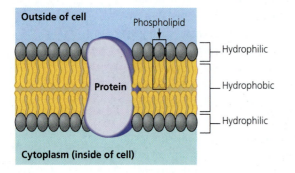

Cell Surfaces

The walls that encase plant cells support plants against the pull of gravity and also prevent cells from absorbing too much water. Animal cells are coated by a sticky extracellular matrix.

The Nucleus and Ribosomes: Genetic Control of the Cell

The Nucleus

An envelope consisting of two membranes encloses the nucleus. Within the nucleus, DNA and proteins make up chromatin fibers; each very long fiber is a single chromosome. The nucleus also contains the nucleolus, which produces components of ribosomes.

Ribosomes

Ribosomes produce proteins in the cytoplasm using messages produced by the DNA.

How DNA Directs Protein Production

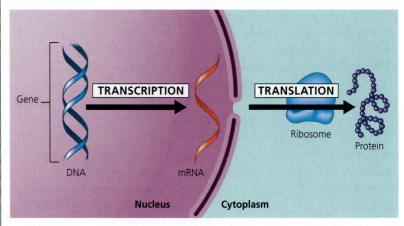

The Endomembrane System: Manufacturing and Distributing Cellular Products

The Endoplasmic Reticulum

The ER consists of membrane-enclosed tubes and sacs within the cytoplasm. Rough ER, named because of the ribosomes attached to its surface, makes membrane and secretory proteins. The functions of smooth ER include lipid synthesis and detoxification.

The Golgi Apparatus

The Golgi apparatus refines certain ER products and packages them in transport vesicles targeted for other organelles or export from the cell.

Lysosomes

Lysosomes, sacs containing digestive enzymes, aid digestion and recycling within the cell.

Vacuoles

Vacuoles include the contractile vacuoles that expel water from certain freshwater protists and the large, multifunctional central vacuoles of plant cells.

Chloroplasts and Mitochondria: Providing Cellular Energy

Chloroplasts and Mitochondria

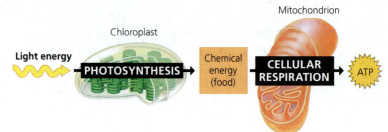

The Cytoskeleton: Cell Shape and Movement

Maintaining Cell Shape

Microtubules are an important component of the cytoskeleton, an organelle that gives support to, and maintains the shape of, cells.

Cilia and Flagella

Cilia and eukaryotic flagella are appendages that aid in movement, and they are made primarily of microtubules. Cilia are short and numerous and move the cell by coordinated beating. Flagella are long, often occur singly, and propel a cell with whiplike movements.

Mastering Biology

For practice quizzes, BioFlix animations, MP3 tutorials, video tutors, and more study tools designed for this textbook, go to Mastering Biology™

SELF-QUIZ

1. You look into a microscope and view an unknown cell. What might you see that would tell you whether the cell is prokaryotic or eukaryotic?
   a. a rigid cell wall
   b. a nucleus
   c. a plasma membrane
   d. ribosomes

2. Explain how each word in the term *fluid mosaic* describes the structure of a membrane.

3. Identify which of the following structures includes all the others in the list: rough ER, smooth ER, endomembrane system, the Golgi apparatus.

4. The ER has two distinct regions that differ in structure and function. Lipids are synthesized within the _____, and proteins are synthesized within the _____.

5. Why do cell walls make good targets for antibiotic drugs?

6. Name two similarities in the structure or function of chloroplasts and mitochondria. Name two differences.

7. Match the following organelles with their functions:

   a. ribosomes   1. movement
   b. microtubules   2. photosynthesis
   c. mitochondria   3. protein synthesis
   d. chloroplasts   4. digestion
   e. lysosomes   5. cellular respiration

8. DNA controls the cell by transmitting genetic messages that result in protein production. Place the following organelles in the order that represents the flow of genetic information from the DNA through the cell: nuclear pores, ribosomes, nucleus, rough ER, Golgi apparatus.

9. Compare and contrast cilia and flagella.

For answers to the Self Quiz, see Appendix D.

IDENTIFYING MAJOR THEMES

For each statement, identify which major theme is evident (the relationship of structure to function, information flow, pathways that transform energy and matter, interactions within biological systems, or evolution) and explain how the statement relates to the theme. If necessary, review the themes (see Chapter 1) and review the examples highlighted in blue in this chapter.

10. The genetic message contained in DNA is used to build proteins.

11. Several different organelles work together to carry out instructions in DNA.

12. Sunlight can be used to drive the photosynthesis of sugars.

For answers to Identifying Major Themes, see Appendix D.

THE PROCESS OF SCIENCE

13. Plant seeds store oils as droplets. An oil droplet membrane is a single layer of phospholipids. Draw a model for such a membrane. Explain why it is more stable than a bilayer.

14. Imagine that you are a pediatrician and one of your patients is a newborn who may have a lysosomal storage disease. You remove some cells from the patient and examine them under the microscope. What would you expect to see? Design a series of tests that could reveal whether the patient is indeed suffering from a lysosomal storage disease.

15. **Interpreting Data** A population of bacteria may evolve resistance to a drug over time. Draw a graph that represents such a change. Label the *x*-axis as "time" and the *y*-axis as "population size." Draw a line on the graph that represents how the size of the population might change over time after the introduction of a new antibiotic. Label the point on your line where the new drug was introduced, and then indicate how the population size might change over time after that introduction.

BIOLOGY AND SOCIETY

16. Doctors at a university medical center removed John Moore's spleen, which is a standard treatment for his type of leukemia. The disease did not recur. Researchers kept the spleen cells alive in a nutrient medium. They found that some cells produced a blood protein that showed promise as a treatment for cancer and AIDS. The researchers patented the cells. Moore sued, claiming a share in profits from any products derived from his cells. The U.S. Supreme Court ruled against Moore, stating that his lawsuit "threatens to destroy the economic incentive to conduct important medical research." Moore argued that the ruling left patients "vulnerable to exploitation at the hands of the state." Do you think Moore was treated fairly? What else would you like to know about this case that might help you decide?

17. Scientists can manipulate living cells, changing their genetic composition and the way they function. Some companies have sought to patent engineered cell lines. Should society allow cells to be patented? Should the same patent rules apply to human and bacterial cells?

5 The Working Cell

CHAPTER CONTENTS
Some Basic Energy Concepts 76
ATP and Cellular Work 79
Enzymes 80
Membrane Function 83

WHAT DO BEEF JERKY AND RAISINS HAVE IN COMMON? THEY ARE BOTH MADE THROUGH OSMOSIS.

Why Cellular Functions Matter

Life begins at the level of the cell: Nothing smaller is alive, and a single cell may contain all of the machinery necessary to carry out life's functions. Living cells are dynamic, performing work constantly. It is therefore important for any student of biology to understand how a working cell goes about the business of life.

BOTH NERVE GAS AND INSECTICIDES WORK BY CRIPPLING A VITAL ENZYME.

GET MOVING! YOU HAVE TO WALK MORE THAN 2 HOURS TO BURN THE CALORIES IN HALF A PEPPERONI PIZZA.

BIOLOGY AND SOCIETY   Harnessing Cellular Structures 75

THE PROCESS OF SCIENCE   Can Enzymes Be Engineered? 81

EVOLUTION CONNECTION   The Origin of Membranes 87

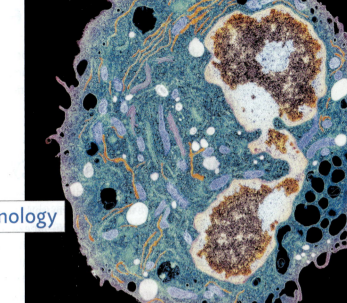

Colorized TEM 4,800×

Cellular structures. Even the smallest cell, such as this human white blood cell, is a miniature machine of startling complexity.

BIOLOGY AND SOCIETY   Nanotechnology

Harnessing Cellular Structures

Imagine a tiny robot with balls of carbon atoms for wheels, or a microchip carved onto an object 1,000 times smaller than a grain of sand. These are real-world examples of nano-technology, the manipulation of materials at the molecular scale. When designing devices of such small size, researchers often turn to living cells for inspiration. After all, you can think of a cell as a machine that continuously and efficiently performs a variety of functions, such as movement, energy processing, and manufacturing a wide variety of products. Let's consider one example of cell-based nanotechnology and see how it relates to working cells.

Researchers at Cornell University are attempting to harvest the energy-producing capability of human sperm cells. Like other cells, a sperm cell generates energy by breaking down sugars and other molecules that pass through its plasma membrane. Enzymes within the cell release energy by breaking down glucose. The released energy is used to produce molecules of ATP. Within a living sperm, such ATP provides the energy that propels the cell through the female reproductive tract. In an attempt to harness this energy-producing system, the Cornell researchers attached three enzymes to a computer chip. The enzymes continued to function in this artificial system, producing energy from sugar. The hope is that a larger set of enzymes can eventually be used to power micro-scopic robots. Such nanorobots could use glucose from the bloodstream to power the delivery of drugs to body tissues, perhaps homing in on cancerous or otherwise abnormal cells. This example is only a glimpse into the incredible potential of new technologies inspired by working cells.

In this chapter, we'll explore three processes common to all living cells: energy metabolism, the use of enzymes to speed chemical reactions, and transport regulation by the plasma mem-brane. Along the way, we'll further consider nanotechnologies that mimic the natural activities of living cells.

melt if the car's radiator did not disperse heat into the atmosphere. That is why high-end performance cars need sophisticated air flow systems to avoid overheating.

Cells also use oxygen in reactions that release energy from fuel molecules. As in a car engine, the "exhaust" from such reactions in cells is mostly carbon dioxide and water. The combustion of fuel in cells is called cellular respiration, which is a more gradual and efficient "burning" of fuel compared with the explosive combustion in an automobile engine. Cellular respiration is the energy-releasing chemical breakdown of fuel molecules and the storage of that energy in a form the cell can use to perform work. (We will discuss the details of cellular respiration in Chapter 6.) You convert about 34% of your food energy to useful work, such as movement of your muscles. The rest of the energy released by the breakdown of fuel molecules generates body heat. Humans and many other animals can use this heat to keep the body at an almost constant temperature (37°C, or 98.6°F, in the case of humans), even when the surrounding air is much colder. You've probably noticed how quickly a crowded room warms up—it's all that released metabolic heat energy! The liberation of heat energy also explains why you feel hot after exercise. Sweating and other cooling mechanisms enable your body to lose the excess heat, much as a car's radiator keeps the engine from overheating.

YOU HAVE TO WALK MORE THAN 2 HOURS TO BURN THE CALORIES IN HALF A PEPPERONI PIZZA.

✅ CHECKPOINT

According to Figure 5.3, about how long would you have to ride your bicycle to burn off the energy in one slice of pizza?

■ *Answer: about 1 hour (riding a bicycle burns about 150 calories each half hour, and a slice of pizza is about 300 calories)*

Food Calories

Read any packaged food label and you'll find the number of calories in each serving of that food. Calories are units of energy. A **calorie** (cal) is the amount of energy that can raise the temperature of 1 gram (g) of water by 1°C. You could actually measure the caloric content of a peanut by burning it under a container of water to convert all of the stored chemical energy to heat and then measuring the temperature increase of the water.

Calories are tiny units of energy, so using them to describe the fuel content of foods is not practical. Instead, it's conventional to use kilocalories (kcal), units of 1,000 calories. In fact, the Calories (capital C) on a food package are actually kilocalories. For example, one peanut has about 5 Calories. That's a lot of energy, enough to increase the temperature of 1 kg (a little more than a quart) of water by 5°C. And just a handful of peanuts contains enough Calories, if converted to heat, to boil 1 kg of water. In living organisms, of course, food isn't used to boil water but instead is used to fuel the activities of life. **Figure 5.3** shows the number of Calories in several foods and how many Calories are burned by some typical activities. ✅

▼ **Figure 5.3** A comparison of the Calories contained in some common foods and burned by some common activities.

ATP and Cellular Work

The carbohydrates, fats, and other fuel molecules we obtain from food cannot be used directly as fuel for our cells. Instead, the chemical energy released by the break-down of organic molecules during cellular respiration is used to generate molecules of ATP. These molecules of ATP then power cellular work. ATP acts like an energy shuttle, storing energy obtained from food and then releasing it as needed at a later time. **Such energy transformations are essential for all life on Earth.**

The Structure of ATP

The abbreviation ATP stands for adenosine triphosphate. **ATP** consists of an organic molecule called adenosine plus a tail of three phosphate groups **(Figure 5.4)**. The triphosphate tail is the "business" end of ATP, the part that provides energy for cellular work. Each phosphate group is negatively charged. Negative charges repel each other. The crowding of negative charges in the triphosphate tail contributes to the potential energy of ATP. It's analogous to storing energy by compressing a spring; if you release the spring, it will relax, and you can use that springiness to do some useful work. For ATP power, it is release of the phosphate at the tip of the triphosphate tail that makes energy available to working cells. What remains is **ADP**, adenosine diphosphate (two phosphate groups instead of three, shown on the right side of Figure 5.4).

Phosphate Transfer

When ATP drives work in cells by being converted to ADP, the released phosphate groups don't just fly off into space. ATP energizes other molecules in cells by transferring phosphate groups to those molecules. When a target molecule accepts the third phosphate group, it becomes energized and can then perform work in the cell. Imagine a bicyclist pedaling up a hill. In the muscle cells of the rider's legs, ATP transfers phosphate groups to motor proteins. The proteins then change shape, causing the muscle cells to contract **(Figure 5.5a)**. This contraction provides the mechanical energy needed to propel the rider. ATP also enables the transport of ions and other dissolved substances across the membranes of the rider's nerve cells **(Figure 5.5b)**, helping them send signals to her legs. And ATP drives the production of a cell's large molecules from smaller molecular building blocks **(Figure 5.5c)**.

▼ **Figure 5.4 ATP power.** Each ⓟ in the triphosphate tail of ATP represents a phosphate group, a phosphorus atom bonded to oxygen atoms. The transfer of a phosphate from the triphosphate tail to other molecules provides energy for cellular work.

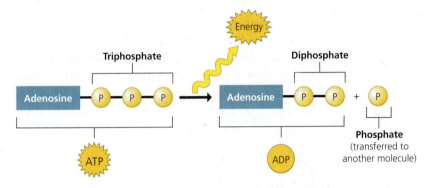

▼ **Figure 5.5 How ATP drives cellular work.** Each type of work shown here is powered when an enzyme transfers phosphate from ATP to a recipient molecule.

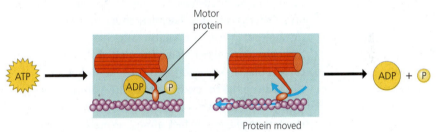

(a) **Motor protein performing mechanical work (moving a muscle fiber)**

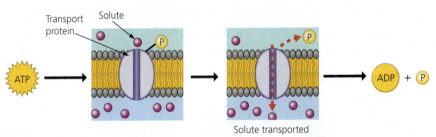

(b) **Transport protein performing transport work (importing a solute)**

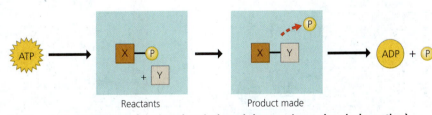

(c) **Chemical reactants performing chemical work (promoting a chemical reaction)**

✓ **CHECKPOINT**

1. Explain how ATP powers cellular work.
2. What is the source of energy for regenerating ATP from ADP?

■ *Answers:* 1. *ATP transfers a phosphate to another molecule, increasing that molecule's energy.* 2. *chemical energy harvested from sugars and other organic fuels via cellular respiration.*

The ATP Cycle

Your cells spend ATP continuously. Fortunately, it is a renewable resource. ATP can be restored by adding a phosphate group back to ADP. That takes energy, like recompressing a spring. And that's where food enters the picture. The chemical energy that cellular respiration harvests from sugars and other organic fuels is put to work regenerating a cell's supply of ATP. Cellular work spends ATP, which is recycled when ADP and phosphate are combined using energy released by cellular respiration **(Figure 5.6)**. Thus, energy from processes that yield energy, such as the breakdown of organic fuels, is transferred to processes that consume energy,

▼ **Figure 5.6** The ATP cycle.

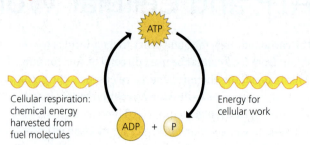

Cellular respiration: chemical energy harvested from fuel molecules

Energy for cellular work

such as muscle contraction and other cellular work. The ATP cycle can run at an astonishing pace: Up to 10 million ATPs are consumed and recycled each second in a working muscle cell. ✓

Enzymes

A living organism contains a vast collection of chemicals, and countless chemical reactions constantly change the organism's molecular makeup. In a sense, a living organism is a complex "chemical square dance," with the molecular "dancers" continually changing partners through chemical reactions. The total of all the chemical reactions in an organism is called **metabolism**. **Illustrating the theme of system interactions, a cell's metabolism depends on the coordination of many molecular players.** In fact, almost no metabolic reactions occur without help. Most require the assistance of **enzymes**, molecules that speed up chemical reactions without being consumed by those reactions. All living cells contain thousands of different enzymes, each promoting a different chemical reaction. Almost all enzymes are proteins, but some RNA molecules also function as enzymes.

✓ **CHECKPOINT**

How does an enzyme affect the activation energy of a chemical reaction?

■ *Answer: An enzyme lowers the activation energy.*

Activation Energy

For a chemical reaction to begin, chemical bonds in the reactant molecules must be broken. (The first step in swapping partners during a square dance is to let go of your current partner's hand.) This process requires that the molecules absorb energy from their surroundings. In other words, for most chemical reactions, a cell has to spend a little energy first. You can easily relate this concept to your own life: It takes effort to clean your room, but this will save you more energy in the long run because you won't have to hunt for your belongings. The energy that must be invested to start a reaction is called **activation energy** because it activates the reactants and triggers the chemical reaction.

Enzymes enable metabolism to occur by reducing the amount of activation energy required to break the bonds of reactant molecules. Without an enzyme, the activation energy barrier might never be breached. For example, lactose (milk sugar) can sit for years without breaking down into its components. But if you add a small amount of the enzyme lactase, all the lactose will be broken down in seconds. If you think of the activation energy as a barrier to a chemical reaction, an enzyme's function is to lower that barrier **(Figure 5.7)**. It does so by binding to reactant molecules and putting them under physical or chemical stress, making it easier to break their bonds and start a reaction. In our analogy of cleaning your room, this is like a friend offering to help you. You start and end in the same place whether solo or assisted, but your friend's help lowers your activation energy, making it more likely that you'll proceed. Next, we'll return to our theme of nanotechnology to see how enzymes can be engineered to be even more efficient. ✓

▼ **Figure 5.7** Enzymes and activation energy.

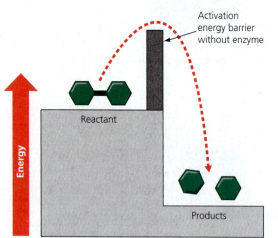

(a) Without enzyme. A reactant molecule must overcome the activation energy barrier before a chemical reaction can break the molecule into products.

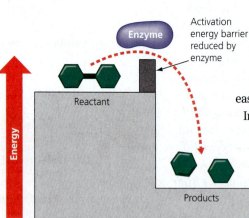

(b) With enzyme. An enzyme speeds the chemical reaction by lowering the activation energy barrier.

Can Enzymes Be Engineered?

BACKGROUND

Like all other proteins, enzymes are encoded by genes. Observations of genetic sequences suggest that many of our genes were formed through a type of molecular evolution: One ancestral gene duplicated, and the two copies diverged over time through random genetic changes, eventually becoming distinct genes.

METHOD

Scientists often use the natural world as inspiration for new experiments. For example, in 2015, a group of researchers from the University of British Columbia used a technique called directed evolution to produce a better version of a particular enzyme. This enzyme cuts the molecular tags that appear on the outside of red blood cells; these tags are responsible for the different blood types (A, B, AB, and O). If a patient receives a transfusion of blood that contains tags that are incompatible with the patient's blood, a potentially deadly rejection could result. The researchers hoped to alleviate such problems by producing an enzyme that efficiently removes the problematic tags from donated blood, thereby making all blood donations safe.

In their attempt to mimic the natural process of evolution, the researchers randomly mutated many copies of the gene that codes for the enzyme **(Figure 5.8)**. Each mutated gene was screened for its ability to code for an enzyme that removes the molecular tags from red blood cells. Any genes that were not efficient were removed from the experimental pool. The genes for the enzymes that were most effective were kept, then subjected to several more rounds of duplication, mutation, and screening.

RESULTS

After many rounds of directed evolution, the researchers isolated a gene that codes for an enzyme that is 170 times more efficient at promoting the reaction than the original enzyme. Testing of this new enzyme is now underway. These results are another example of how scientists can mimic natural processes to modify cellular components into useful medical tools.

Thinking Like a Scientist

In what ways does directed evolution mimic natural selection? In what ways does it differ?

For the answer, see Appendix D.

▶ **Figure 5.8 Directed evolution of an enzyme.** By mutating the gene for glycoside hydrolase, which can remove compatibility proteins from the outside of red blood cells, researchers isolated a gene that codes for a much more effective enzyme.

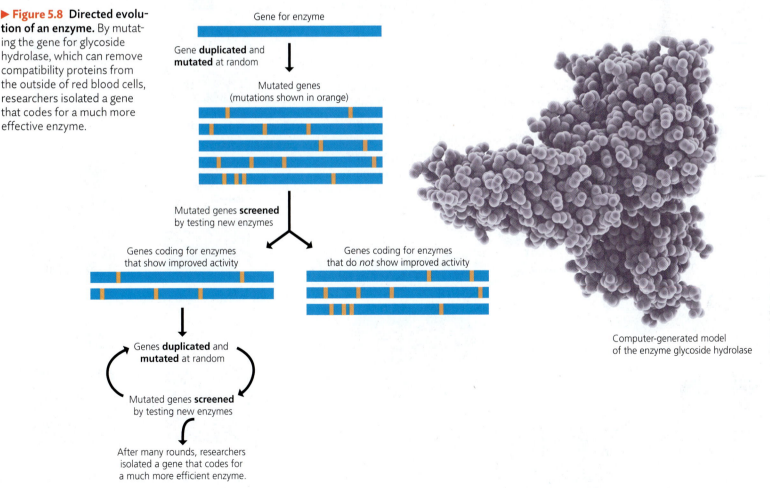

Gene for enzyme

Gene **duplicated** and **mutated** at random

Mutated genes
(mutations shown in orange)

Mutated genes **screened** by testing new enzymes

Genes coding for enzymes that show improved activity

Genes coding for enzymes that do *not* show improved activity

Genes **duplicated** and **mutated** at random

Mutated genes **screened** by testing new enzymes

After many rounds, researchers isolated a gene that codes for a much more efficient enzyme.

Computer-generated model of the enzyme glycoside hydrolase

Enzyme Activity

An enzyme is very selective in the reaction it catalyzes. The specific molecule that an enzyme acts on is called the enzyme's **substrate**. Enzymes illustrate the close relationship between structure and function. Each kind of enzyme has a unique three-dimensional shape that determines what specific chemical reaction the enzyme promotes. A region of the enzyme called the **active site** has a shape and chemistry that fit the substrate molecule. The active site is typically a pocket or groove on the surface of the enzyme. When a substrate slips into this docking station, the active site changes shape slightly to embrace the substrate and catalyze the reaction. This interaction is called **induced fit** because the entry of the substrate induces the enzyme to change shape slightly, making the fit between substrate and active site snugger. Think of a handshake: As your hand makes contact with another hand, it changes shape slightly to make a better fit.

After the products are released from the active site, the enzyme can accept another molecule of substrate. In fact, the ability to function repeatedly is a key characteristic of enzymes. **Figure 5.9** follows the action of the enzyme lactase, which breaks down the disaccharide lactose (the substrate). This enzyme is underproduced or defective in lactose-intolerant people. Like lactase, many enzymes are named for their substrates, with an *-ase* ending. ✓

Enzyme Inhibitors

Certain molecules called **enzyme inhibitors** can inhibit a metabolic reaction by binding to an enzyme and disrupting its function **(Figure 5.10)**. Some of these inhibitors are substrate imposters that plug up the active site. (You can't shake a person's hand if someone else puts a banana in it first!) Other inhibitors bind to the enzyme at a site remote from the active site, but the binding changes the enzyme's shape. (Imagine trying to shake hands when someone is tickling your ribs, causing you to clench your hand.) In each case, an inhibitor disrupts the enzyme by altering its shape.

In some cases, the binding of an inhibitor is reversible. For example, if a cell is producing more of a certain product than it needs, that product may reversibly inhibit an enzyme required for its production. This feedback regulation keeps the cell from wasting resources by building an unneeded product.

Many beneficial drugs work by inhibiting enzymes. Penicillin blocks the active site of an enzyme that bacteria use in making cell walls. Ibuprofen inhibits an enzyme involved in sending pain signals. Many cancer drugs inhibit enzymes that promote cell division. Many toxins and poisons also work as inhibitors. Nerve gases (a form of chemical warfare) irreversibly bind to the active site of an enzyme vital to transmitting nerve impulses, leading to rapid paralysis and death. Many pesticides are toxic to insects because they inhibit this same enzyme.

BOTH NERVE GAS AND INSECTICIDES WORK BY CRIPPLING A VITAL ENZYME.

☑ **CHECKPOINT**

How does an enzyme recognize its substrate?

■ *Answer: The substrate and the enzyme's active site are complementary in shape and chemistry.*

▼ **Figure 5.9 How an enzyme works.** Our example is the enzyme lactase, named for its substrate, lactose.

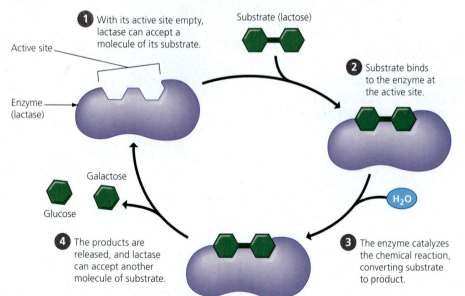

1 With its active site empty, lactase can accept a molecule of its substrate.

Active site

Enzyme (lactase)

Substrate (lactose)

2 Substrate binds to the enzyme at the active site.

Galactose

Glucose

H_2O

3 The enzyme catalyzes the chemical reaction, converting substrate to product.

4 The products are released, and lactase can accept another molecule of substrate.

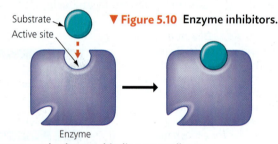

Substrate
Active site

▼ **Figure 5.10 Enzyme inhibitors.**

Enzyme

(a) Enzyme and substrate binding normally

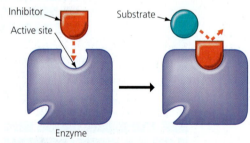

Inhibitor
Active site

Substrate

Enzyme

(b) Enzyme inhibition by a substrate imposter

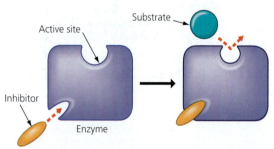

Active site

Substrate

Inhibitor

Enzyme

(c) Inhibition of an enzyme by a molecule that causes the active site to change shape

Membrane Function

So far, we have discussed how cells control the flow of energy and how enzymes affect the pace of chemical reactions. In addition to these vital processes, cells must also regulate the flow of materials to and from the environment. The plasma membrane consists of a double layer of fat with embedded proteins—a phospholipid bilayer (see Figure 4.5). **Figure 5.11** describes the major functions of these membrane proteins. Of all the functions shown in the figure, one of the most important is the regulation of transport in and out of the cell. A steady traffic of small molecules moves across a cell's plasma membrane in both directions. But this traffic flow is never willy-nilly. Instead, all biological membranes are selectively permeable—that is, they only allow certain molecules to pass. Let's explore this in more detail.

Passive Transport: Diffusion across Membranes

Molecules are restless. They constantly vibrate and wander randomly. One result of this motion is **diffusion**, the movement of molecules spreading out evenly into the available space. Each molecule moves randomly, but the overall diffusion of a population of molecules is usually directional, from a region where the molecules are more concentrated to a region where they are less concentrated. For example, imagine many molecules of perfume inside a bottle. If you remove the bottle top, every molecule of perfume will move randomly about, but the overall movement will be out of the bottle, and the whole room will eventually smell of the perfume. You could, with great effort, return the perfume molecules to its bottle, but the molecules would never all return spontaneously.

▼ **Figure 5.11 Primary functions of membrane proteins.** An actual cell may have just a few of the types of proteins shown here, and many copies of each particular protein may be present.

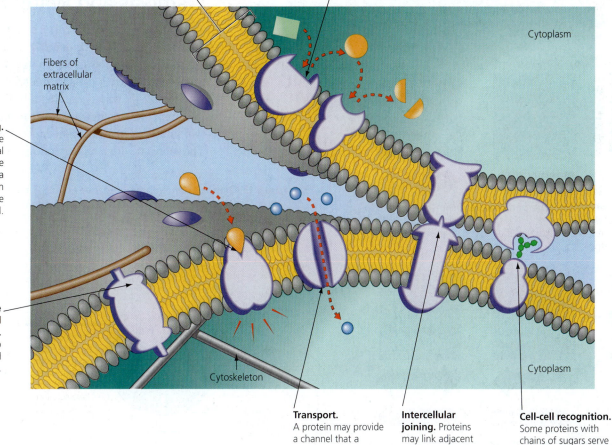

Phospholipid bilayer

Enzymatic activity. This protein and the one next to it are enzymes, having an active site that fits a substrate. Enzymes may form an assembly line that carries out steps of a pathway.

Cytoplasm

Fibers of extracellular matrix

Cell signaling. A binding site fits the shape of a chemical messenger. The messenger may cause a change in the protein that relays the message to the inside of the cell.

Attachment to the cytoskeleton and extracellular matrix. Such proteins help maintain cell shape and coordinate changes.

Cytoskeleton

Cytoplasm

Transport. A protein may provide a channel that a chemical substance can pass through.

Intercellular joining. Proteins may link adjacent cells.

Cell-cell recognition. Some proteins with chains of sugars serve as identification tags recognized by other cells.

For an example closer to a living cell, imagine a membrane separating pure water from a mixture of dye dissolved in water (Figure 5.12). Assume that this membrane has tiny holes that allow dye molecules to pass. Although each dye molecule moves randomly, there will be a net migration across the membrane to the side that began as pure water. Movement of the dye will continue until both solutions have equal concentrations. After that, there will be a dynamic equilibrium: Molecules will still be moving, but at that point as many dye molecules move in one direction as in the other, so the concentration of dye molecules in water remains steady.

Diffusion of dye across a membrane is an example of **passive transport**—*passive* because no energy is needed for the diffusion to happen. In passive transport, a substance diffuses down its **concentration gradient**, from where the substance is more concentrated to where it is less concentrated. In our lungs, for example, there is more oxygen gas (O_2) in the air than in the blood. Therefore, oxygen moves by passive transport from the air into the bloodstream. But remember that the cell membrane is selectively permeable. For example, small molecules such as oxygen (O_2) generally pass through more readily than larger molecules such as amino acids. But the membrane

is relatively impermeable to even some very small substances, such as most ions, which are too hydrophilic to pass through the fatty phospholipid bilayer.

Substances that do not cross membranes spontaneously—or otherwise cross very slowly—can be transported by proteins that act as corridors for specific molecules (see Figure 5.11). This assisted transport is called **facilitated diffusion**. For example, water molecules can move through the plasma membrane of some cells by way of transport proteins—each of which can help 3 billion water molecules per second pass through! People with a rare mutation in the gene that encodes these water-transport proteins have defective kidneys that cannot reabsorb water; such people must drink 20 liters of water every day to prevent dehydration. On the flip side, a common complication of pregnancy is fluid retention, the culprit responsible for swollen ankles and feet, often caused by increased synthesis of water channel proteins. Other specific transport proteins move glucose across cell membranes 50,000 times faster than diffusion. Even at this rate, facilitated diffusion is a type of passive transport because it does not require the cell to expend energy. As in all passive transport, the driving force is the concentration gradient. ✓

☑ CHECKPOINT

1. What does it mean to say that molecules move "down the concentration gradient"?

2. Why is facilitated diffusion a form of passive transport?

■ Answer: 1. The molecule moves from where it is more concentrated to where it is less concentrated. 2. It uses proteins to transport materials down a concentration gradient without expending energy.

▼ **Figure 5.12 Passive transport: diffusion across a membrane.**
A substance will diffuse from where it is more concentrated to where it is less concentrated. Put another way, a substance tends to diffuse down its concentration gradient.

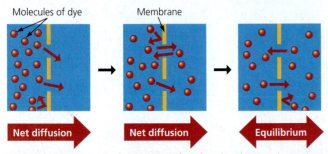

Molecules of dye   Membrane

Net diffusion   Net diffusion   Equilibrium

(a) Passive transport of one type of molecule. The membrane is permeable to these dye molecules, which diffuse down the concentration gradient. At equilibrium, the molecules are still restless, but the rate of transport is equal in both directions.

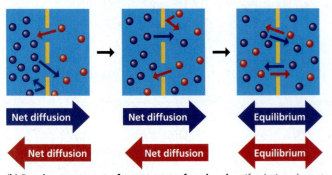

Net diffusion   Net diffusion   Equilibrium

Net diffusion   Net diffusion   Equilibrium

(b) Passive transport of two types of molecules. If solutions have two or more solutes, each will diffuse down its own concentration gradient.

Osmosis and Water Balance

The diffusion of water across a selectively permeable membrane is called **osmosis (Figure 5.13)**. A **solute** is a substance that is dissolved in a liquid solvent, and the resulting mixture is called a solution. For example, a solution of salt water contains salt (the solute) dissolved in water (the solvent). Imagine a membrane separating two solutions with different concentrations of a solute. The solution with a higher concentration of solute is

▼ **Figure 5.13 Osmosis.** A membrane separates two solutions with different sugar concentrations. Water molecules can pass through the membrane, but the sugar molecules cannot.

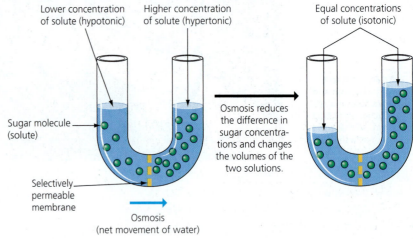

Lower concentration of solute (hypotonic)

Higher concentration of solute (hypertonic)

Equal concentrations of solute (isotonic)

Sugar molecule (solute)

Osmosis reduces the difference in sugar concentrations and changes the volumes of the two solutions.

Selectively permeable membrane

Osmosis (net movement of water)

said to be **hypertonic** to the other solution. The solution with the lower solute concentration is said to be **hypotonic** to the other. Note that the hypotonic solution, by having the lower solute concentration, has the higher water concentration (less solute = more water). Therefore, water will diffuse across the membrane along its concentration gradient from an area of higher water concentration (hypotonic solution) to one of lower water concentration (hypertonic solution). This reduces the difference in solute concentrations and changes the volumes of the two solutions.

People can take advantage of osmosis to preserve foods. Salt is often applied to meats—like beef and fish—to cure them into jerky; the salt causes water to move out of food-spoiling bacteria and fungi. Food can also be preserved in honey because a high sugar concentration draws water out of food.

When the solute concentrations are the same on both sides of a membrane, water molecules will move at the same rate in both directions, so there will be no net change in solute concentration. Solutions of equal solute concentration are said to be **isotonic**. For example, many marine animals, such as sea stars and crabs, are isotonic to seawater, so that overall they neither gain nor lose water from the environment. In your body, red blood cells are isotonic to the bloodstream in which they flow.

Water Balance in Animal Cells

The survival of a cell depends on its ability to balance water uptake and loss. When an animal cell is immersed in an isotonic solution, the cell's volume remains constant because the cell gains water at the same rate that it loses water (**Figure 5.14a**, top). But what happens if an animal cell is in contact with a hypotonic solution, which has a lower solute concentration than the cell? Due to osmosis, the cell would gain water, swell, and possibly burst (lyse) like an overfilled water balloon (**Figure 5.14b**, top). A hypertonic environment is also harsh on an animal cell; the cell shrivels from water loss (**Figure 5.14c**, top).

For an animal to survive a hypotonic or hypertonic environment, the animal must have a way to balance the uptake and loss of water. The control of water balance is called **osmoregulation**. For example, a freshwater fish has kidneys and gills that work constantly to prevent an excessive buildup of water in the body. Humans can suffer consequences of osmoregulation failure. Dehydration (consumption of too little water) can cause fatigue and even death. Drinking too much water—called hyponatremia, or "water intoxication"—can also cause death by overdiluting necessary ions.

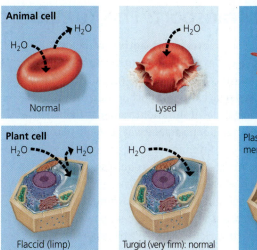

◀ **Figure 5.14 Osmotic environments.** Animal cells (such as a red blood cell) and plant cells behave differently in different osmotic environments.

Animal cell

Normal — Lysed — Shriveled

Plant cell

Plasma membrane

Flaccid (limp) — Turgid (very firm): normal — Shriveled

(a) **Isotonic solution** (b) **Hypotonic solution** (c) **Hypertonic solution**

Water Balance in Plant Cells

Problems of water balance are somewhat different for cells that have rigid cell walls, such as those from plants, fungi, many prokaryotes, and some protists. A plant cell immersed in an isotonic solution is flaccid (floppy), and the plant wilts (Figure 5.14a, bottom). In contrast, a plant cell is turgid (very firm) and healthiest in a hypotonic environment, with a net inflow of water (Figure 5.14b, bottom). Although the elastic cell wall expands a bit, the back pressure it exerts prevents the cell from taking in too much water and bursting. Turgor is necessary for plants to retain their upright posture and the extended state of their leaves (**Figure 5.15**). However, in a hypertonic environment, a plant cell is no better off than an animal cell. As a plant cell loses water, it shrivels, and its plasma membrane pulls away from the cell wall (Figure 5.14c, bottom). This usually kills the cell. Thus, plant cells thrive in a hypotonic environment, whereas animal cells thrive in an isotonic one. ☑

BEEF JERKY AND RAISINS ARE BOTH MADE THROUGH OSMOSIS.

▼ **Figure 5.15 Plant turgor.** Watering a wilted plant will make it regain its turgor.

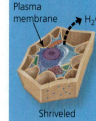

Wilted — Turgid

☑ CHECKPOINT

1. An animal cell shrivels when it is _____ compared with its environment.

2. What would happen if you placed one of your cells in pure water?

◾ Answers: **1.** hypotonic **2.** Water would rush in (because it is more concentrated outside the cell and less concentrated inside) and would burst the cell.

Active Transport: The Pumping of Molecules across Membranes

In contrast to passive transport, **active transport** requires that a cell expend energy to move molecules across a membrane. Cellular energy (usually provided by ATP) is used to drive a transport protein that pumps a solute *against* the concentration gradient—that is, in the direction that is opposite the way it would naturally flow **(Figure 5.16)**. Movement against a force, like rolling a boulder uphill against gravity, requires a considerable expenditure of energy. Consider this analogy: During a storm, water spontaneously flows downhill into a basement (passive transport) but requires a powered sump pump to move it back uphill (active transport).

Active transport allows cells to maintain internal concentrations of small solutes that differ from environmental concentrations. For example, compared with its surroundings, an animal nerve cell has a much higher concentration of potassium ions and a much lower concentration of sodium ions. The plasma membrane helps maintain these differences by pumping sodium out of the

▼ **Figure 5.16 Active transport.** Transport proteins are specific in their recognition of atoms or molecules. This transport protein (purple) has a binding site that accepts only a certain solute. Using energy from ATP, the protein pumps the solute against its concentration gradient.

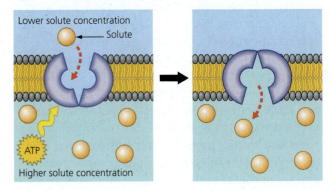

cell and potassium into the cell. This particular case of active transport (called the sodium-potassium pump) is vital to the nervous system of most animals. ☑

Exocytosis and Endocytosis: Traffic of Large Molecules

So far, we've focused on how water and small solutes enter and leave cells by moving through the plasma membrane. The story is different for large molecules such as proteins, which are much too big to fit through the membrane. Their traffic into and out of the cell depends on the ability of the cell to package large molecules inside sacs called vesicles. You have already seen an example of this: During protein production by the cell, secretory proteins exit the cell from transport vesicles that fuse with the plasma membrane, spilling the contents outside the cell (see Figures 4.13 and 4.17). That process is called **exocytosis (Figure 5.17)**. When you cry, for example, cells in your tear glands use exocytosis to export the salty tears. In your brain, the exocytosis of neurotransmitter chemicals such as dopamine helps neurons communicate.

In **endocytosis**, a cell takes in material by vesicles that bud inward **(Figure 5.18)**. For example, in a process called

phagocytosis ("cellular eating"), a cell engulfs a particle and packages it within a food vacuole. Other times, a cell "gulps" droplets of fluid into vesicles. Endocytosis can also be triggered by the binding of certain external molecules to specific receptor proteins built into the plasma membrane. This binding causes the local region of the membrane to form a vesicle that transports the specific substance into the cell. In human liver cells, this process is used to take up cholesterol from the blood. An inherited defect in the receptors on liver cells can lead to an inability to process cholesterol, which can lead to heart attacks at ages as young as 5. Cells of your immune system use endocytosis to engulf and destroy invading bacteria and viruses.

Because all cells have a plasma membrane, it is logical to infer that membranes first formed early in the evolution of life on Earth. In the final section of this chapter, we'll consider the evolution of membranes.

▼ **Figure 5.17 Exocytosis.**

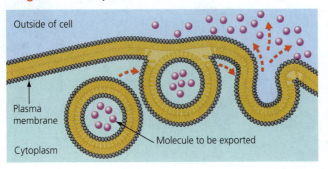

▼ **Figure 5.18 Endocytosis.**

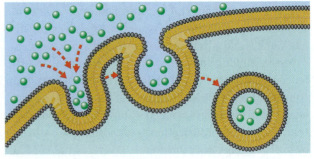

The Origin of Membranes

By simulating conditions found on the early Earth, scientists have been able to demonstrate that many of the molecules important to life can form spontaneously. (See Figure 15.3 and the accompanying text for a description of one such experiment.) Such results suggest that phospholipids, the key ingredients in all membranes, were probably among the first organic compounds that formed from chemical reactions on the early Earth. Once formed, they could self-assemble into simple membranes. When a mixture of phospholipids and water is shaken, for example, the phospholipids organize into bilayers, forming water-filled bubbles of membrane (Figure 5.19). This assembly requires neither genes nor other information beyond the properties of the phospholipids themselves.

The tendency of lipids in water to spontaneously form membranes has led biomedical engineers to produce liposomes (a type of artificial vesicle) that can encase specific chemicals. In the future, these engineered liposomes may be used to deliver nutrients or medications to specific sites within the body. In fact, over a dozen drugs have been approved for delivery by liposomes, including ones that target fungal infections, influenza, and hepatitis. Thus, membranes—like the other cellular components discussed in the Biology and Society and the Process of Science sections—have inspired novel nanotechnologies.

The formation of membrane-enclosed collections of molecules would have been a critical step in the evolution of the first cells. A membrane can enclose a solution that is different in composition from its surroundings. A plasma membrane that allows cells to regulate their chemical exchanges with the environment is a basic requirement for life. Indeed, all cells are enclosed by a plasma membrane that is similar in structure and function—illustrating the evolutionary unity of life.

▼ **Figure 5.19** The spontaneous formation of membranes: a key step in the origin of life.

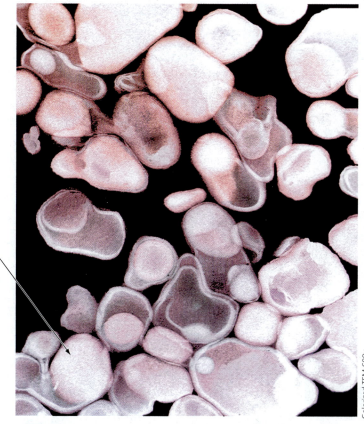

Water-filled bubble made of phospholipids

Colorized TEM 600×

Chapter Review

SUMMARY OF KEY CONCEPTS

Some Basic Energy Concepts

Conservation of Energy

Machines and organisms can transform kinetic energy (energy of motion) to potential energy (stored energy) and vice versa. In all such energy transformations, total energy is conserved. Energy cannot be created or destroyed.

Heat

Every energy transformation releases some randomized energy in the form of heat. Entropy is a measure of disorder, or randomness.

Chemical Energy

Molecules store varying amounts of potential energy in the arrangement of their atoms. Organic compounds are relatively rich in such chemical energy. The combustion of gasoline within a car's engine and the breakdown of glucose by cellular respiration within living cells are both examples of how the chemical energy stored in molecules can be converted to useful work.

Food Calories

Food Calories, actually kilocalories, are units used to measure the amount of energy in our foods and the amount of energy we expend in various activities.

ATP and Cellular Work

Your cells recycle ATP: As ATP is broken down to ADP to drive cellular work, new molecules of ATP are built from ADP using energy obtained from food.

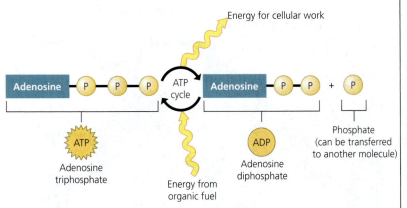

Enzymes

Activation Energy

Enzymes are biological catalysts that speed up metabolic reactions by lowering the activation energy required to break the bonds of reactant molecules.

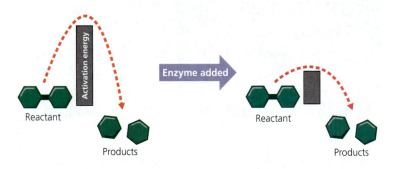

Enzyme Activity

The entry of a substrate into the active site of an enzyme causes the enzyme to change shape slightly, allowing for a better fit and thereby promoting the interaction of enzyme with substrate.

Enzyme Inhibitors

Enzyme inhibitors are molecules that can disrupt metabolic reactions by binding to enzymes, either at the active site or elsewhere.

Membrane Function

Proteins embedded in the plasma membrane perform a wide variety of functions, including regulating transport, anchoring to other cells or substances, promoting enzymatic reactions, and recognizing other cells.

Passive Transport, Osmosis, and Active Transport

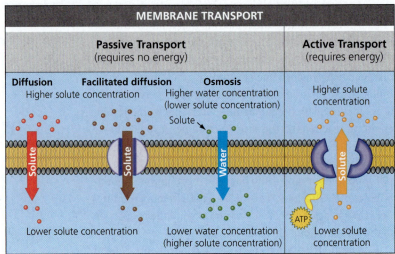

Most animal cells require an isotonic environment, with equal concentrations of water within and outside the cell. Plant cells need a hypotonic environment, which causes water to flow inward, keeping walled cells turgid.

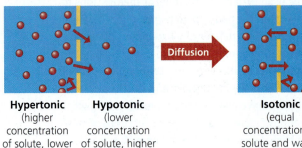

Exocytosis and Endocytosis: Traffic of Large Molecules

Exocytosis is the secretion of large molecules within vesicles. Endocytosis is the import of large substances by vesicles into the cell.

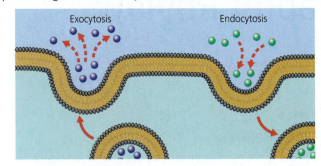

Mastering Biology

For practice quizzes, BioFlix animations, MP3 tutorials, video tutors, and more study tools designed for this textbook, go to Mastering Biology™

SELF-QUIZ

1. Describe the energy transformations that occur when you climb to the top of a stairway.

2. _____ is the capacity to perform work, while _____ is a measure of randomness.

3. The label on a candy bar says that it contains 150 Calories. If you could convert all of that energy to heat, you could raise the temperature of how much water by 15°C?

4. Why does removing a phosphate group from the triphosphate tail in a molecule of ATP release energy?

5. Your digestive system uses a variety of enzymes to break down large food molecules into smaller ones that your cells can assimilate. A generic name for a digestive enzyme is hydrolase. What is the chemical basis for that name? (*Hint:* Review Figure 3.4.)

6. Explain how an inhibitor can disrupt an enzyme's action without binding to the active site.

7. If someone at the other end of a room smokes a cigarette, you may breathe in some smoke. The movement of smoke is similar to what type of transport?
   a. osmosis
   b. diffusion
   c. facilitated diffusion
   d. active transport

8. Explain why it is not enough to say that a solution is "hypertonic."

9. What is the primary difference between passive and active transport in terms of concentration gradients?

10. Which of these types of cellular transport require(s) energy?
    a. facilitated diffusion
    b. active transport
    c. osmosis
    d. a and b

For answers to the Self Quiz, see Appendix D.

IDENTIFYING MAJOR THEMES

For each statement, identify which major theme is evident (the relationship of structure to function, information flow, pathways that transform energy and matter, interactions within biological systems, or evolution) and explain how the statement relates to the theme. If necessary, review the themes (see Chapter 1) and review the examples highlighted in blue in this chapter.

11. You turn the crank of a well, raising a bucket of water from below ground to the surface.

12. Your ability to walk depends upon the coordination of many different enzymes and other cellular structures.

13. Enzymes unravel and stop functioning if the environment gets too hot.

For answers to Identifying Major Themes, see Appendix D.

THE PROCESS OF SCIENCE

14. HIV, the virus that causes AIDS, depends on an enzyme called reverse transcriptase to multiply. Reverse transcriptase reads a molecule of RNA and creates a molecule of DNA from it. A molecule of AZT, an anti-AIDS drug, has a shape very similar to (but a bit different than) that of the DNA base thymine. Propose a model for how AZT inhibits HIV.

15. Gaining and losing weight are matters of caloric accounting: Calories in food minus Calories spent in activity. One pound of human body fat contains approximately 3,500 Calories. Using Figure 5.3, compare ways you could burn off those Calories. How far would you have to run, swim, or walk to burn 1 pound of fat, and how long would it take? Which method of burning Calories appeals the most to you? The least? How much of each food would you have to consume in order to gain a single pound? How does 1 pound's worth of food compare with 1 pound's worth of exercise? Does it seem like an even trade-off?

16. **Interpreting Data** The graph illustrates two chemical reactions. Which curve represents the reaction in the presence of the enzyme? What energy changes are represented by lines a, b, and c?

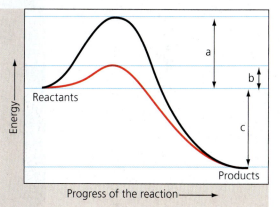

BIOLOGY AND SOCIETY

17. Obesity is a serious health problem for many Americans. Several popular diet plans advocate low-carbohydrate diets. Most low-carb dieters compensate by eating more protein and fat. What are the advantages and disadvantages of such a diet? Should the government regulate the claims of diet books? How should the claims be tested? Should diet proponents be required to obtain and publish data before making claims?

18. Nanotechnology devices can improve human health. But can they do harm or be abused? What regulations would allow for beneficial uses of nanorobots without allowing for harm?

19. Lead inhibits enzymes and can interfere with development of the nervous system. One battery manufacturer banned female employees of childbearing age from working in areas where they might be exposed to high levels of lead. The Supreme Court ruled the policy illegal. But many people are uncomfortable about the "right" to work in an unsafe environment. What rights and responsibilities of employers, employees, and government agencies are in conflict? What criteria should be used to decide who can work in a particular environment?

6 Cellular Respiration: Obtaining Energy from Food

CHAPTER CONTENTS

Energy Flow and Chemical Cycling in the Biosphere 92
Cellular Respiration: Aerobic Harvest of Food Energy 94
Fermentation: Anaerobic Harvest of Food Energy 101

THE METABOLIC PROCESSES THAT PRODUCE ACID IN YOUR MUSCLES AFTER A HARD WORKOUT ARE SIMILAR TO THE PROCESSES THAT PRODUCE PEPPERONI, SOY SAUCE, YOGURT, AND BREAD.

Why Cellular Respiration Matters

You can survive for weeks without eating and days without drinking, but you can only live for minutes without breathing. Why? Because every cell in your body relies on the energy created by using oxygen to break down glucose during the process of cellular respiration.

YOU HAVE SOMETHING IN COMMON WITH A SPORTS CAR: YOU BOTH REQUIRE AN AIR INTAKE SYSTEM TO BURN FUEL EFFICIENTLY.

THE CELLS OF YOUR BRAIN BURN THROUGH A QUARTER POUND OF GLUCOSE EACH DAY.

BIOLOGY AND SOCIETY   Getting the Most Out of Your Muscles 91

THE PROCESS OF SCIENCE   What Causes Muscle Burn? 102

EVOLUTION CONNECTION   The Importance of Oxygen 103

BIOLOGY AND SOCIETY   Exercise Science

Getting the Most Out of Your Muscles

Serious athletes train extensively to reach the peak of their physical potential. A key aspect of athletic conditioning involves increasing aerobic capacity, the ability of the heart and lungs to deliver oxygen to body cells. For many endurance athletes, such as long-distance runners or cyclists, the rate at which oxygen is provided to working muscles is the limiting factor in their performance.

Why is oxygen so important? Whether you are exercising or just going about your daily tasks, your muscles need a continuous supply of energy to perform work. Muscle cells obtain this energy from the sugar glucose through a series of chemical reactions that depend upon a constant input of oxygen (O_2). Therefore, to keep moving, your body needs a steady supply of O_2. When there is enough oxygen reaching your cells to support their energy needs, metabolism is said to be aerobic. As your muscles work harder, you breathe faster and more deeply to inhale O_2. If you continue to pick up the pace, you will approach your aerobic capacity, the maximum rate at which O_2 can be taken in and used by your muscle cells and therefore the most strenuous exercise that your body can maintain aerobically. Exercise scientists can use oxygen-monitoring equipment to precisely determine the maximum possible aerobic output for any given person. Such data allow a well-trained athlete to stay within aerobic limits, ensuring the maximum possible output—in other words, his or her best effort.

If you work even harder and exceed your aerobic capacity, the demand for oxygen in your muscles will outstrip your body's ability to deliver it; metabolism then becomes anaerobic. With insufficient O_2, your muscle cells switch to an "emergency mode" in which they break down glucose very inefficiently and produce lactic acid as a by-product. As lactic acid and other wastes accumulate, muscle activity is impaired. Your muscles can work under these conditions for only a few minutes before they give out. When this happens (sometimes called "hitting the wall"), your muscles cannot function, and you will likely collapse, unable to even stand.

Every living organism depends on processes that provide energy. In fact, we need energy to walk, talk, and think—in short, to stay alive. The human body has trillions of cells, all hard at work, all demanding fuel continuously. In this chapter, you'll learn how cells harvest food energy and put it to work with the help of oxygen. Along the way, we'll consider the implications of how the body responds to exercise.

The science of exercise. Endurance athletes must carefully monitor their efforts so that they maintain an aerobic pace over the long term.

Energy Flow and Chemical Cycling in the Biosphere

All life requires energy. In almost all ecosystems on Earth, this energy originates with the sun. During **photosynthesis**, the energy of sunlight is converted to the chemical energy of sugars and other organic molecules (as we'll discuss in Chapter 7). Photosynthesis takes place in the chloroplasts of plants and algae, as well as in some prokaryotes. All animals depend on this conversion for food and more. You're probably wearing clothing made of a product of photosynthesis—cotton. Most of our homes are framed with lumber, which is wood produced by photosynthetic trees. Even textbooks are printed on a material (paper) that can be traced to photosynthesis in plants. But from an animal's point of view, photosynthesis is primarily about providing food.

✓ CHECKPOINT

What chemical ingredients do plants require from the environment to synthesize their own food?

■ *Answer: CO₂, H₂O, and soil minerals*

Producers and Consumers

Plants and other **autotrophs** ("self-feeders") are organisms that make all their own organic matter—including carbohydrates, lipids, proteins, and nucleic acids—from nutrients that are entirely inorganic: carbon dioxide from the air and water and minerals from the soil. In other words, autotrophs make their own food; they don't need to eat to gain energy to power their cellular processes. In contrast, humans and other animals are **heterotrophs** ("other-feeders"), organisms that cannot make organic molecules from inorganic

ones. Therefore, we must eat organic material to get our nutrients and provide energy for life's processes.

Most ecosystems depend entirely on photosynthesis for food. For this reason, biologists refer to plants and other autotrophs as **producers**. Heterotrophs, in contrast, are **consumers** because they obtain their food by eating plants or by eating animals that have eaten plants **(Figure 6.1)**. We animals and other heterotrophs depend on autotrophs for organic fuel and for the raw organic materials we need to build our cells and tissues. ✓

Chemical Cycling between Photosynthesis and Cellular Respiration

Within a plant, the chemical ingredients for photosynthesis are carbon dioxide (CO_2), a gas that passes from the air into a plant through tiny pores, and water (H_2O), absorbed from the soil by the plant's roots. Inside leaf cells, organelles called chloroplasts use light energy to rearrange the atoms of these ingredients to produce sugars—most importantly glucose ($C_6H_{12}O_6$)—and other organic molecules **(Figure 6.2)**. You can think of chloroplasts as tiny solar-powered sugar factories. A by-product of photosynthesis is oxygen gas (O_2) that is released through pores into the atmosphere.

▶ **Figure 6.1 Producer and consumer.** A koala (consumer) eating leaves produced by a photosynthetic plant (producer).

A chemical process called cellular respiration uses O_2 to convert the energy stored in the chemical bonds of sugars to another source of chemical energy called ATP. Cells expend ATP for almost all their work. In both plants and animals, the production of ATP during cellular respiration occurs mainly in the organelles called mitochondria (see Figure 4.19).

You might notice in Figure 6.2 that energy takes a one-way trip through an ecosystem, entering as sunlight and exiting as heat. Chemicals, in contrast, are recycled. Notice also in Figure 6.2 that the waste products of cellular respiration are CO_2 and H_2O—the very same ingredients used as inputs for photosynthesis. Plants store chemical energy through photosynthesis and then harvest this energy through cellular respiration. (Note that plants perform *both* photosynthesis to produce fuel molecules

and cellular respiration to burn them, while animals perform *only* cellular respiration.) Plants usually make more organic molecules than they need for fuel. This photosynthetic surplus provides material for the plant to grow or can be stored (as starch in potatoes, for example). Thus, when you consume a carrot, potato, or turnip, you are eating the energy reservoir that plants (if unharvested) would have used to grow the following spring.

People have always taken advantage of plants' photosynthetic abilities by eating them. More recently, engineers have managed to tap into this energy reserve to produce liquid biofuels, primarily ethanol (see Chapter 7 for a discussion of biofuels). But no matter the end product, you can trace the energy and raw materials for growth back to solar-powered photosynthesis. ✓

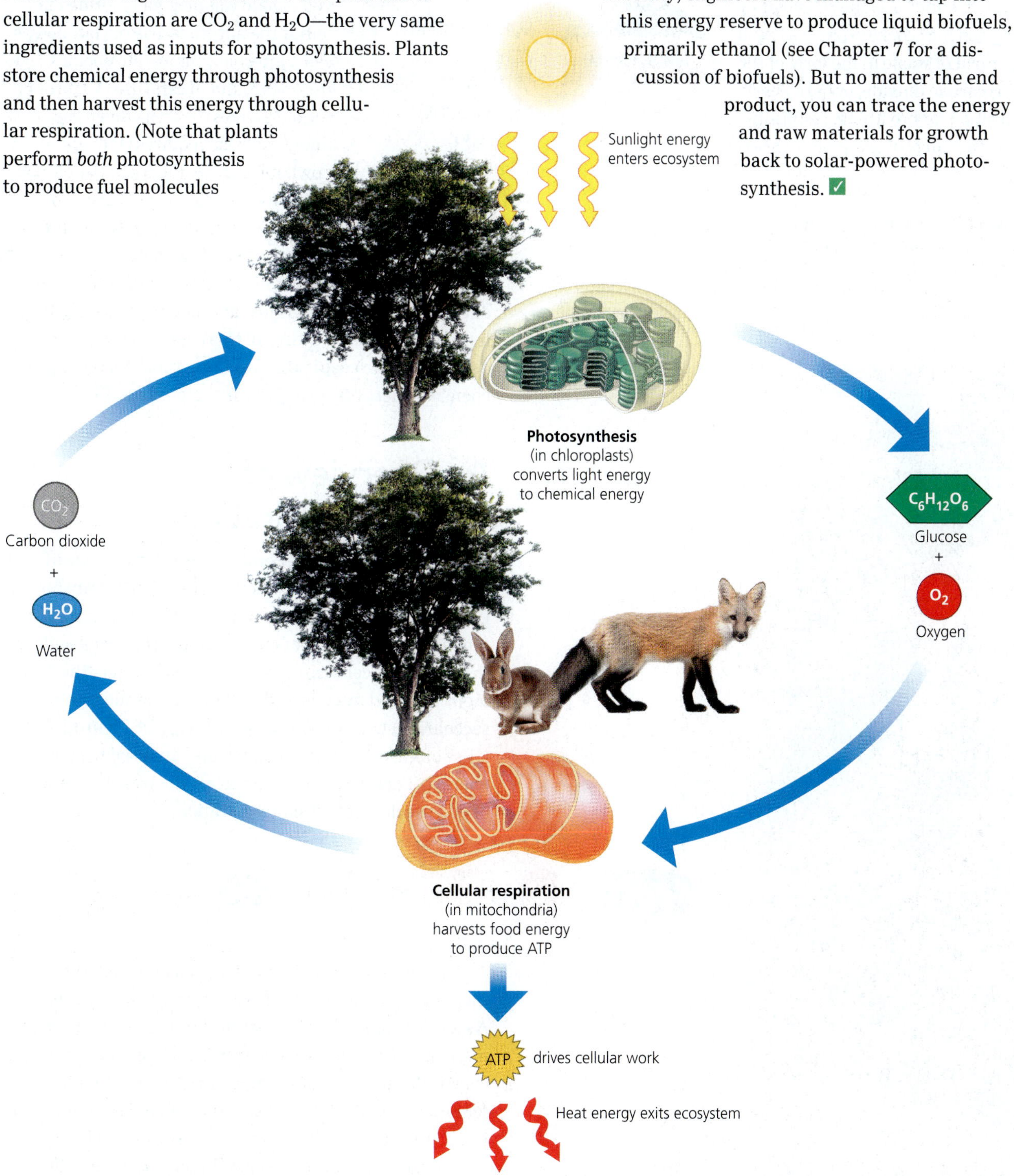

✅ CHECKPOINT

What is misleading about the following statement? "Plants perform photosynthesis, whereas animals perform cellular respiration."

■ *Answer: It implies that only animals perform cellular respiration, when in fact all life does.*

Sunlight energy enters ecosystem

Photosynthesis
(in chloroplasts)
converts light energy
to chemical energy

$C_6H_{12}O_6$

Glucose
+

O_2

Oxygen

CO_2

Carbon dioxide
+

H_2O

Water

Cellular respiration
(in mitochondria)
harvests food energy
to produce ATP

ATP drives cellular work

Heat energy exits ecosystem

◀ **Figure 6.2 Energy flow and chemical cycling in ecosystems.** Energy flows through an ecosystem, entering as sunlight and exiting as heat. In contrast, chemical elements are recycled within an ecosystem.

Figure Walkthrough

Mastering **Biology**
goo.gl/SvLydh

Cellular Respiration: Aerobic Harvest of Food Energy

We usually use the word *respiration* to mean breathing. Although respiration on the organismal level should not be confused with cellular respiration, the two processes are closely related (**Figure 6.3**). Cellular respiration requires a cell to exchange two gases with its surroundings. The cell takes in oxygen in the form of the gas O_2. It gets rid of waste in the form of the gas carbon dioxide, or CO_2. Respiration, or breathing, results in the exchange of these same gases between your blood and the outside air. Oxygen present in the air you inhale diffuses across the lining of your lungs and into your bloodstream. And the CO_2 in your bloodstream diffuses into your lungs and exits your body when you exhale. Every molecule of CO_2 that you exhale was originally formed in one of the mitochondria of your body's cells. Internal combustion engines, like the ones found in cars, use O_2 (through the air intakes) to break down gasoline. A cell also requires O_2 to break down its fuel (see Figure 5.2). Cellular respiration—a biological version of internal combustion—is the main way that chemical energy is harvested from food and converted to ATP energy (see Figure 5.6). Cellular respiration is an **aerobic** process, which is just another way of saying that it requires oxygen. Putting all this together, we can now define **cellular respiration** as the aerobic harvesting of chemical energy from organic fuel molecules. ✅

YOU HAVE SOMETHING IN COMMON WITH A SPORTS CAR: YOU BOTH REQUIRE AN AIR INTAKE SYSTEM TO BURN FUEL EFFICIENTLY.

✅ CHECKPOINT

At both the organismal and cellular levels, respiration involves taking in the gas _____ and expelling the gas _____.

■ *Answer: O_2; CO_2*

▼ **Figure 6.3 How breathing is related to cellular respiration.** When you inhale, you breathe in O_2. The O_2 is delivered to your cells, where it is used in cellular respiration. Carbon dioxide, a waste product of cellular respiration, diffuses from your cells to your blood and travels to your lungs, where it is exhaled.

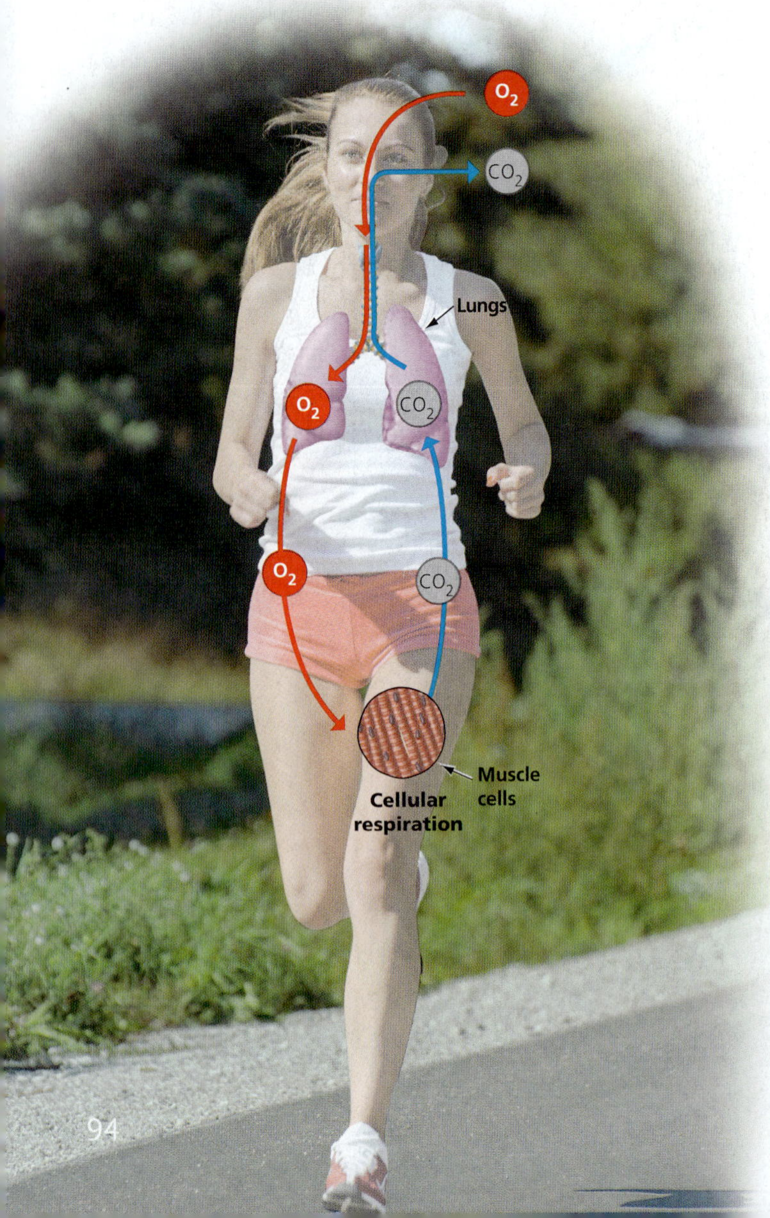

O_2

CO_2

Lungs

O_2

CO_2

O_2

CO_2

Cellular respiration

Muscle cells

An Overview of Cellular Respiration

All living organisms depend on transformations of energy and matter. We see examples of such transformations throughout the study of life, but few are as important as the conversion of energy in fuel (food molecules) to a form that cells can use directly. Most often, the fuel molecule used by cells is glucose, a simple sugar (monosaccharide) with the formula $C_6H_{12}O_6$ (see Figure 3.6). (Less often, other organic molecules are used to gain energy.) This equation summarizes the transformation of glucose during cellular respiration:

$$C_6H_{12}O_6 + 6\ O_2 \xrightarrow{\text{Many steps}} 6\ CO_2 + 6\ H_2O + \text{approx. } 32\ ATP + \text{heat}$$

The series of arrows in this formula represents the fact that cellular respiration consists of many chemical steps. A specific enzyme catalyzes each reaction in the pathway, more than two dozen reactions in all. In fact, these reactions constitute one of the most important metabolic pathways for nearly every eukaryotic cell: those found in plants, fungi, protists, and animals. This pathway provides the energy these cells need to maintain the functions of life.

The many chemical reactions that make up cellular respiration can be grouped into three main stages: glycolysis, the citric acid cycle, and electron transport. **Figure 6.4** is a road map that will help you follow the three stages of respiration and see where each stage occurs in your cells. During **glycolysis**, a molecule of glucose is split into two molecules of a compound called pyruvic acid. The enzymes for glycolysis are located in the cytoplasm. The **citric acid cycle** (also called the Krebs cycle) completes the breakdown of glucose all the way to CO_2, which is then released as a waste product. The enzymes for the citric acid cycle are dissolved in the fluid within mitochondria. Glycolysis and the citric acid cycle generate a small amount of ATP directly. They generate much more ATP indirectly, by reactions that transfer electrons from fuel molecules to a molecule called NAD^+ (<u>n</u>icotinamide <u>a</u>denine <u>d</u>inucleotide) that cells make from niacin, a B vitamin. The electron transfer forms a molecule called **NADH** (the H represents the transfer of

hydrogen along with the electrons) that acts as a shuttle carrying high-energy electrons from one area of the cell to another. The third stage of cellular respiration is **electron transport**. Electrons captured from food by the NADH formed in the first two stages are stripped of their energy, a little bit at a time, until they are finally combined with oxygen to form water. The proteins and other molecules that make up electron transport chains are embedded within the inner membrane of the mitochondria. The transport of electrons from NADH to oxygen releases the energy your cells use to make most of their ATP.

The overall equation for cellular respiration shows that the atoms of the reactant molecules glucose and oxygen are rearranged to form the products carbon dioxide and water. But don't lose track of why this process occurs: The main function of cellular respiration is to generate ATP for cellular work. In fact, the process can produce around 32 ATP molecules for each glucose molecule consumed. ✓

☑ **CHECKPOINT**

Which stages of cellular respiration take place in the mitochondria? Which stage takes place outside the mitochondria?

■ *Answer: the citric acid cycle and electron transport; glycolysis*

▶ **Figure 6.4** A road map for cellular respiration.

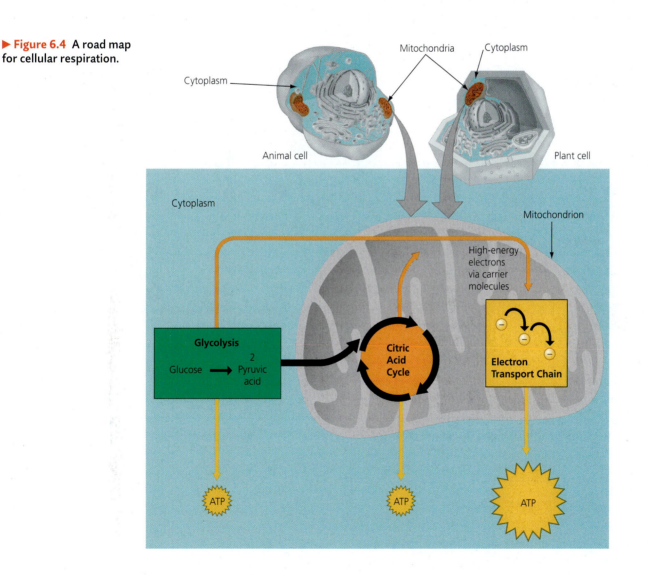

Cytoplasm

Mitochondria   Cytoplasm

Animal cell   Plant cell

Cytoplasm   Mitochondrion

High-energy electrons via carrier molecules

Glycolysis
2
Glucose → Pyruvic acid

Citric Acid Cycle

Electron Transport Chain

ATP   ATP   ATP

The Three Stages of Cellular Respiration

Now that you have a big-picture view of cellular respiration, let's examine the process in more detail. A small version of Figure 6.4 will help you keep the overall process of cellular respiration in plain view as we take a closer look at its three stages.

Stage 1: Glycolysis

The word *glycolysis* means "splitting of sugar" **(Figure 6.5)**, and that's just what happens. ❶ During glycolysis, a six-carbon glucose molecule is broken in

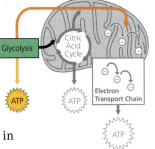

half, forming two three-carbon molecules. Notice in Figure 6.5 that the initial split requires an energy "investment" of two ATP molecules per glucose. ❷ The three-carbon molecules then donate high-energy electrons to NAD$^+$, forming NADH. ❸ In addition to NADH, glycolysis also "banks" four ATP molecules directly when enzymes transfer phosphate groups from fuel molecules to ADP. Glycolysis thus produces a "profit" of two molecules of ATP per molecule of glucose (two invested, but four banked; this fact will become important during our discussion of fermentation later). What remains of the fractured glucose at the end of glycolysis are two molecules of pyruvic acid. The pyruvic acid still holds most of the energy of glucose, and that energy is harvested in the second stage of cellular respiration, the citric acid cycle.

▼ **Figure 6.5 Glycolysis.** In glycolysis, a team of enzymes splits glucose, eventually forming two molecules of pyruvic acid. After investing 2 ATP at the start, glycolysis generates 4 ATP directly. More energy will be harvested later from high-energy electrons used to form NADH and from the two molecules of pyruvic acid.

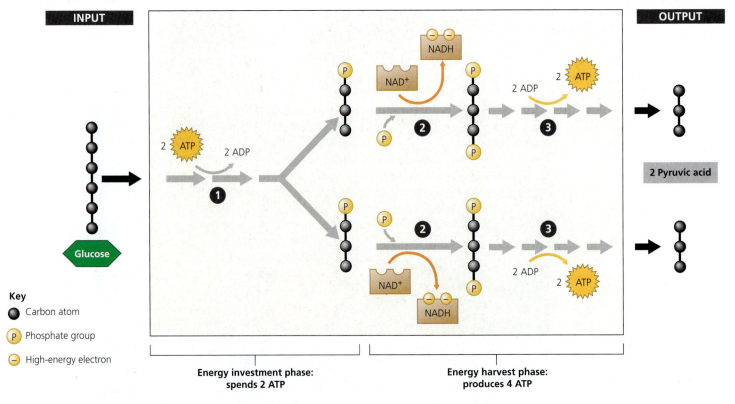

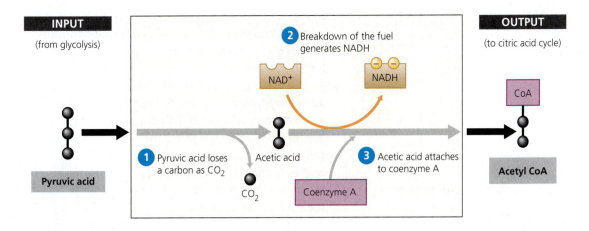

◀ Figure 6.6 The link between glycolysis and the citric acid cycle: the conversion of pyruvic acid to acetyl CoA. Remember that one molecule of glucose is split into two molecules of pyruvic acid. Therefore, the process shown here occurs twice for each starting glucose molecule.

Stage 2:
The Citric Acid Cycle

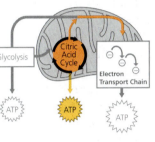

During glycolysis, one molecule of glucose is split into two molecules of pyruvic acid. But before pyruvic acid can be used by the citric acid cycle, it must be "groomed"—converted to a form the citric acid cycle can use **(Figure 6.6)**. **1** First, each pyruvic acid loses a carbon as CO_2. This is the first of this waste product we've seen so far in the breakdown of glucose. The remaining fuel molecules, each with only two carbons left, are called acetic acid (the acid that's in vinegar). **2** Electrons are stripped from these molecules and transferred to another molecule of NAD^+, forming more NADH. **3** Finally, each acetic acid is attached to a molecule called coenzyme A (CoA), an enzyme derived from the B vitamin pantothenic acid, to form acetyl CoA. The

CoA escorts the acetic acid into the first reaction of the citric acid cycle. The CoA is then stripped and recycled.

The citric acid cycle finishes extracting the energy of sugar by dismantling the acetic acid molecules all the way down to CO_2 **(Figure 6.7)**. **1** Acetic acid joins a four-carbon acceptor molecule to form a six-carbon product called citric acid (for which the cycle is named). For every acetic acid molecule that enters the cycle as fuel, **2** two CO_2 molecules eventually exit as a waste product. Along the way, the citric acid cycle harvests energy from the fuel. **3** Some of the energy is used to produce ATP directly. However, the cycle captures much more energy in the form of **4** NADH and **5** a second, closely related electron carrier called $FADH_2$. **6** All the carbon atoms that entered the cycle as fuel are accounted for as CO_2 exhaust, and the four-carbon acceptor molecule is recycled. We have tracked only one acetic acid molecule through the citric acid cycle here. But because glycolysis splits glucose in two, the citric acid cycle occurs twice for each glucose molecule that fuels a cell. ☑

☑ CHECKPOINT

1. Two molecules of what compound are produced by glycolysis? Does this molecule enter the citric acid cycle?

2. Pyruvic acid must travel from the _____, where glycolysis takes place, to the _____, where the citric acid cycle takes place.

Answers: 1. Pyruvic acid. No; it is first converted to acetic acid. 2. cytoplasm; mitochondria

▶ Figure 6.7
The citric acid cycle.

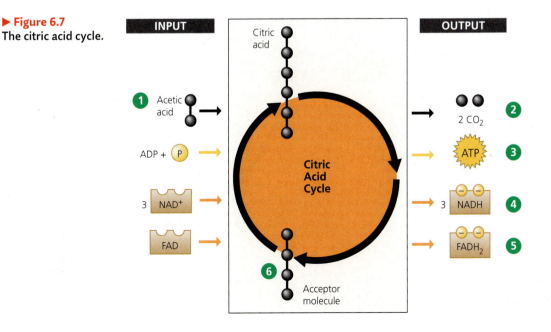

Stage 3: Electron Transport

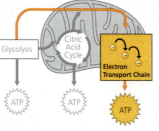

During cellular respiration, the electrons gathered from food molecules gradually "fall," losing energy at each step. In this way, cellular respiration unlocks chemical energy in small amounts, bit by bit, that cells can put to productive use.

Electrons are transferred from glucose in food molecules to NAD$^+$. This electron transfer converts NAD$^+$ to NADH. Then NADH releases two electrons that enter an **electron transport chain**, a series of electron carrier molecules. This chain is like a bucket brigade, with each molecule passing an electron to the next molecule. With each exchange, the electron gives up a bit of energy. This downward cascade releases energy from the electron and uses it to make ATP **(Figure 6.8)**.

Let's take a closer look at the path that electrons take **(Figure 6.9)**. Each link in an electron transport chain is a molecule, usually a protein (shown as purple circles in Figure 6.9). In a series of reactions, each member of the chain transfers electrons. With each transfer, the electrons give up a small amount of energy that can then be used indirectly to generate ATP. The first molecule of the chain accepts electrons from NADH. Thus, NADH carries electrons from glucose and other fuel molecules and deposits them at the top of an electron transport chain. The molecule at the bottom of the chain finally "drops" the electrons to oxygen. At the same time, oxygen picks up hydrogen, forming water.

The overall effect of all this transfer of electrons during cellular respiration is a "downward" trip for electrons from glucose to NADH to an electron transport chain to oxygen. During the stepwise release of chemical energy in the electron transport chain, our cells make most of their ATP. It is actually oxygen, the "electron grabber," at the end, that makes it all possible. By pulling electrons down the transport chain from fuel molecules, oxygen functions somewhat like gravity pulling objects downhill. Because oxygen is the final electron acceptor, we cannot survive more than a few minutes without breathing. Viewed this way, drowning is deadly because it deprives cells of the final "electron grabbers" (oxygen) needed to drive cellular respiration.

▶ **Figure 6.8 Cellular respiration illustrated using a hard-hat analogy.** Splitting the bonds of a molecule of glucose provides the energy that boosts an electron from a low-energy state to a high-energy state.

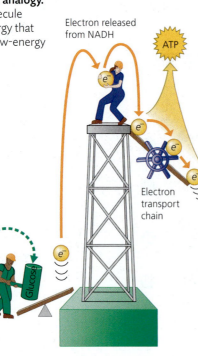

▼ **Figure 6.9 The role of oxygen in harvesting food energy.** In cellular respiration, electrons "fall" in small steps from food to oxygen, producing water. NADH transfers electrons from food to an electron transport chain. The attraction of oxygen to electrons "pulls" the electrons down the chain.

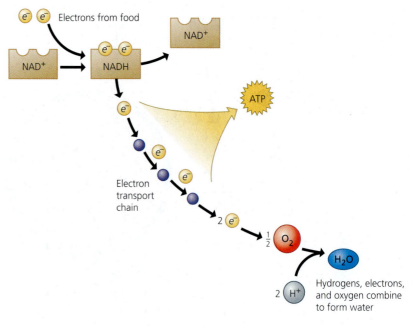

The molecules of electron transport chains are built into the inner membranes of mitochondria (see Figure 4.19). **Because these membranes are highly folded, their large surface area can accommodate thousands of copies of the electron transport chain—a good example of how biological structure fits function.** Each chain acts as a chemical pump that uses the energy released by the "fall" of electrons to move hydrogen ions (H⁺) across the inner mitochondrial membrane. This pumping causes ions to become more concen-

trated on one side of the membrane than on the other. Such a difference in concentration stores potential energy, similar to the way water can be stored behind a dam. There is a tendency for hydrogen ions to gush back to where they are less concentrated, just as there is a tendency for water to flow downhill. The inner membrane temporarily "dams" hydrogen ions.

The energy of dammed water can be harnessed to perform work. Gates in a dam allow the water to rush downhill, turning giant turbines, and this work can be used to generate electricity. Your mitochondria have structures that act like turbines. Each of these miniature machines, called an **ATP synthase**, is constructed from proteins built into the inner mitochondrial membrane, adjacent to the proteins of the electron transport chains. **Figure 6.10** shows a simplified view of how the energy previously stored in NADH and FADH₂ can now be used to generate ATP. ❶ NADH and ❷ FADH₂ transfer electrons to an electron transport chain. ❸ The electron transport chain uses this energy supply to pump H⁺ across the inner mitochondrial membrane. ❹ Oxygen pulls electrons down the transport chain. ❺ The H⁺ concentrated on one side of the membrane rushes back "downhill" through an ATP synthase. This action spins a component of the ATP synthase, just as water turns the turbines in a dam. ❻ The rotation activates parts of the synthase molecule that attach phosphate groups to ADP molecules to generate ATP.

The poison cyanide produces its deadly effect by binding to one of the protein complexes in the electron transport chain (marked with a skull-and-crossbones symbol in Figure 6.10). When bound there, cyanide blocks the passage of electrons to oxygen. This blockage is like clogging the outflow channel of a dam. As a result, no H⁺ gradient is generated, and no ATP is made. Cells stop working, and the organism dies. ✔

☑ **CHECKPOINT**

What is the potential energy source that drives ATP production by ATP synthase?

■ *Answer: a concentration gradient of H⁺ across the inner membrane of a mitochondrion*

▼ **Figure 6.10 How electron transport drives ATP synthase machines.**

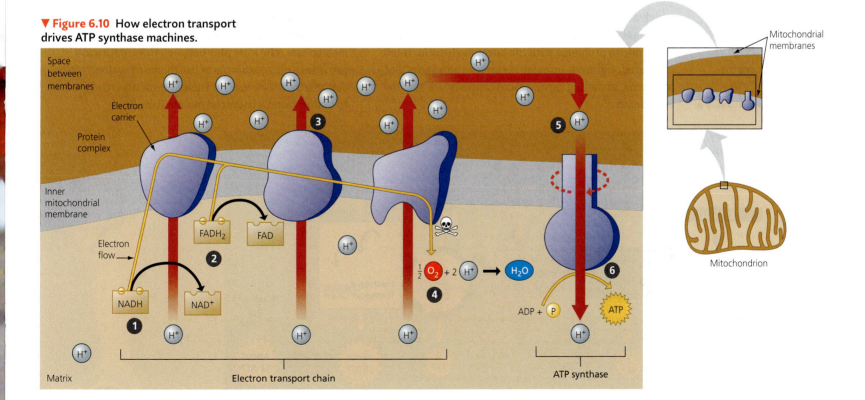

THE PROCESS OF SCIENCE | Exercise Science

What Causes Muscle Burn?

BACKGROUND

You probably know that your muscles burn after hard exercise ("Feel the burn!"). But what causes the burn? This question was investigated by one of the founders of the field of exercise science, a British biologist named A.V. Hill. In fact, Hill won a 1922 Nobel Prize for his investigations of muscle contraction.

Hill knew that muscles produce lactic acid under anaerobic conditions. In 1929, Hill developed a technique for electrically stimulating dissected frog muscles in a laboratory solution. He wondered if a buildup of lactic acid would cause muscle activity to stop.

METHOD

Hill's experiment tested frog muscles under two different sets of conditions **(Figure 6.14)**. First, he showed that muscle performance declined when lactic acid could not diffuse away from the muscle tissue. Next, he showed that when lactic acid was allowed to diffuse away, performance improved significantly. These results led Hill to the conclusion that the buildup of lactic acid is the primary cause of muscle failure under anaerobic conditions.

RESULTS

Given his scientific stature—he was considered the world's leading authority on muscle activity—Hill's conclusion went unchallenged for many decades. Gradually, however, evidence that contradicted Hill's results began to accumulate. For example, the effect that Hill demonstrated did not appear to occur at human body temperature. And certain people who are unable to accumulate lactic acid have muscles that

fatigue *more* rapidly, which is the opposite of what you would expect. Recent experiments have directly refuted Hill's conclusions. Some research indicates that increased levels of other ions may be to blame. Therefore, the cause of muscle fatigue remains hotly debated.

The changing view of lactic acid's role in muscle fatigue illustrates an important point about the process of science: Scientific belief is dynamic and subject to constant adjustment as new evidence is uncovered. This would not have surprised Hill, who himself observed that all scientific hypotheses may become obsolete, and that changing conclusions in light of new evidence is necessary for the advancement of science.

Thinking Like a Scientist

The explanation for the cause of muscle burn has changed over time as new evidence has accumulated. Does this represent a failure of the scientific method?

For the answer, see Appendix D.

▼ **Figure 6.14** A. V. Hill's 1929 apparatus for measuring muscle fatigue.

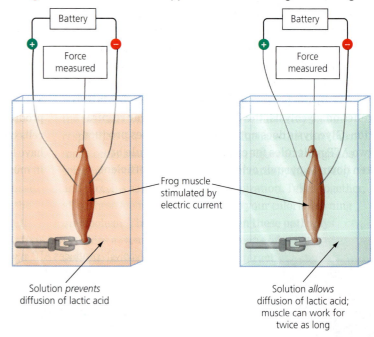

Frog muscle stimulated by electric current

Solution *prevents* diffusion of lactic acid

Solution *allows* diffusion of lactic acid; muscle can work for twice as long

Fermentation in Microorganisms

Our muscles cannot rely on lactic acid fermentation for very long. However, the two ATP molecules produced per glucose molecule during fermentation are enough to sustain many microorganisms. We have domesticated such microbes to transform milk into cheese, sour cream, and yogurt. These foods owe their

THE METABOLIC PROCESSES THAT PRODUCE ACID IN YOUR MUSCLES AFTER A HARD WORKOUT ARE SIMILAR TO THE PROCESSES THAT PRODUCE PEPPERONI, SOY SAUCE, YOGURT, AND BREAD.

sharp or sour flavor mainly to lactic acid. The food industry also uses fermentation to produce soy sauce from soybeans, to pickle cucumbers, olives, and cabbage, and to produce meat products like sausage, pepperoni, and salami.

Yeast, a microscopic fungus, is capable of both cellular respiration and fermentation. When kept in an anaerobic environment, yeast cells ferment sugars and other foods to stay alive. As they do, the yeast produce ethyl alcohol as a waste product instead of lactic acid

(Figure 6.15). This alcoholic fermentation also releases CO_2. For thousands of years, people have put yeast to work producing alcoholic beverages such as beer and wine using airtight barrels and vats, where the lack of oxygen forces the yeast to ferment glucose into ethanol. And as every baker knows, the CO_2 bubbles from fermenting yeast also cause bread dough to rise. (The alcohol produced in fermenting bread is burned off during baking.) ✓

✓ CHECKPOINT

What kind of acid builds up in human muscle during strenuous activity?

■ Answer: lactic acid

▼ **Figure 6.15 Fermentation: producing ethyl alcohol.** The alcohol produced by yeast as bread rises is burned off during baking.

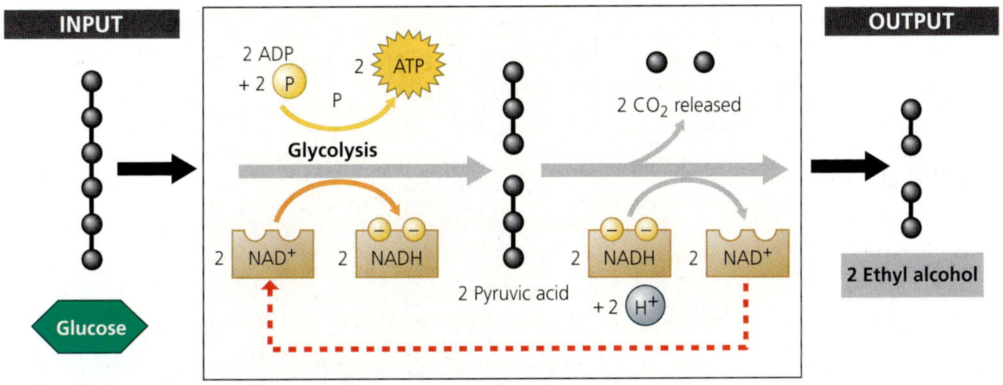

EVOLUTION CONNECTION Exercise Science

The Importance of Oxygen

In the Biology and Society and the Process of Science sections, we were reminded of the important role that oxygen plays during aerobic exercise. But in the last section on fermentation, we learned that exercise can continue on a limited basis even under anaerobic (oxygen-free) conditions. Both aerobic and anaerobic respiration start with glycolysis, the splitting of glucose to form pyruvic acid. Glycolysis is thus the universal energy-harvesting process of life. If you looked inside a bacterial cell, one of your body cells, or any other living cell, you would find the metabolic machinery of glycolysis.

The universality of glycolysis in life on Earth has an evolutionary basis. Ancient prokaryotes probably used glycolysis to make ATP long before oxygen was present in Earth's atmosphere. The oldest known fossils of bacteria date back more than 3.5 billion years and resemble some living photosynthetic bacteria. Evidence indicates, however, that significant levels of O_2, formed as a by-product of bacterial photosynthesis, did not accumulate in the atmosphere until about 2.7 billion years ago (Figure 6.16). For almost 1 billion years, prokaryotes must have generated ATP exclusively from glycolysis, a process that does not require oxygen.

The fact that glycolysis occurs in almost all organisms suggests that it evolved very early in ancestors common to all the domains of life. The location of glycolysis within the cell also implies great antiquity; the pathway does not require any of the membrane-enclosed organelles of the eukaryotic cell, which evolved about a billion years after the prokaryotic cell. Glycolysis is a metabolic heirloom from early cells that continues to function in fermentation and as the first stage in the breakdown of organic molecules by cellular respiration. The ability of our muscles to function anaerobically can therefore be viewed as a vestige of our ancient ancestors, who relied exclusively on this metabolic pathway.

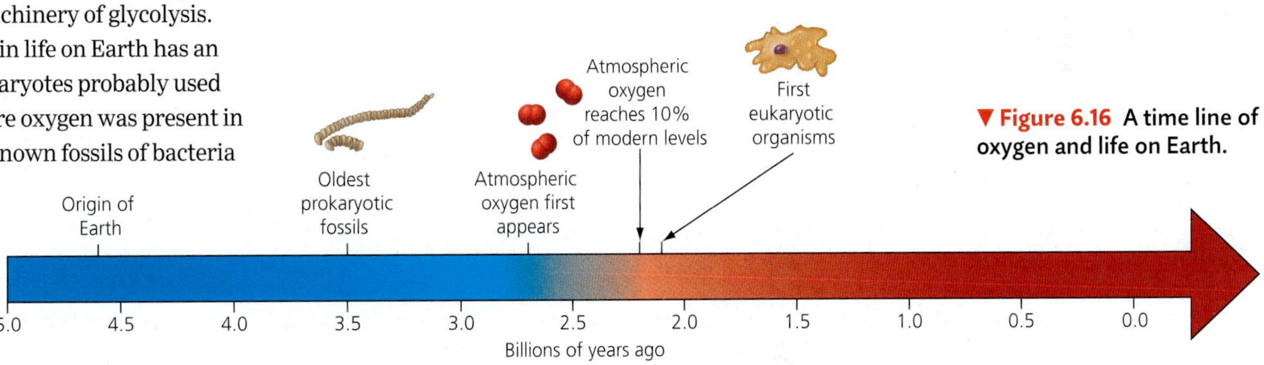

▼ **Figure 6.16 A time line of oxygen and life on Earth.**

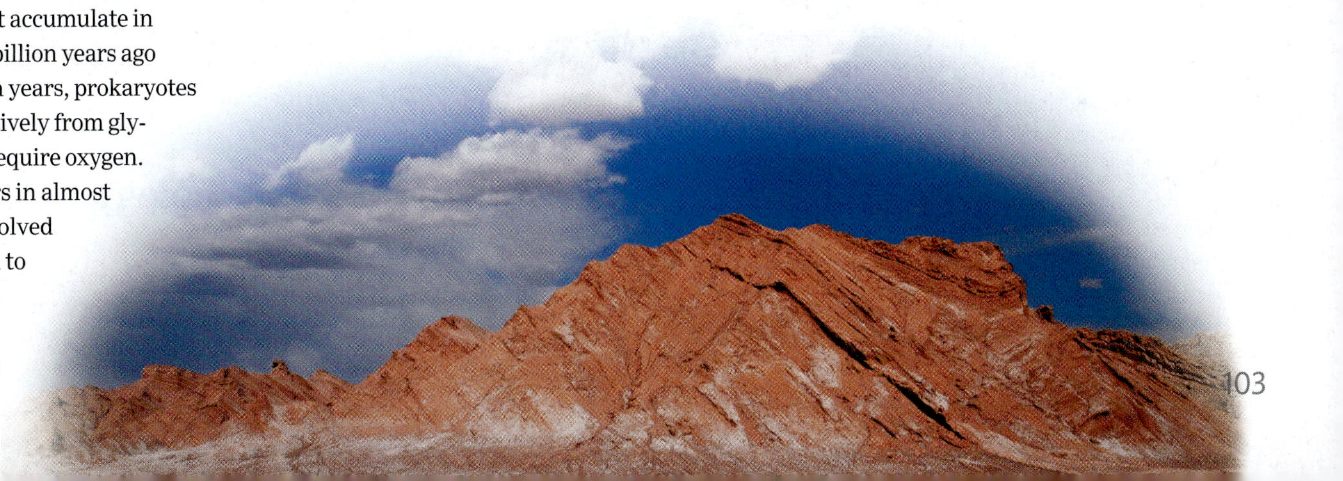

Chapter Review

SUMMARY OF KEY CONCEPTS

Energy Flow and Chemical Cycling in the Biosphere

Producers and Consumers

Autotrophs (producers) make organic molecules from inorganic nutrients by photosynthesis. Heterotrophs (consumers) must consume organic material and obtain energy by cellular respiration.

Chemical Cycling between Photosynthesis and Cellular Respiration

The molecular outputs of cellular respiration—CO_2 and H_2O—are the molecular inputs of photosynthesis, and vice versa. While these chemicals cycle through an ecosystem, energy flows through, entering as sunlight and exiting as heat.

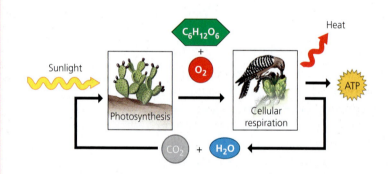

Cellular Respiration: Aerobic Harvest of Food Energy

An Overview of Cellular Respiration

The overall equation of cellular respiration simplifies a great many chemical steps into one formula:

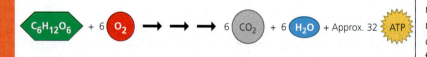

The Three Stages of Cellular Respiration

Cellular respiration occurs in three stages. During glycolysis, a molecule of glucose is split into two molecules of pyruvic acid, producing two molecules of ATP and two high-energy electrons stored in NADH. During the citric acid cycle, what remains of glucose is completely broken down to CO_2, producing a bit of ATP and a lot of high-energy electrons stored in NADH and $FADH_2$. The electron transport chain uses the high-energy electrons to pump H^+ across the inner mitochondrial membrane, eventually handing them off to O_2, producing H_2O. Backflow of H^+ across the membrane powers the ATP synthases, which produce ATP from ADP.

The Results of Cellular Respiration

You can follow the flow of molecules through the process of cellular respiration in the diagram. Notice that the first two stages primarily produce high-energy electrons carried by NADH and that it is the final stage that uses these high-energy electrons to produce the bulk of the ATP molecules produced during cellular respiration.

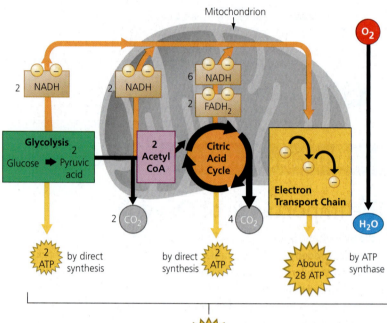

Fermentation: Anaerobic Harvest of Food Energy

Fermentation in Human Muscle Cells

When muscle cells consume ATP faster than O_2 can be supplied for cellular respiration, the conditions become anaerobic, and muscle cells will begin to regenerate ATP by fermentation. The waste product under these anaerobic conditions is lactic acid. The ATP yield per glucose is much lower during fermentation (2 ATP) than during cellular respiration (about 32 ATP).

Fermentation in Microorganisms

Yeast and some other organisms can survive with or without O_2. Wastes from fermentation can be ethyl alcohol, lactic acid, or other compounds, depending on the species.

Mastering Biology

For practice quizzes, BioFlix animations, MP3 tutorials, video tutors, and more study tools designed for this textbook, go to Mastering Biology™

SELF-QUIZ

1. Which of the following statements is a correct distinction between autotrophs and heterotrophs?
    a. Only heterotrophs require chemical compounds from the environment.
    b. Cellular respiration is unique to heterotrophs.
    c. Only heterotrophs have mitochondria.
    d. Only autotrophs can live on nutrients that are entirely inorganic.
2. Why are plants called producers? Why are animals called consumers?
3. How is your breathing related to your cellular respiration?
4. Of the three stages of cellular respiration, which produces the most ATP molecules per glucose?
5. The final electron acceptor of electron transport chains in mitochondria is _____.
6. The poison cyanide acts by blocking a key step in the electron transport chain. Knowing this, explain why cyanide kills so quickly.
7. Cells can harvest the most chemical energy from which of the following?
    a. an NADH molecule
    c. six CO_2 molecules
    b. a glucose molecule
    d. two pyruvic acid molecules
8. _____ is a metabolic pathway common to both fermentation and cellular respiration.
9. Exercise scientists at an Olympic training center want to monitor athletes to determine at what point their muscles are functioning anaerobically. They can do this by checking for a buildup of
    a. ADP.
    c. carbon dioxide.
    b. lactic acid.
    d. oxygen.
10. A glucose-fed yeast cell is moved from an aerobic environment to an anaerobic one. For the cell to continue to generate ATP at the same rate, approximately how much glucose must it consume in the anaerobic environment compared with the aerobic environment?

For answers to the Self-Quiz, see Appendix D.

IDENTIFYING MAJOR THEMES

For each statement, identify which major theme is evident (the relationship of structure to function, information flow, pathways that transform energy and matter, interactions within biological systems, or evolution) and explain how the statement relates to the theme. If necessary, review the themes (see Chapter 1) and review the examples highlighted in blue in this chapter.

11. The highly folded membranes of the mitochondria make these organelles well suited to carry out the huge number of chemical reactions required for cellular respiration to proceed.
12. Cellular respiration and photosynthesis are linked, with each process using inputs created by the other.
13. Your body uses many different intersecting chemical pathways that, all together, constitute your metabolism.

For answers to Identifying Major Themes, see Appendix D.

THE PROCESS OF SCIENCE

14. Your body makes NAD^+ from the vitamins niacin and riboflavin. The recommended daily allowances are 20 mg for niacin and 1.7 mg for riboflavin, thousands of times less than the amount of glucose needed each day. How many NAD^+ molecules are needed to break down each glucose molecule? Why are the requirements for niacin and riboflavin so small?
15. **Interpreting Data** Basal metabolic rate (BMR) is the amount of energy that must be consumed by a person at rest to maintain his or her body weight. BMR depends on several factors, including sex, age, height, and weight. The following graph shows the BMR for a 6'0" 45-year-old male. For this person, how does BMR correlate with weight? How many more calories must a 250-pound man consume to maintain his weight than a 200-pound man? Why does BMR depend on weight?

BIOLOGY AND SOCIETY

16. Delivery of oxygen to muscles is the limiting factor for many athletes. Some try blood doping (injecting oxygenated blood). Others train at high altitude. Both approaches increase the number of red blood cells. Why is doping considered cheating? How would you enforce antidoping rules?
17. The technology of fermentation dates back to the earliest civilizations. Suggest a hypothesis for how it was discovered.
18. Alcohol consumption during pregnancy can cause a series of birth defects called fetal alcohol syndrome (FAS). Symptoms include head and facial irregularities, heart defects, intellectual disability, and behavioral problems. The U.S. Surgeon General's Office recommends that pregnant women abstain from alcohol and liquor bottles have a warning label. If you were a server in a restaurant and a pregnant woman ordered a daiquiri, how would you respond? Is it her right to make those decisions about her unborn child's health? Do you bear any responsibility? Is a restaurant responsible for monitoring dietary habits of customers?

The Basics of Photosynthesis

The process of photosynthesis is the ultimate source of energy for nearly every ecosystem on Earth. **Photosynthesis** is a process whereby plants, algae (which are protists), and certain bacteria transform light energy into chemical energy, using carbon dioxide and water as starting materials and releasing oxygen gas as a by-product. The chemical energy produced through photosynthesis is stored in the bonds of sugar molecules. Organisms that generate their own food are called autotrophs (see Chapter 6). Plants and other organisms that do this by photosynthesis—photoautotrophs—are the producers for most ecosystems **(Figure 7.1)**. Photoautotrophs not only feed us but also clothe us (as the source of cotton fibers), house us (wood), and provide energy for warmth, light, and transportation (biofuels). **The fact that nearly all organisms—including you—can trace their source of energy back to the sun clearly illustrates that the ability to transform energy and matter is vital to the existence of life on Earth.**

NEARLY ALL LIFE ON EARTH CAN TRACE ITS SOURCE OF ENERGY BACK TO THE SUN.

Chloroplasts: Sites of Photosynthesis

Photosynthesis in plants and algae occurs within light-absorbing organelles called **chloroplasts** (see Figure 4.18). All green parts of a plant have chloroplasts and thus can carry out photosynthesis. In most plants, however, the leaves have the most chloroplasts (about 300 million chloroplasts in a piece of leaf the size of a nickel). Their green color is from **chlorophyll**, a pigment (light-absorbing molecule) in the chloroplasts that plays a central role in converting solar energy to chemical energy.

Chloroplasts are concentrated in the interior cells of leaves **(Figure 7.2)**, with a typical cell containing 30—40 chloroplasts. Carbon dioxide (CO_2) enters a leaf, and oxygen (O_2) exits, by way of tiny pores called **stomata** (singular, *stoma*, meaning "mouth"). The carbon dioxide that enters the leaf is the source of carbon for much of the body of the plant, including the sugars and starches that we eat. So the bulk of the body of a plant derives from the air, not the soil. As proof of this idea, consider hydroponics, a means of growing plants using only air and water; no soil whatsoever is involved. In addition to carbon dioxide, photosynthesis requires water, which is absorbed by the plant's roots and transported to the leaves, where veins carry it to the photosynthetic cells.

Membranes within the chloroplast form the framework where many of the reactions of photosynthesis occur. Like a mitochondrion, a chloroplast has a double-membrane envelope. The chloroplast's inner membrane encloses a compartment filled with **stroma**, a thick fluid. (It's easy to confuse two terms associated with photosynthesis: *Stomata* are pores through which gases are exchanged,

▼ **Figure 7.1** A diversity of photoautotrophs.

PHOTOSYNTHETIC AUTOTROPHS

Plants
(mostly on land)

Photosynthetic Protists
(aquatic)

Photosynthetic Bacteria
(aquatic)

Forest plants

Kelp, a large, multicellular alga

Micrograph of cyanobacteria

LM 375×

and *stroma* is the fluid within the chloroplast.) Suspended in the stroma are interconnected membranous sacs called **thylakoids**. The thylakoids are concentrated in stacks called **grana** (singular, *granum*). The chlorophyll molecules that capture light energy are built into the thylakoid membranes. The structure of a chloroplast—with its stacks of disks—aids its function by providing a large surface area for the reactions of photosynthesis. ☑

▼ **Figure 7.2 Journey into a leaf.** This series of blowups takes you into a leaf's interior, then into a plant cell, and finally into a chloroplast, the site of photosynthesis.

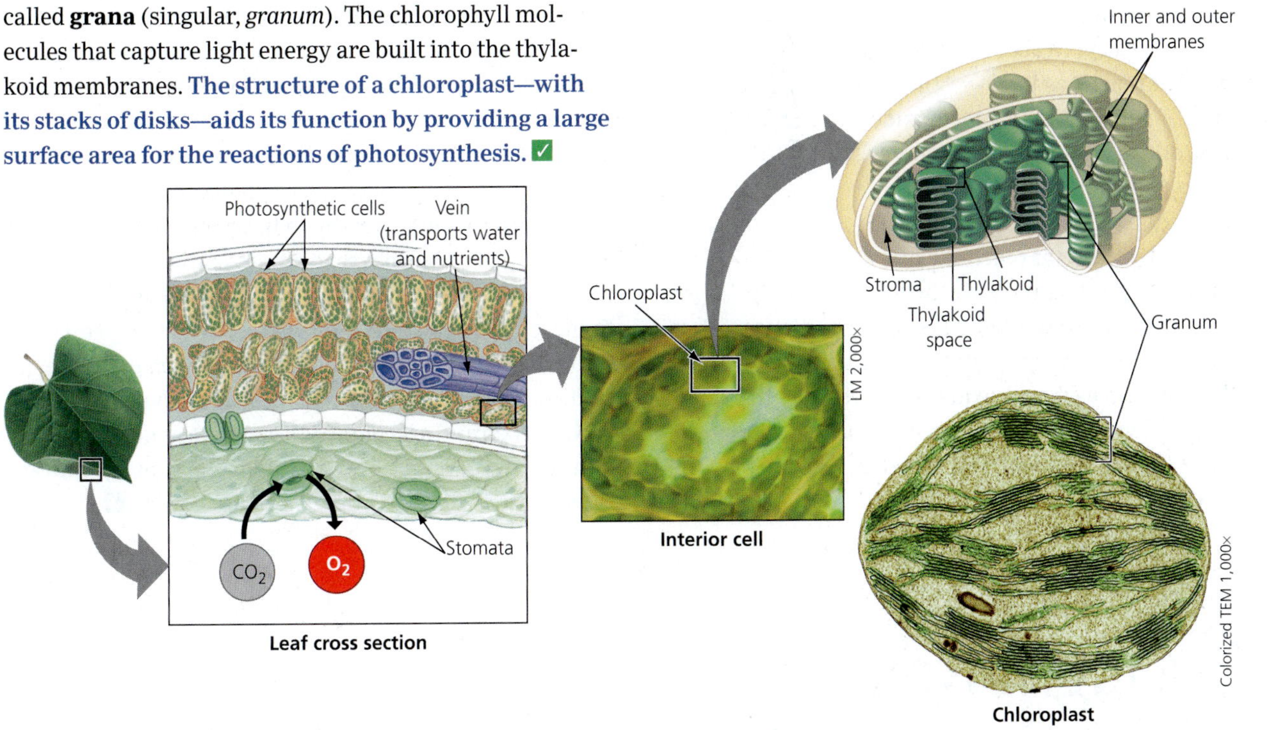

Leaf cross section

Interior cell

Chloroplast

☑ **CHECKPOINT**

Photosynthesis takes place within organelles called _____ using gases that are exchanged through pores called _____.

Answer: chloroplasts; stomata

An Overview of Photosynthesis

The following chemical equation, simplified to highlight the relationship between photosynthesis and cellular respiration, provides a summary of the reactants and products of photosynthesis:

$$6\ CO_2 + 6\ H_2O \longrightarrow\longrightarrow\longrightarrow C_6H_{12}O_6 + 6\ O_2$$

(Light energy)

Notice that the reactants of photosynthesis—carbon dioxide (CO_2) and water (H_2O)—are the same as the waste products of cellular respiration (see Figure 6.2). Also notice that photosynthesis produces what respiration uses—glucose ($C_6H_{12}O_6$) and oxygen (O_2). In other words, photosynthesis recycles the "exhaust" of cellular respiration and rearranges its atoms to produce food and oxygen. Photosynthesis is a chemical transformation that requires a lot of energy, and sunlight absorbed by chlorophyll provides that energy.

Recall that cellular respiration is a process of electron transfer (see Chapter 6). A "fall" of electrons from food molecules to oxygen to form water releases the energy that mitochondria can use to make ATP (see Figure 6.9). The opposite occurs in photosynthesis: Electrons are boosted "uphill" and added to carbon dioxide to produce sugar. Hydrogen is moved along with the electrons being

transferred from water to carbon dioxide. This transfer of hydrogen requires the chloroplast to split water molecules into hydrogen and oxygen. The hydrogen is transferred along with electrons to carbon dioxide to form sugar. The oxygen escapes through stomata in leaves into the atmosphere as O_2, a waste product of photosynthesis.

The overall equation for photosynthesis is a simple summary of a complex process. Like many energy-producing processes within cells, photosynthesis is a multistep chemical pathway, with each step in the path producing products that are used as reactants in the next step. For a better overview, let's look at the two stages of photosynthesis: the light reactions and the Calvin cycle **(Figure 7.3)**.

In the **light reactions**, chlorophyll in the thylakoid membranes absorbs solar energy (the "photo" part of photosynthesis), which is then converted to the chemical energy of ATP (the molecule

Figure Walkthrough

Mastering Biology
goo.gl/3rETcw

▼ **Figure 7.3**
A road map for photosynthesis.

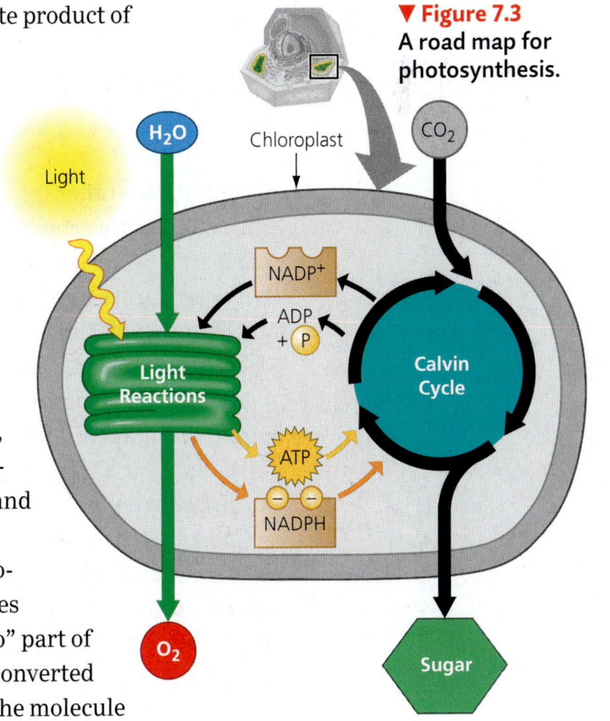

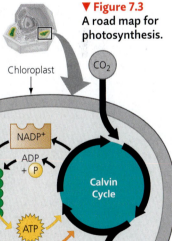

that drives most cellular work) and **NADPH** (an electron carrier). During the light reactions, water is split, providing a source of electrons and giving off O_2 gas as a by-product.

The **Calvin cycle** uses the products of the light reactions to power the production of sugar from carbon dioxide (the "synthesis" part of photosynthesis). The enzymes that drive the Calvin cycle are dissolved in the stroma. ATP generated by the light reactions provides the energy for sugar synthesis. And the NADPH produced by the light reactions provides the high-energy electrons that drive the synthesis of glucose from carbon dioxide. Thus, the Calvin cycle indirectly depends on light to produce sugar because it requires the supply of ATP and NADPH produced by the light reactions.

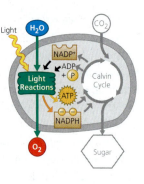

WANT TO DO SOMETHING SIMPLE TO COMBAT GLOBAL CLIMATE CHANGE? PLANT A TREE— YOU'LL BE GLAD YOU DID!

The initial incorporation of carbon from CO_2 into organic compounds is called **carbon fixation**. This process has important implications for global climate, because the removal of carbon from the air and its incorporation into plant material can help reduce the concentration of carbon dioxide in the atmosphere. Deforestation, which removes a lot of photosynthetic plant life, thereby reduces the ability of the biosphere to absorb carbon. Planting new forests can have the opposite effect of fixing carbon from the atmosphere, potentially reducing the effect of the gases that contribute to global climate change. **The relationship between photosynthesis, carbon fixation, and global climate is a good example of how interactions between biological components at many different levels affect all life on Earth.** ✓

The Light Reactions: Converting Solar Energy to Chemical Energy

Chloroplasts are solar-powered sugar factories. Let's look at how they use the energy in sunlight to store chemical energy.

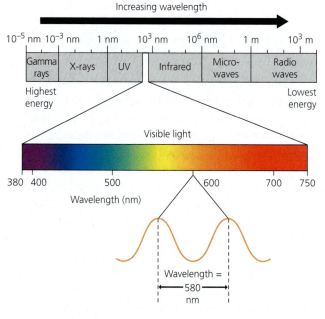

The Nature of Sunlight

Sunlight, like other forms of radiation, travels through space as rhythmic waves, like the ripples made by a pebble dropped into a pond. The distance between the crests of two adjacent waves is called a **wavelength**. The full range of radiation, from the very short wavelengths of gamma rays to the very long wavelengths of radio signals, is called the **electromagnetic spectrum (Figure 7.4)**. Visible light is the small fraction of the spectrum that our eyes see as different colors.

When sunlight shines on a pigmented material, certain wavelengths (colors) of the visible light are absorbed. The colors that are not absorbed are reflected by the material. For example, we see a pair of jeans as blue because pigments in the fabric absorb the other colors, leaving only blue light to be reflected to our

▼ **Figure 7.4 The electromagnetic spectrum.** The middle of the figure expands the thin slice of the spectrum that is visible to us as different colors of light, from about 380 nanometers (nm) to about 750 nm in wavelength. The bottom of the figure shows waves of one particular wavelength of visible light.

eyes. In the 1800s, plant biologists discovered that only certain wavelengths of light are used by plants, as we'll see next.

What Colors of Light Drive Photosynthesis?

BACKGROUND
Scientific breakthroughs often occur after careful observations of natural phenomena. In 1883, German biologist Theodor Engelmann noticed that certain aquatic bacteria tend to cluster in areas with higher oxygen concentrations. He formed a hypothesis that oxygen-seeking bacteria would congregate near algae that were performing the most photosynthesis and therefore producing the most oxygen. He conducted a simple but elegant experiment to determine which wavelengths (colors) of light are used during photosynthesis.

METHOD
Engelmann placed a string of freshwater algal cells within a drop of water on a microscope slide and used a prism to expose the cells to the color spectrum. He then added oxygen-seeking bacteria to the water **(Figure 7.5)**.

RESULTS
The experiment showed that most bacteria congregated around algae exposed to blue-violet and red-orange light. Other experiments have since verified that those wavelengths are mainly responsible for photosynthesis. Variations of this classic experiment are still performed. For example, biofuel researchers test different species of algae to determine which wavelengths result in optimal fuel production. Biofuel facilities of the future may use a variety of species that take advantage of the full spectrum of light that shines down on them.

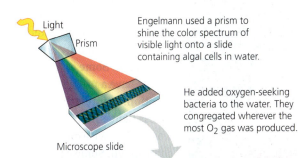

Engelmann used a prism to shine the color spectrum of visible light onto a slide containing algal cells in water.

He added oxygen-seeking bacteria to the water. They congregated wherever the most O_2 gas was produced.

Microscope slide

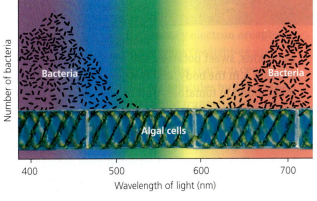

◄ **Figure 7.5** Investigating how light wavelength affects photosynthesis.

Most bacteria congregated around algal cells exposed to blue-violet and red-orange light.

Thinking Like a Scientist

Why was Engelmann's use of a prism with natural sunlight a key to his discovery?

For the answer, see Appendix D.

Chloroplast Pigments

The selective absorption of light by leaves explains why they appear green to us; light of that color is poorly absorbed by chloroplasts and is thus reflected or transmitted toward the observer **(Figure 7.6)**. Energy cannot be destroyed, so the absorbed energy must be converted to other forms. Chloroplasts contain several different pigments that absorb light of different wavelengths. Chlorophyll *a*,

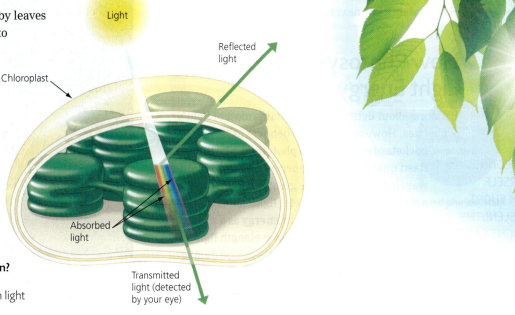

▶ **Figure 7.6 Why are leaves green?** Chlorophyll and other pigments in chloroplasts reflect or transmit green light while absorbing other colors.

8 Cellular Reproduction: Cells from Cells

CHAPTER CONTENTS

What Cell Reproduction Accomplishes 122
The Cell Cycle and Mitosis 123
Meiosis, the Basis of Sexual Reproduction 130

Why Cellular Reproduction Matters

About 9 months before you were born, you were just a single cell. From that microscopic beginning, every cell in your body was created through cell reproduction. Even today, millions of cells are reproducing themselves in your body every second—and errors in the process can be deadly.

IF STRETCHED OUT, THE DNA IN ANY ONE OF YOUR CELLS WOULD BE TALLER THAN YOU!

KEEPING TRACK OF CHROMOSOMES DURING CELL DIVISION IS VITAL: DUPLICATING THE WRONG NUMBER OF CHROMOSOMES IS ALMOST ALWAYS FATAL.

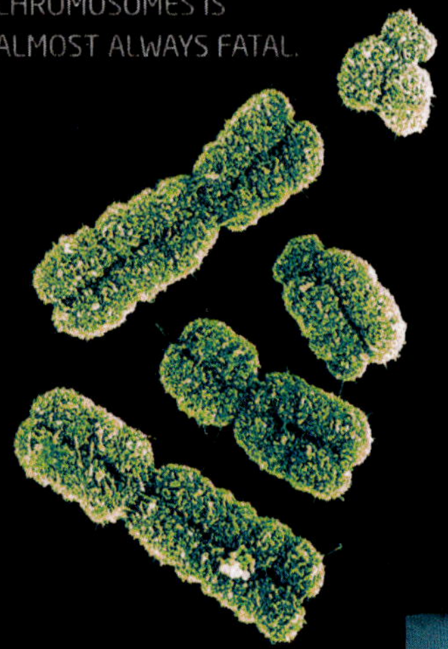

A TUMOR RESULTS FROM AN ERROR IN THE DIVISION OF ONE OF THE BODY'S OWN CELLS.

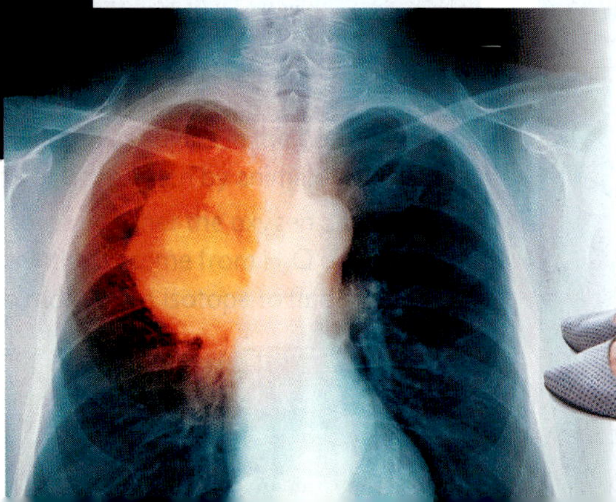

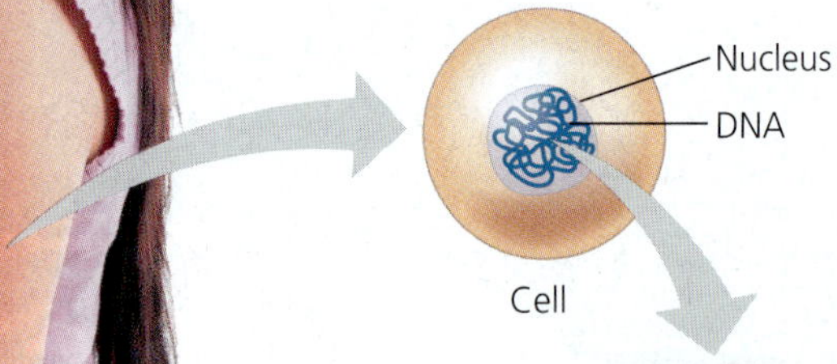

Nucleus

DNA

Cell

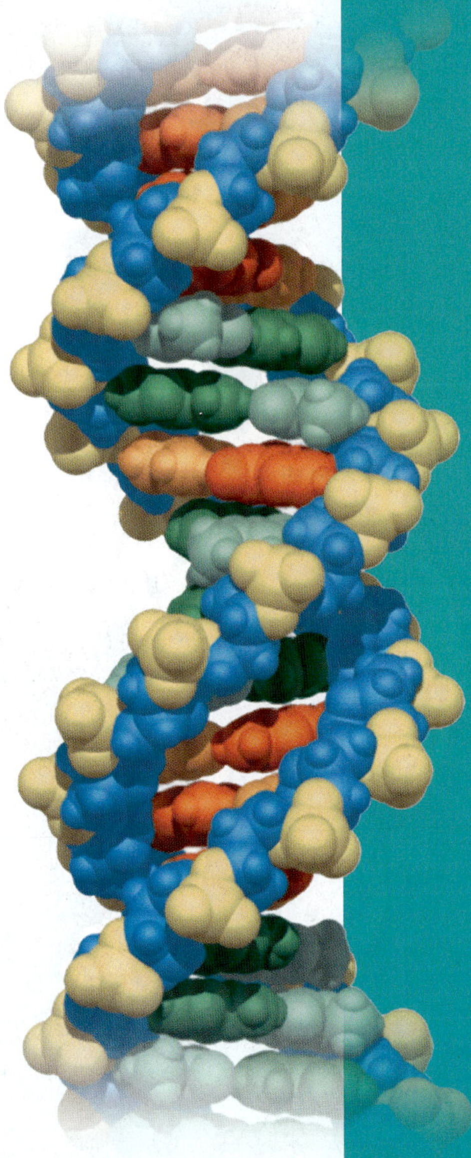

CHAPTER THREAD

Life with and without Sex

BIOLOGY AND SOCIETY Virgin Birth of a Shark 121

THE PROCESS OF SCIENCE Do All Animals Have Sex? 137

EVOLUTION CONNECTION The Advantages of Sex 140

BIOLOGY AND SOCIETY   Life with and without Sex

Zebra shark. The zebra shark can be found living around coral reefs and sandy bottoms in the Indian and Pacific Oceans.

Virgin Birth of a Shark

Biologists at the Reef HQ Aquarium in Queensland, Australia helped care for Leonie, a female zebra shark (*Stegostoma fasciatum*) captured from the wild. To promote a breeding program, a male shark was introduced to Leonie's tank, resulting in several litters of offspring. When the aquarium decided to end the breeding program, the male shark was removed, and Leonie lived in the tank with a daughter.

No one was surprised that Leonie continued to lay eggs. After all, sharks, like chickens (or, for that matter, humans), produce infertile eggs even in the absence of sex. But Leonie's keepers were shocked when, in 2015, she laid eggs with viable embryos in them. One of those eggs hatched into Cleo, a healthy daughter. This is surprising because the vast majority of animal species create off-spring only through sexual reproduction, involving the union of a male's sperm with a female's egg. But, despite the fact that Leonie was last in the company of a male three years prior, she produced a normal daughter. Cleo remains on display at the aquarium today.

DNA analysis confirmed that Cleo's genes came solely from her mother. The birth had resulted from parthenogenesis, the production of offspring by a female without involvement of a male. Parthenogenesis is rare among vertebrates (animals with backbones), although it has been documented in about 70 species as diverse as lizards (including the Komodo dragon, the world's largest lizard), domesticated birds (such as chickens and turkeys), and frogs. (In case you were wondering, there are no documented cases of parthenogenesis among humans or other mammals.) It is extremely rare, however, for a single female to produce offspring via both sexual reproduction and parthenogenesis. Leonie's case proved that zebra sharks can switch between two reproduc-tive modes—the first documented case of this ability among sharks. Biologists are investigating the evolutionary basis of this phenomenon and considering what implications it may have on efforts to repopulate endangered species such as the zebra sharks and Komodo dragons.

The ability of organisms to procreate best distinguishes living things from nonliving matter. All organisms—from bacteria to lizards to you—are the result of repeated cell divisions. Therefore, the perpetuation of life depends on cell division, the production of new cells. In this chapter, we'll look at how individual cells are copied and then see how cell reproduction underlies the process of sexual reproduction. Throughout our discussion, we'll consider examples of asexual and sexual reproduction among both plants and animals.

What Cell Reproduction Accomplishes

When you hear the word *reproduction*, you probably think of the birth of new organisms. But reproduction actually occurs much more often at the cellular level. Consider the skin on your arm. Skin cells are constantly reproducing themselves and moving outward toward the surface, replacing dead cells that have rubbed off. This renewal of your skin goes on throughout your life. And when your skin is injured, reproduction of cells helps heal the wound.

When a cell undergoes reproduction, the process is called **cell division**. The two "daughter" cells that result from cell division are genetically identical to each other and to the original "parent" cell. (Biologists traditionally use the word *daughter* in this context to refer to offspring cells even though cells lack gender.) Before the parent cell splits into two, it duplicates its **chromosomes**, the structures that contain most of the cell's DNA. Then, during cell division, each daughter cell receives one identical set of chromosomes from the original parent cell.

As summarized in **Figure 8.1**, cell division plays several important roles in the lives of organisms. For example, within your body, millions of cells must divide every second to replace damaged or lost cells. Another function of cell division is growth. All of the trillions of cells in your body are the result of repeated cell divisions that began in your mother's body with a single fertilized egg cell.

Another vital function of cell division is reproduction. Many single-celled organisms, such as amoebas, reproduce by dividing in half, and the offspring are genetic replicas of the parent. This is an example of

asexual reproduction, the creation of genetically identical offspring by a single parent, without the participation of sperm and egg. Offspring produced by asexual reproduction inherit all their chromosomes from a single parent and are thus genetic duplicates. An individual that reproduces asexually gives rise to a **clone**, a group of genetically identical individuals.

Many multicellular organisms can reproduce asexually as well. For example, some sea star species can grow new individuals from fragmented pieces. If you've ever grown a houseplant from a clipping, you've observed asexual reproduction in plants. In asexual reproduction, there is one simple principle of inheritance: The lone parent and each of its offspring have identical genes. The type of cell division responsible for asexual reproduction and for the growth and maintenance of multicellular organisms is called mitosis.

Sexual reproduction is different; it requires fertilization of an egg by a sperm. The production of **gametes**—egg and sperm—involves a special type of cell division called meiosis, which occurs only in reproductive organs. As we'll discuss later, a gamete has only half as many chromosomes as the parent cell that gave rise to it.

In summary, two kinds of cell division are involved in sexually reproducing organisms: mitosis for growth and maintenance and meiosis for reproduction. To define and distinguish these processes, the remainder of the chapter is divided into two main sections: one on mitosis and one on meiosis. ✓

✓ CHECKPOINT

Ordinary cell division produces two daughter cells that are genetically identical. Name three functions of this type of cell division. Which of these functions occur in your body?

■ Answer: Cell replacement, growth of an organism, and asexual reproduction of an organism. Only the first two occur in your body.

▶ **Figure 8.1** Three functions of cell division by mitosis.

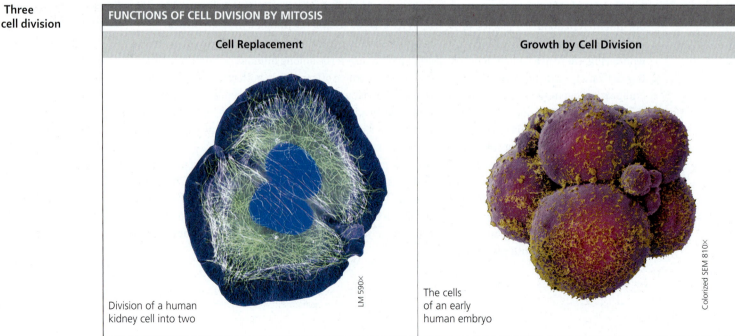

FUNCTIONS OF CELL DIVISION BY MITOSIS

Cell Replacement	Growth by Cell Division

Division of a human kidney cell into two

LM 590×

The cells of an early human embryo

Colorized SEM 810×

The Cell Cycle and Mitosis

Almost all the genes of a eukaryotic cell—around 21,000 in humans—are located on chromosomes in the cell nucleus. (The main exceptions are genes on small DNA molecules found in mitochondria and chloroplasts.) Because chromosomes are the lead players in cell division, we'll focus on them before turning our attention to the cell as a whole.

IF STRETCHED OUT, THE DNA IN ANY ONE OF YOUR CELLS WOULD BE TALLER THAN YOU!

▼ **Figure 8.2 The number of chromosomes in the cells of selected mammals.** Notice that humans have 46 chromosomes and that the number of chromosomes does not correspond to the size or complexity of an organism.

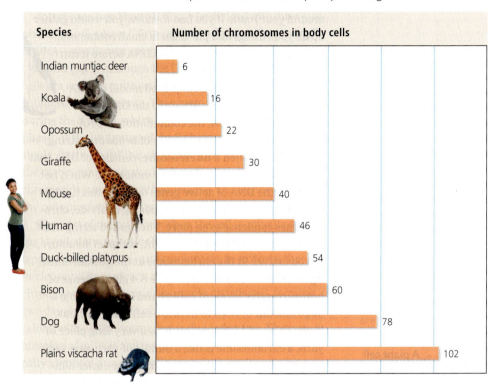

Species	Number of chromosomes in body cells
Indian muntjac deer	6
Koala	16
Opossum	22
Giraffe	30
Mouse	40
Human	46
Duck-billed platypus	54
Bison	60
Dog	78
Plains viscacha rat	102

Eukaryotic Chromosomes

Each eukaryotic chromosome contains one very long DNA molecule, typically bearing thousands of genes. The number of chromosomes in a eukaryotic cell depends on the species **(Figure 8.2)**. For example, human body cells have 46 chromosomes, while the body cells of a dog have 78 and those of a koala have 16. Chromosomes are made up of a material called **chromatin**, fibers composed of roughly equal amounts of DNA and protein molecules. The protein molecules help organize the chromatin and help control the activity of its genes.

Most of the time, the chromosomes exist as thin fibers that are much longer than the nucleus they are stored in. In fact, if fully extended, the DNA in just one of your cells would be more than 6 feet long! Chromatin in this state is too thin to be seen using a light

Asexual Reproduction

LM 250×

Reproduction of an amoeba

Fragmentation and regeneration of a sea star. The sea star on the right lost and replaced an arm. The severed arm grew into the new sea star on the left.

Reproduction of an African violet from a clipping (large leaf)

Mitosis and Cytokinesis

Figure 8.7 illustrates the cell cycle for an animal cell using drawings, descriptions, and fluorescent light micrographs. The micrographs running along the bottom row of the page show dividing cells from a salamander, with chromosomes depicted in blue. The drawings in the top row include details that are not visible in the micrographs. In these cells, we illustrate just four chromosomes to keep the process a bit simpler to follow; remember that one of your cells actually contains 46 chromosomes. The text within the figure describes the events occurring

▼ **Figure 8.7 Cell reproduction: a dance of the chromosomes.** After the chromosomes duplicate during interphase, the elaborately choreographed stages of mitosis—prophase, metaphase, anaphase, and telophase—distribute the duplicate sets of chromosomes to two separate nuclei. Cytokinesis then divides the cytoplasm, yielding two genetically identical daughter cells.

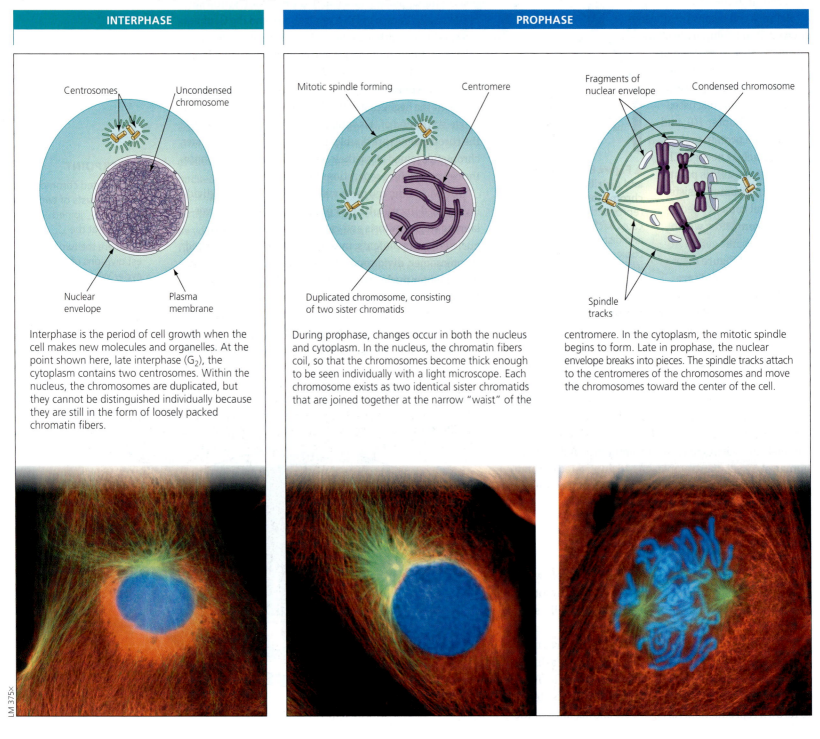

INTERPHASE

Centrosomes

Uncondensed chromosome

Nuclear envelope

Plasma membrane

Interphase is the period of cell growth when the cell makes new molecules and organelles. At the point shown here, late interphase (G₂), the cytoplasm contains two centrosomes. Within the nucleus, the chromosomes are duplicated, but they cannot be distinguished individually because they are still in the form of loosely packed chromatin fibers.

PROPHASE

Mitotic spindle forming

Centromere

Duplicated chromosome, consisting of two sister chromatids

During prophase, changes occur in both the nucleus and cytoplasm. In the nucleus, the chromatin fibers coil, so that the chromosomes become thick enough to be seen individually with a light microscope. Each chromosome exists as two identical sister chromatids that are joined together at the narrow "waist" of the

Fragments of nuclear envelope

Condensed chromosome

Spindle tracks

centromere. In the cytoplasm, the mitotic spindle begins to form. Late in prophase, the nuclear envelope breaks into pieces. The spindle tracks attach to the centromeres of the chromosomes and move the chromosomes toward the center of the cell.

LM 375×

at each stage. Study this figure carefully (it has a lot of information and it's important!) and notice the striking changes in the nucleus and other cellular structures.

Biologists distinguish four main stages of mitosis: **prophase**, **metaphase**, **anaphase**, and **telophase**. The timing of these stages is not precise, and they overlap a bit. Think of stages in your own life—infancy, childhood, adulthood, old age—and you'll realize that the division between stages isn't always clear, and the timings of the stages vary from person to person; so it is with the stages of mitosis.

The chromosomes are the stars of the mitotic drama, and their movements depend on the **mitotic spindle**, a football-shaped structure of microtubule tracks (colored green in the figure) that guides the separation of the two sets of daughter chromosomes. The tracks of spindle microtubules grow from structures within the cytoplasm called centrosomes.

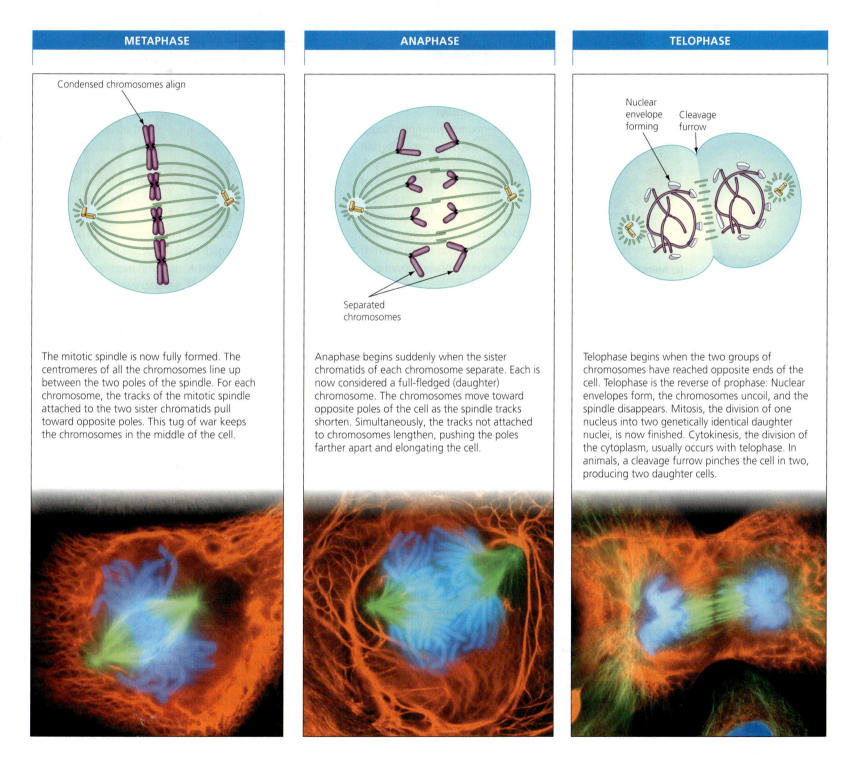

METAPHASE

Condensed chromosomes align

The mitotic spindle is now fully formed. The centromeres of all the chromosomes line up between the two poles of the spindle. For each chromosome, the tracks of the mitotic spindle attached to the two sister chromatids pull toward opposite poles. This tug of war keeps the chromosomes in the middle of the cell.

ANAPHASE

Separated chromosomes

Anaphase begins suddenly when the sister chromatids of each chromosome separate. Each is now considered a full-fledged (daughter) chromosome. The chromosomes move toward opposite poles of the cell as the spindle tracks shorten. Simultaneously, the tracks not attached to chromosomes lengthen, pushing the poles farther apart and elongating the cell.

TELOPHASE

Nuclear envelope forming   Cleavage furrow

Telophase begins when the two groups of chromosomes have reached opposite ends of the cell. Telophase is the reverse of prophase: Nuclear envelopes form, the chromosomes uncoil, and the spindle disappears. Mitosis, the division of one nucleus into two genetically identical daughter nuclei, is now finished. Cytokinesis, the division of the cytoplasm, usually occurs with telophase. In animals, a cleavage furrow pinches the cell in two, producing two daughter cells.

Review: Comparing Mitosis and Meiosis

You have now learned the two ways that cells of eukaryotic organisms divide **(Figure 8.15)**. Mitosis—which provides for growth, tissue repair, and asexual reproduction—produces daughter cells that are genetically identical to the parent cell. Meiosis, needed for sexual reproduction, yields genetically unique haploid daughter cells—cells with only one member of each homologous chromosome pair.

For both mitosis and meiosis, the chromosomes duplicate only once, in the preceding interphase. Mitosis involves one division of the nucleus and cytoplasm (duplication, then division in half), producing two diploid cells. Meiosis involves two nuclear and cytoplasmic divisions

▶ **Figure 8.15** **Comparing mitosis and meiosis.** The events unique to meiosis occur during meiosis I: In prophase I, duplicated homologous chromosomes pair along their lengths, and crossing over occurs between homologous (nonsister) chromatids. In metaphase I, pairs of homologous chromosomes (rather than individual chromosomes) are aligned at the center of the cell. During anaphase I, sister chromatids of each chromosome stay together and go to the same pole of the cell as homologous chromosomes separate. At the end of meiosis I, there are two haploid cells, but each chromosome still has two sister chromatids.

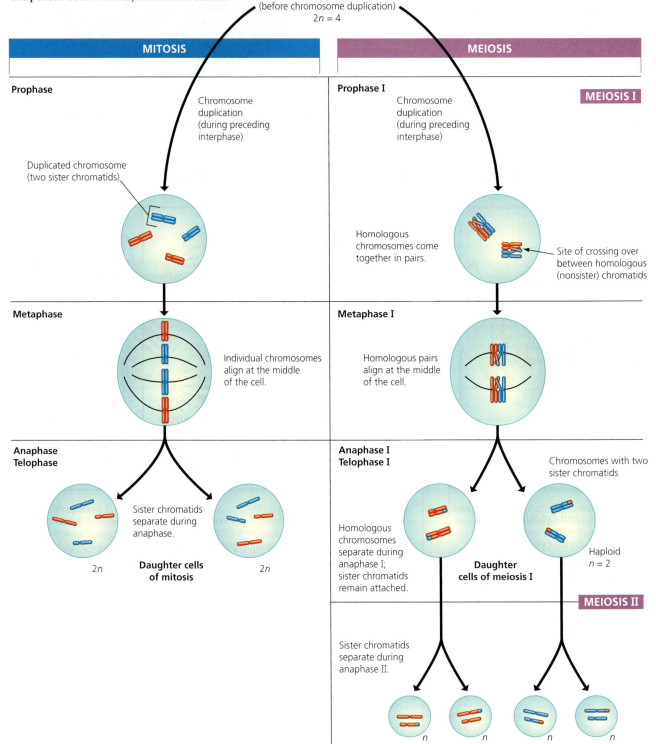

(duplication, division in half, then division in half again), yielding four haploid cells.

In comparing mitosis and meiosis, Figure 8.15 traces these two processes for a diploid parent cell with four chromosomes. As before, homologous chromosomes are those matching in size. Imagine that the red chromosomes were inherited from the mother and the blue chromosomes from the father. Notice that all the events unique to meiosis occur during meiosis I. Meiosis II is virtually identical to mitosis in that it separates sister chromatids. But unlike mitosis, meiosis II yields daughter cells with a haploid set of chromosomes. ✅

The Origins of Genetic Variation

As discussed earlier, offspring that result from sexual reproduction are genetically different from their parents and from one another. How does meiosis produce such genetic variation?

Independent Assortment of Chromosomes

Figure 8.16 illustrates one way in which meiosis contributes to genetic variety. The figure shows how the arrangement of homologous chromosomes at metaphase of meiosis I affects the resulting gametes. Once again, our example is from a hypothetical diploid organism with four chromosomes (two pairs of homologous chromosomes), with colors used to differentiate homologous chromosomes (red for chromosomes inherited from the mother and blue for chromosomes from the father).

When aligned during metaphase I, the side-by-side orientation of each homologous pair of chromosomes is a matter of chance. Either the red or blue chromosome may be on the left or right. Thus, in this example, there are two possible ways that the chromosome pairs can align during metaphase I. In possibility 1, the chromosome pairs are oriented with both red chromosomes on the same side (blue/red and blue/red). In this case, each of the gametes produced at the end of meiosis II has only red or only blue chromosomes (combinations a and b). In possibility 2, the chromosome pairs are oriented differently (blue/red and red/blue). This arrangement produces gametes with one red and one blue chromosome (combinations c and d). Thus, with the two possible arrangements shown in this example, the organism will produce gametes with four different combinations of chromosomes. For a species with more than two pairs of chromosomes, such as humans, every chromosome pair orients independently of all the others at metaphase I. (Chromosomes X and Y behave as a homologous pair in meiosis.)

✅ CHECKPOINT

True or false: Both mitosis and meiosis are preceded by chromosome duplication.

■ *Answer: true*

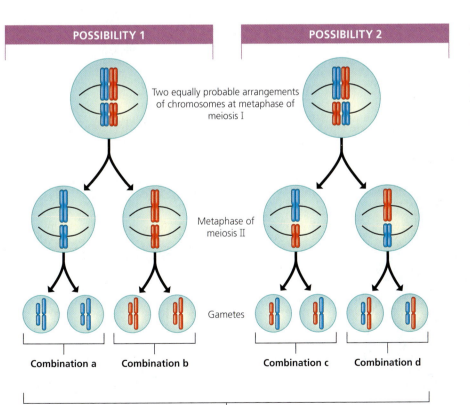

POSSIBILITY 1   **POSSIBILITY 2**

Two equally probable arrangements of chromosomes at metaphase of meiosis I

Metaphase of meiosis II

Gametes

Combination a   **Combination b**   **Combination c**   **Combination d**

Because possibilities 1 and 2 are equally likely, the four possible types of gametes will be made in approximately equal numbers.

◀ **Figure 8.16 Results of alternative arrangements of chromosomes at metaphase of meiosis I.** The arrangement of chromosomes at metaphase I determines which chromosomes will be packaged together in the haploid gametes.

For any species, the total number of chromosome combinations that can appear in gametes is 2^n, where n represents the haploid number. For the hypothetical organism in Figure 8.16, $n = 2$, so the number of chromosome combinations is 2^2, or 4. For a human ($n = 23$), there are 2^{23}, or about 8 million, possible chromosome combinations! This means that every gamete a person produces contains one of about 8 million possible combinations of maternal and paternal chromosomes. When you consider that a human egg cell with about 8 million possibilities is fertilized at random by a human sperm cell with about 8 million possibilities **(Figure 8.17)**, you can see that a single man and a single woman can produce zygotes with 64 trillion combinations of chromosomes!

Crossing Over

So far, we have focused on genetic variety in gametes and zygotes at the whole-chromosome level. We'll now take a closer look at **crossing over**, the exchange of corresponding segments between nonsister chromatids of homologous chromosomes, which occurs during prophase I of meiosis. **Figure 8.18** shows crossing over between two homologous chromosomes and the resulting gametes. At the time that crossing over begins, very early in prophase I, homologous chromosomes are closely paired all along their lengths, with a precise gene-by-gene alignment.

The exchange of segments between nonsister chromatids—one maternal chromatid and one paternal chromatid of a homologous pair—adds to the genetic variety resulting from sexual reproduction. In Figure 8.18, if there were no crossing over, meiosis could produce only two types of gametes: the ones ending up with chromosomes that exactly match the parents' chromosomes, either all blue or all red (as in Possibility 1 in Figure 8.16). With crossing over, gametes arise with chromosomes that are partly from the mother and partly from the father. These chromosomes are called "recombinant" because they result from genetic recombination, the production

▼ **Figure 8.17 The process of fertilization: a close-up view.** Here you see many human sperm contacting an egg. Only one sperm can add its chromosomes to produce a zygote.

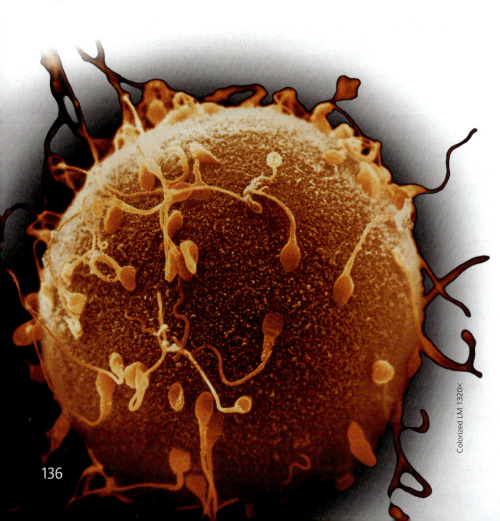

Colorized LM 1320×

▼ **Figure 8.18 The results of crossing over during meiosis for a single pair of homologous chromosomes.** A real cell has multiple pairs of homologous chromosomes that produce a huge variety of recombinant chromosomes in the gametes.

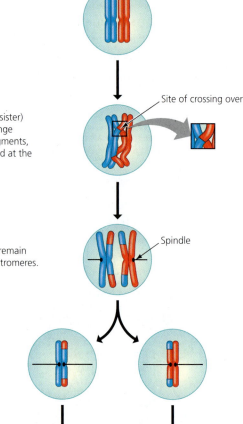

Meiosis I

Duplicated pair of homologous chromosomes

Homologous (nonsister) chromatids exchange corresponding segments, remaining attached at the crossover points.

Site of crossing over

Sister chromatids remain joined at their centromeres.

Spindle

Meiosis II

Recombinant chromosomes combine genetic information originally derived from different parents.

Recombinant chromosomes

of gene combinations different from those carried by the parental chromosomes.

Because most chromosomes contain thousands of genes, a single crossover event can affect many genes.

When we also consider that multiple crossovers can occur in each pair of homologous chromosomes, it's not surprising that gametes and the offspring that result from them are so incredibly varied. ✓

✓ **CHECKPOINT**

Name two events during meiosis that contribute to genetic variety among gametes. During what stages of meiosis does each occur?

■ *Answer: crossing over between homologous chromosomes during prophase I and independent orientation/assortment of the pairs of homologous chromosomes at metaphase I*

THE PROCESS OF SCIENCE | **Life with and without Sex**

Do All Animals Have Sex?

BACKGROUND

As discussed in the Biology and Society section, some species such as zebra sharks can reproduce through both sexual and asexual methods. Although some animal species can reproduce asexually, very few animals reproduce *only* asexually. In fact, biologists have traditionally considered asexual reproduction an evolutionary dead end (for reasons we'll discuss in the Evolution Connection section at the end of the chapter).

To investigate a case in which asexual reproduction seemed to be the norm, researchers from Harvard University studied a group of animals called bdelloid rotifers **(Figure 8.19a)**. This class of nearly microscopic freshwater invertebrates includes about 460 species. Despite hundreds of years of observations, no one has ever found bdelloid rotifer males or evidence of sexual reproduction. Has this entire class of animals reproduced solely by asexual means for tens of millions of years?

METHOD

In most species, the two versions of a gene in a pair of homologous chromosomes are very similar due to the constant trading of genes during sexual reproduction. If a species has survived without sex for millions of years, the researchers reasoned, then changes in the DNA sequences of homologous genes should accumulate independently,

and the two versions of the genes should have significantly diverged from each other over time. The researchers therefore compared the sequences of a series of genes in bdelloid rotifers and in other closely related rotifers who were known to reproduce sexually.

RESULTS

Their results were striking **(Figure 8.19b)**. Among non-bdelloid rotifers that reproduce sexually, homologous versions of the genes were 99.5% identical, differing by only 0.5% on average. In contrast, the versions of the same gene in bdelloid rotifers were only 46 to 96% identical, differing by between 4 and 54%.

These data provided strong evidence that bdelloid rotifers have evolved for millions of years without sex. Recent research has suggested that these fascinating creatures may indeed rarely undergo some type of previously undiscovered sexual reproduction, but precisely what it is and how it works remain unknown. The study of bdelloid rotifers illustrates a general principle: Exploration of unusual cases can provide general insights into nature. In this case, studies of a tiny, obscure creature are shedding light on one of the biggest questions in all of animal biology: Why have sex?

Thinking Like a Scientist

A botanist noticed that a tree growing in a monastery in Brazil produced seedless oranges. Every navel orange tree alive today is derived from this single mutant. What common principle do navel oranges and bdelloid rotifers illustrate?

For the answer, see Appendix D.

▼ **Figure 8.19** A study of bdelloid rotifers.

(a) One of over 460 species of bdelloid rotifers

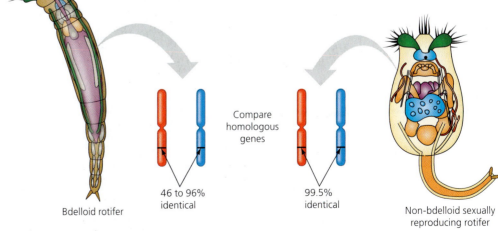

Compare homologous genes

46 to 96% identical

99.5% identical

Bdelloid rotifer

Non-bdelloid sexually reproducing rotifer

(b) An experiment to determine whether bdelloid rotifer chromosomes behave like those of sexually reproducing rotifers

When Meiosis Goes Wrong

So far, our discussion of meiosis has focused on the process as it normally and correctly occurs. But what happens when there is an error in the process? Such a mistake can result in genetic abnormalities that range from mild to severe to fatal.

How Accidents During Meiosis Can Alter Chromosome Number

Within the human body, meiosis occurs repeatedly as the testes or ovaries produce gametes. Almost always, chromosomes are distributed to daughter cells without any errors. But occasionally there is a mishap, called a **nondisjunction**, in which the members of a chromosome pair fail to separate at anaphase. Nondisjunction can occur during meiosis I or II **(Figure 8.20)**. In either case, gametes with abnormal numbers of chromosomes are the result.

Figure 8.21 shows what can happen when an abnormal gamete produced by nondisjunction unites with a normal gamete during fertilization. When a normal sperm fuses with an egg cell with an extra chromosome, the result is a zygote with a total of $2n + 1$ chromosomes. Because mitosis duplicates the

chromosomes as they are, the abnormality will be passed to all embryonic cells. If the organism survives, it will have an abnormal karyotype and probably a medical disorder caused by the abnormal number of genes. Nondisjunction is estimated to be involved in 10–30% of human conceptions and is the main reason for pregnancy loss. Next, we'll examine one particular case of survival nondisjunction. ✓

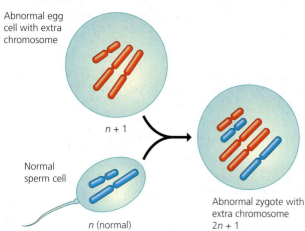

▼ **Figure 8.21** Fertilization after nondisjunction in the mother.

Abnormal egg cell with extra chromosome

$n + 1$

Normal sperm cell

n (normal)

Abnormal zygote with extra chromosome

$2n + 1$

▶ **Figure 8.20 Two types of nondisjunction.** In both examples in the figure, the cell at the top is diploid (2n), with two pairs of homologous chromosomes.

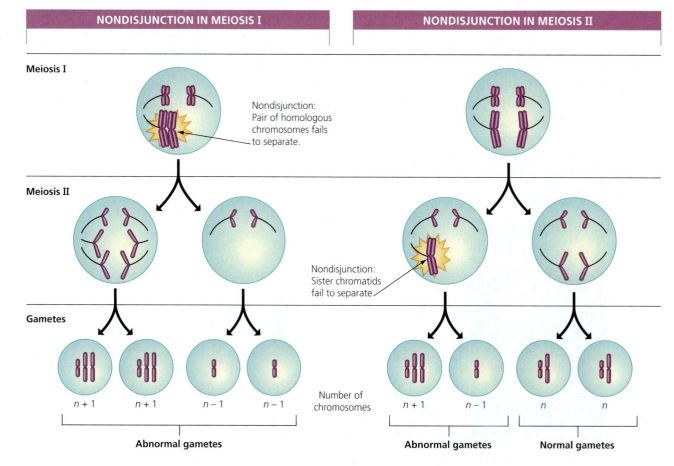

| NONDISJUNCTION IN MEIOSIS I | NONDISJUNCTION IN MEIOSIS II |

Meiosis I

Nondisjunction: Pair of homologous chromosomes fails to separate.

Meiosis II

Nondisjunction: Sister chromatids fail to separate.

Gametes

Number of chromosomes

$n + 1$ $n + 1$ $n - 1$ $n - 1$

$n + 1$ $n - 1$ n n

Abnormal gametes

Abnormal gametes   Normal gametes

Down Syndrome: An Extra Chromosome 21

Figure 8.11 showed a normal human complement of 23 pairs of chromosomes. Compare it with the karyotype in **Figure 8.22**. Besides having two X chromosomes (because it's from a female), the karyotype in Figure 8.21 has three number 21 chromosomes. This person is triploid for chromosome 21 (instead of having the usual diploid condition) and therefore has 47 chromosomes. This condition is called **trisomy 21**.

▼ **Figure 8.22 Trisomy 21 and Down syndrome.** This child displays the characteristic facial features of Down syndrome. The karyotype (bottom) shows trisomy 21; notice the three copies of chromosome 21.

LM 4,000×

Trisomy 21

In most cases, a human embryo with an atypical number of chromosomes is spontaneously aborted (miscarried) long before birth, often before the woman is even aware that she is pregnant. In fact, some doctors speculate that miscarriages due to genetic defects occur in nearly one-quarter of all pregnancies, although this number is difficult to verify. However, some aberrations in chromosome number seem to upset the genetic balance less drastically, and people with such abnormalities can survive. These people usually have a characteristic set of symptoms, called a syndrome. A person with trisomy 21 has a condition called **Down syndrome** (named after John Langdon Down, an English physician who first described this condition in 1866).

Trisomy 21 affects about 1 out of every 850 children and is the most common chromosome number abnormality and most common serious birth defect in the United States. Down syndrome includes characteristic facial features—frequently a fold of skin at the inner corner of the eye, a round face, and a flattened nose—as well as short stature, heart defects, and susceptibility to leukemia and Alzheimer's disease. People with Down syndrome usually have a life span shorter than normal. They also exhibit varying degrees of developmental delays. However, with proper care, many people with Down syndrome live to middle age or beyond, and many are socially adept, live independently, and hold jobs. Although no one is sure why, the risk of Down syndrome increases with the age of the mother, climbing to about 1% risk for mothers at age 40. Therefore, the fetuses of pregnant women age 35 and older are candidates for chromosomal prenatal screenings (see Chapter 9). ✅

Abnormal Numbers of Sex Chromosomes

Trisomy 21 is an autosomal nondisjunction. Nondisjunction in meiosis can also lead to abnormal numbers of sex chromosomes, X and Y. Unusual numbers of sex chromosomes seem to upset the genetic balance less than unusual numbers of autosomes. This may be because the Y chromosome is very small and carries relatively few genes. Furthermore, mammalian cells normally operate with only one functioning X chromosome because other copies of the chromosome become inactivated in each cell (see Chapter 11).

DUPLICATING THE WRONG NUMBER OF CHROMOSOMES IS ALMOST ALWAYS FATAL.

Table 8.1	Abnormalities of Sex Chromosome Number in Humans	
Sex Chromosomes	Syndrome	Symptoms
XXY	Klinefelter syndrome (male)	Sterile; underdeveloped testes; secondary female characteristics
XYY	None (normal male)	Slightly taller than average
XXX	None (normal female)	Slightly taller than average; slight risk of learning disabilities
XO	Turner syndrome (female)	Sterile; immature sex organs

Table 8.1 lists the most common human sex chromosome abnormalities. An extra X chromosome in a male, making him XXY, produces a condition called Klinefelter syndrome. If untreated, men with this disorder have abnormally small testes, are sterile, and often have breast enlargement and other feminine body contours. These symptoms can be reduced through administration of the sex hormone testosterone. Klinefelter syndrome is also found in men with more than three sex chromosomes, such as XXYY, XXXY, or XXXXY. These abnormal numbers of sex chromosomes result from multiple nondisjunctions; such men are more likely to have developmental disabilities than XY or XXY men.

Human males with a single extra Y chromosome (XYY) do not have any well-defined syndrome, although they tend to be taller than average. Females with an extra X chromosome (XXX) cannot be distinguished from XX females except by karyotype. Such women tend to be slightly taller than average and have a higher risk of learning disabilities.

Females who are lacking an X chromosome are designated XO; the O indicates the absence of a second sex chromosome. These women have Turner syndrome. They have a characteristic appearance, including short stature and often a web of skin extending between the neck and shoulders. Women with Turner syndrome are of normal intelligence but are sterile. If left untreated, they have poor development of breasts and other secondary sex characteristics. Administration of estrogen can alleviate those symptoms. The XO condition is the sole known case where having only 45 chromosomes is not fatal in humans.

Notice the crucial role of the Y chromosome in determining a person's sex. In general, having at least one Y chromosome produces biological maleness, regardless of the number of X chromosomes. The absence of a Y chromosome results in biological femaleness. ✓

☑ **CHECKPOINT**

Why is a person more likely to survive with an abnormal number of sex chromosomes than an abnormal number of autosomes?

■ *Answer: because the Y chromosome is very small and extra X chromosomes are inactivated*

EVOLUTION CONNECTION　Life with and without Sex

The Advantages of Sex

▼ **Figure 8.23 Sexual and asexual reproduction.** Many plants, such as this strawberry, have the ability to reproduce both sexually (through flowers that produce fruit) and asexually (through runners).

Throughout this chapter, we've examined cell division within the context of reproduction. Like the zebra shark discussed in the Biology and Society section, many species (including a few dozen animal species, but many more within the plant kingdom) can reproduce both sexually and asexually **(Figure 8.23)**. An important advantage of asexual reproduction is that there is no need for a partner. Asexual reproduction may thus confer an evolutionary advantage when organisms are sparsely distributed (on an isolated island, for example) and unlikely to meet a mate. Furthermore, if an organism is superbly suited to a stable environment, asexual reproduction has the advantage of passing on its entire genetic legacy intact. Asexual reproduction also eliminates the need to expend energy forming gametes and copulating with a partner.

In contrast to plants, the vast majority of animals reproduce by sexual means. There are exceptions, such as the few species that can reproduce by parthenogenesis and the bdelloid rotifers discussed in the Process of Science section. But most animals reproduce only through sex. In fact, asexual reproduction seems to be an evolutionary "dead end" in the animal kingdom. Therefore, sex must enhance evolutionary fitness. But how? The answer remains elusive. Most hypotheses focus on the unique combinations of genes formed during meiosis and fertilization. By producing offspring of varied genetic makeup, sexual reproduction may enhance survival by speeding adaptation to a changing environment. Another idea is that shuffling genes during sexual reproduction might reduce the incidence of harmful genes more rapidly and allow for disadvantageous mutations to be more quickly eliminated from the gene pool. But for now, one of biology's most basic questions—Why have sex?—remains a hotly debated topic that is the focus of much ongoing research.

Runner ⟶

Chapter Review

SUMMARY OF KEY CONCEPTS

What Cell Reproduction Accomplishes

Cell reproduction, also called cell division, produces genetically identical cells:

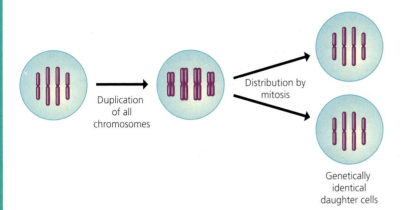

Duplication of all chromosomes

Distribution by mitosis

Genetically identical daughter cells

Some organisms use mitosis (ordinary cell division) to reproduce. This is called asexual reproduction, and it results in offspring that are genetically identical to the lone parent and to each other. Mitosis also enables multicellular organisms to grow and develop and to replace damaged or lost cells. Organisms that reproduce sexually, by the union of a sperm with an egg cell, carry out meiosis, a type of cell division that yields gametes with only half as many chromosomes as body (somatic) cells.

The Cell Cycle and Mitosis

Eukaryotic Chromosomes

The genes of a eukaryotic genome are grouped into multiple chromosomes in the nucleus. Each chromosome contains one very long DNA molecule, with many genes, that is wrapped around histone proteins. Individual chromosomes are coiled up and therefore visible with a light microscope only when the cell is in the process of dividing; otherwise, they are in the form of thin, loosely packed chromatin fibers.

Duplicating Chromosomes

Because chromosomes contain the information needed to control cellular processes, they must be copied and distributed to daughter cells. Before a cell starts dividing, the chromosomes duplicate, producing sister chromatids (containing identical DNA) joined together at the centromere.

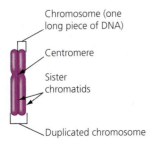

Chromosome (one long piece of DNA)

Centromere

Sister chromatids

Duplicated chromosome

The Cell Cycle

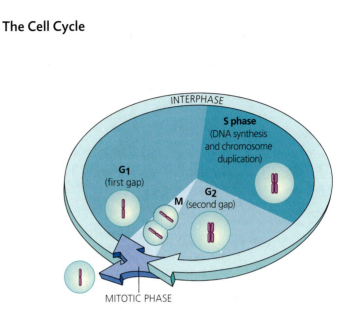

INTERPHASE

S phase (DNA synthesis and chromosome duplication)

G₁ (first gap)

G₂ (second gap)

M

MITOTIC PHASE

Mitosis and Cytokinesis

Mitosis is divided into four phases: prophase, metaphase, anaphase, and telophase. At the start of mitosis, the chromosomes coil up and the nuclear envelope breaks down (prophase). Next, a mitotic spindle made of microtubule tracks moves the chromosomes to the middle of the cell (metaphase). The sister chromatids then separate and are moved to opposite poles of the cell (anaphase), where two new nuclei form (telophase). Cytokinesis overlaps the end of mitosis. In animals, cytokinesis occurs by cleavage, which pinches the cell in two. In plants, a membranous cell plate divides the cell in two. Mitosis and cytokinesis produce genetically identical cells.

Cancer Cells: Dividing Out of Control

When the cell cycle control system malfunctions, a cell may divide excessively and form a tumor. Cancer cells may grow to form malignant tumors, invade other tissues (metastasize), and even kill the host. Surgery can remove tumors, and radiation and chemotherapy are effective as treatments because they interfere with cell division. You can increase the likelihood of surviving some forms of cancer through lifestyle changes and regular screenings.

Meiosis, the Basis of Sexual Reproduction

Homologous Chromosomes

The somatic cells (body cells) of each species contain a specific number of chromosomes; human cells have 46, made up of 23 pairs of homologous chromosomes. The chromosomes of a homologous pair carry genes for the same characteristics at the same places. Mammalian males have X and Y sex chromosomes (only partly homologous), and females have two X chromosomes.

Gametes and the Life Cycle of a Sexual Organism

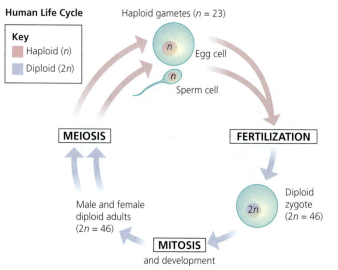

Human Life Cycle

Haploid gametes (*n* = 23)

Key
- Haploid (*n*)
- Diploid (2*n*)

n Egg cell

n Sperm cell

MEIOSIS

FERTILIZATION

Male and female diploid adults (2*n* = 46)

Diploid zygote (2*n* = 46)

2*n*

MITOSIS
and development

The Process of Meiosis

Meiosis, like mitosis, is preceded by chromosome duplication. But in meiosis, the cell divides twice to form four daughter cells. The first division, meiosis I, starts with the pairing of homologous chromosomes. In crossing over, homologous chromosomes exchange corresponding segments. Meiosis I separates the members of the homologous pairs and produces two daughter cells, each with one set of (duplicated) chromosomes. Meiosis II is essentially the same as mitosis; in each of the cells, the sister chromatids of each chromosome separate.

Review: Comparing Mitosis and Meiosis

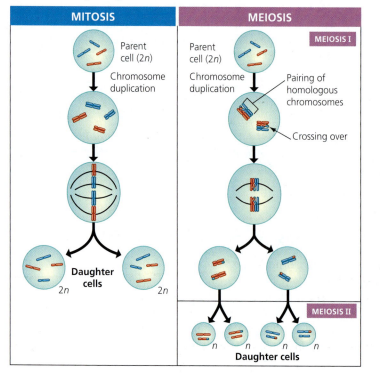

| MITOSIS | MEIOSIS |

MEIOSIS I

Parent cell (2*n*)

Parent cell (2*n*)

Chromosome duplication

Chromosome duplication

Pairing of homologous chromosomes

Crossing over

Daughter cells

2*n*   2*n*

MEIOSIS II

n   *n*   *n*   *n*

Daughter cells

The Origins of Genetic Variation

Because the chromosomes of a homologous pair come from different parents, they carry different versions of many of their genes. The large number of possible arrangements of chromosome pairs at metaphase of meiosis I leads to many different combinations of chromosomes in eggs and sperm. Random fertilization of eggs by sperm greatly increases the variation. Crossing over during prophase of meiosis I increases variation still further.

When Meiosis Goes Wrong

An abnormal number of chromosomes can cause problems. Down syndrome is caused by an extra copy of chromosome 21, a result of nondisjunction, the failure of homologous chromosomes to separate during meiosis I or of sister chromatids to separate during meiosis II. Nondisjunction can also produce gametes with extra or missing sex chromosomes, which lead to varying degrees of malfunction but do not usually affect survival.

Mastering Biology

For practice quizzes, BioFlix animations, MP3 tutorials, video tutors, and more study tools designed for this textbook, go to Mastering Biology™

SELF-QUIZ

1. Which of the following is not a function of mitosis in humans?
   a. repair of wounds
   b. growth
   c. production of gametes from diploid cells
   d. replacement of lost or damaged cells

2. In what sense are the two daughter cells produced by mitosis identical?

3. Why is it hard to observe individual chromosomes in interphase?

4. A biochemist measures the amount of DNA in cells growing in the laboratory. The quantity of DNA in a cell would be found to double
   a. between prophase and anaphase of mitosis.
   b. between the G_1 and G_2 phases of the cell cycle.
   c. during the M phase of the cell cycle.
   d. between prophase I and prophase II of meiosis.

5. What phases of mitosis are opposites in terms of changes in the nucleus?

6. Complete the following table to compare mitosis and meiosis.

	Mitosis	Meiosis
a. Number of chromosomal duplications		
b. Number of cell divisions		
c. Number of daughter cells produced		
d. Number of chromosomes in daughter cells		
e. How chromosomes line up during metaphase		
f. Genetic relationship of daughter cells to parent cells		
g. Functions performed in the human body		

7. If an intestinal cell in a dog contains 78 chromosomes, a dog sperm cell would contain _____ chromosomes.

8. A micrograph of a dividing cell from a mouse shows 19 chromosomes, each consisting of two sister chromatids. During which stage of meiosis could this micrograph have been taken? (Explain your answer.)

9. Tumors that remain at their site of origin are called _____, and tumors that can migrate to other body tissues are called _____.

10. A diploid body (somatic) cell from a fruit fly contains eight chromosomes. This means that _____ different combinations of chromosomes are possible in its gametes.

11. Although nondisjunction is a random event, there are many more people with an extra chromosome 21, which causes Down syndrome, than people with an extra chromosome 3 or chromosome 16. Explain.

For answers to the Self Quiz, see Appendix D.

IDENTIFYING MAJOR THEMES

For each statement, identify which major theme is evident (the relationship of structure to function, information flow, pathways that transform energy and matter, interactions within biological systems, or evolution) and explain how the statement relates to the theme. If necessary, review the themes (see Chapter 1) and review the examples highlighted in blue in this chapter.

12. Passing along genetic instructions from one generation to the next requires a precise duplication of the chromosomes.

13. The combined actions of many different proteins determine whether a cell will divide or not.

14. By examining the compactness of a chromosome region (loose versus tightly wound), you can gain insight into whether the genes in that region are actively being used.

For answers to Identifying Major Themes, see Appendix D.

THE PROCESS OF SCIENCE

15. A mule is the offspring of a horse and a donkey. A donkey sperm contains 31 chromosomes and a horse egg has 32, so the zygote has 63. The zygote develops normally. However, a mule is sterile; meiosis cannot occur normally in its testes or ovaries. Explain why mitosis is normal in cells containing both horse and donkey chromosomes, but the mixed set interferes with meiosis.

16. You prepare a slide with a thin slice of an onion root tip. You see the following view in a light microscope. Identify the stage of mitosis for each of the outlined cells, a–d.

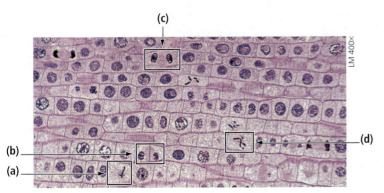

17. **Interpreting Data** The graph shows the incidence of Down syndrome in the offspring of normal parents as the age of the mother increases. For women under the age of 30, how many infants with Down syndrome are born per 1,000 births? At age 40? At age 50? How many times more likely is a 50-year-old woman to give birth to a baby with Down syndrome than a 30-year-old woman?

BIOLOGY AND SOCIETY

18. If an endangered species can reproduce by parthenogenesis, what implications might this have on efforts to repopulate that species? How might a parthenogenesis program harm such a species?

19. Every year, about a million Americans are diagnosed with cancer. This means that about 75 million Americans now living will eventually have cancer, and one in five will die of the disease. There are many kinds of cancers and many causes, such as smoking, overexposure to ultraviolet rays, a high-fat and low-fiber diet, and some workplace chemicals. Hundreds of millions of dollars are spent each year on searching for treatments, yet far less money is spent on prevention. Should we devote more resources to treating cancer or to preventing it? Explain.

20. The practice of buying and selling gametes, particularly eggs from fertile women, is becoming common in the United States and some other industrialized nations. Do you object to this type of transaction? Would you sell your gametes? At any price? Whether you would do so or not, should this practice be restricted?

21. The anticancer drug Taxol freezes the mitotic spindle after it forms, which may stop a tumor from growing. It is made from a chemical in the bark of the Pacific yew, a tree found mainly in the northwestern United States. It has fewer side effects than many other anticancer drugs and seems effective against some hard-to-treat cancers of the ovary and breast. Another drug, vinblastine, prevents the mitotic spindle from forming. It was first obtained from the periwinkle plant, native to the tropical rain forests of Madagascar. Given these examples, preserving biodiversity may be the key to discovering lifesaving anticancer drugs. Such drugs are often discovered in developing areas, but because they are expensive, they primarily benefit people in developed regions. Do you see a conflict in this? Explain.

Genetics and Heredity

Heredity is the transmission of traits from one generation to the next. **Genetics**, the scientific study of heredity, began in the 1860s, when a monk named Gregor Mendel **(Figure 9.1)** deduced its fundamental principles by breeding garden peas. Mendel lived and worked in an abbey in Brunn, Austria (now Brno, in the Czech Republic). Strongly influenced by his study of physics, mathematics, and chemistry at the University of Vienna, his research was both experimentally and mathematically rigorous, and these qualities were largely responsible for his success.

In a paper published in 1866, Mendel correctly argued that parents pass on to their offspring discrete genes (which he called "heritable factors") that are responsible for inherited traits, such as purple flowers or round seeds in pea plants. (It is interesting to note that Mendel's publication came just seven years after Darwin's 1859 publication of *On the Origin of Species*, making the 1860s a banner decade in the advent of modern biology.) In his paper, one of the most influential in the history of biology, Mendel stressed that genes retain their individual identities generation after generation. That is, genes are like playing cards: A deck may be shuffled, but all the cards retain their identities, and no card is ever blended with another. Similarly, genes may be sorted, but each gene retains its intact identity.

▲ **Figure 9.1** Gregor Mendel.

wanted to fertilize one plant with pollen from a different plant, he pollinated the plants by hand, as shown in **Figure 9.3**. Thus, Mendel was always sure of the parentage of his new plants.

Each of the characters Mendel chose to study, such as flower color, occurred in two distinct traits. Mendel worked with his plants until he was sure he had purebred varieties—that is, varieties for which self-fertilization produced offspring all identical to the parent. For instance, he identified a purple-flowered variety that, when self-fertilized, always produced offspring plants that had all purple flowers.

Next Mendel was ready to ask what would happen when he crossed different purebred varieties with each other. For example, what offspring would result if plants

In an Abbey Garden

Mendel chose to study garden peas because they were easy to grow and they came in many readily distinguishable varieties. For example, one variety has purple flowers and another variety has white flowers. A heritable feature that varies among individuals, such as flower color, is called a **character**. Each variant of a character, such as purple or white flowers, is called a **trait**.

Perhaps the most important advantage of pea plants as an experimental model was that Mendel could strictly control their reproduction. The petals of a pea flower **(Figure 9.2)** almost completely enclose the egg-producing organ (the carpel) and the sperm-producing organs (the stamens). Consequently, in nature, pea plants usually self-fertilize because sperm-carrying pollen grains released from the stamens land on the tip of the egg-containing carpel of the same flower. Mendel could ensure self-fertilization by covering a flower with a small bag so that no pollen from another plant could reach the carpel. When he

▼ **Figure 9.2 The structure of a pea flower.** To reveal the reproductive organs—the stamens and carpel—one of the petals is not shown in this drawing.

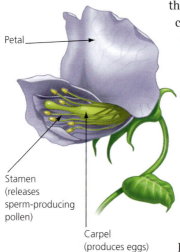

Petal

Stamen (releases sperm-producing pollen)

Carpel (produces eggs)

▼ **Figure 9.3** Mendel's technique for cross-fertilizing pea plants.

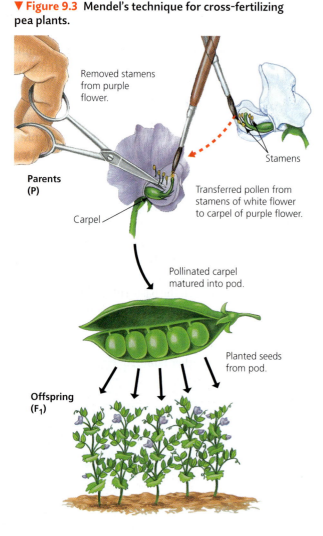

Removed stamens from purple flower.

Stamens

Parents (P)

Carpel

Transferred pollen from stamens of white flower to carpel of purple flower.

Pollinated carpel matured into pod.

Planted seeds from pod.

Offspring (F₁)

with purple flowers and plants with white flowers were cross-fertilized as shown in Figure 9.3? The offspring of two different purebred varieties are called **hybrids**, and the cross-fertilization itself is referred to as a genetic **cross**. The purebred parents are called the **P generation**, and their hybrid offspring are the **F₁ generation** (F for *filial*, from the Latin for "son" or "daughter"). When F₁ plants self-fertilize or fertilize each other, their offspring are the **F₂ generation**. ✅

Mendel's Law of Segregation

Mendel performed many experiments in which he tracked the inheritance of characters, such as flower color, that occur as two alternative traits **(Figure 9.4)**. The results led him to formulate several hypotheses about inheritance. Let's look at some of his experiments and follow the reasoning that led to his hypotheses.

▼ **Figure 9.4 The seven characters of pea plants studied by Mendel.** Each character comes in the two alternative traits shown here.

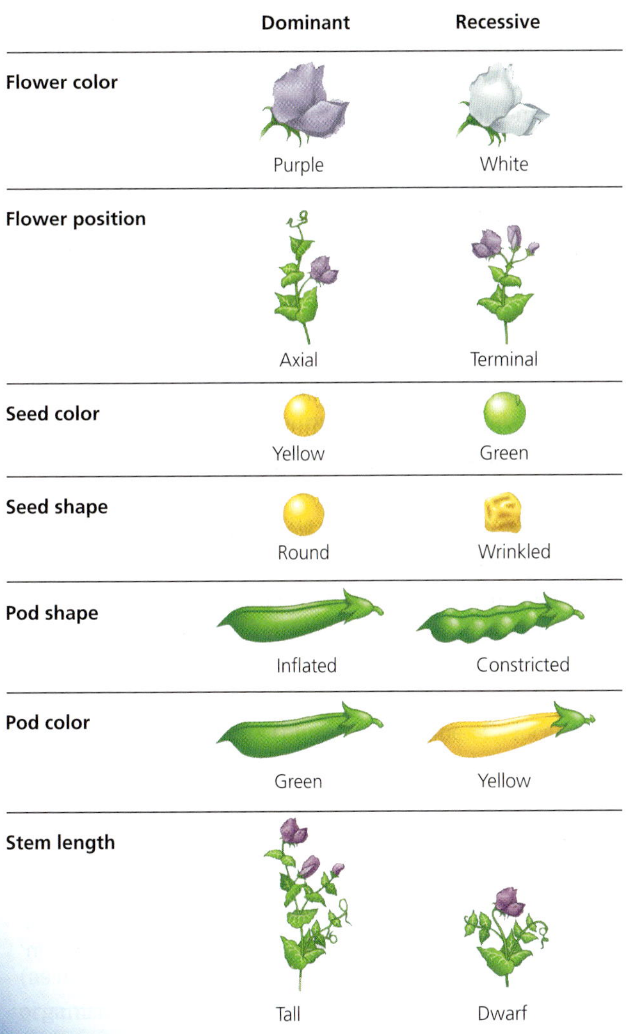

	Dominant	Recessive
Flower color	Purple	White
Flower position	Axial	Terminal
Seed color	Yellow	Green
Seed shape	Round	Wrinkled
Pod shape	Inflated	Constricted
Pod color	Green	Yellow
Stem length	Tall	Dwarf

Imagine you have two purebred dogs, one male and one female. Use the correct genetic names to describe the next two possible generations.

■ Answer: *The purebred parents are the P generation. Their puppies are the F₁ generation. If those puppies are bred with each other, they would produce the F₂ generation.*

Variations on Mendel's Laws

Mendel's two laws explain inheritance in terms of genes that are passed along from generation to generation according to simple rules of probability. These laws are valid for all sexually reproducing organisms, including garden peas, Labrador retrievers, and people. But just as the basic rules of musical harmony cannot account for all the rich sounds of a symphony, Mendel's laws stop short of explaining some patterns of genetic inheritance. In fact, for most sexually reproducing organisms, cases in which Mendel's rules can strictly account for the patterns of inheritance are relatively rare. More often, the observed inheritance patterns are more complex. Next, we'll look at several extensions to Mendel's laws that help account for this complexity.

Incomplete Dominance in Plants and People

The F_1 offspring of Mendel's pea crosses always looked like one of the two parent plants. In such situations,

the dominant allele has the same effect on the phenotype whether present in one or two copies. But for some characters, the appearance of F_1 hybrids falls between the phenotypes of the two parents, an effect called **incomplete dominance**. For instance, when red snapdragons are crossed with white snapdragons, all the F_1 hybrids have pink flowers **(Figure 9.18)**. And in the F_2 generation, the genotypic ratio and the phenotypic ratio are the same: 1:2:1.

We also see examples of incomplete dominance in people. One case involves a recessive allele (h) that causes hypercholesterolemia, dangerously high levels of cholesterol in the blood. Normal individuals are homozygous dominant, HH. Heterozygotes (Hh) have blood cholesterol levels about twice what is normal. Such heterozygotes are very prone to cholesterol buildup in artery walls and may have heart attacks from blocked heart arteries by their mid-30s. Hypercholesterolemia is even more serious in homozygous individuals (hh). Homozygotes have about five times the normal amount of blood cholesterol and may have heart attacks as early as age 2. If we look at the molecular basis for hypercholesterolemia, we can understand the intermediate phenotype of heterozygotes **(Figure 9.19)**. The H allele specifies a cell-surface receptor protein that liver cells use to mop up excess low-density lipoprotein (LDL, or "bad cholesterol") from the blood. With only half as many receptors as HH individuals, heterozygotes can remove much less excess cholesterol.

▼ **Figure 9.18 Incomplete dominance in snapdragons.**
Compare this diagram with Figure 9.6, where one of the alleles displays complete dominance.

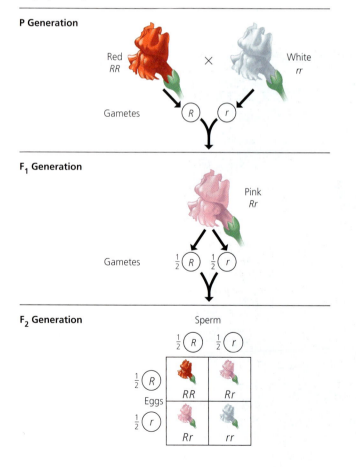

▼ **Figure 9.19 Incomplete dominance in human hypercholesterolemia.**
LDL (low-density lipoprotein) receptors on liver cells promote the breakdown of cholesterol carried in the bloodstream by LDL. This process helps prevent the accumulation of cholesterol in the arteries. Having too few receptors allows dangerous levels of LDL to build up in the blood.

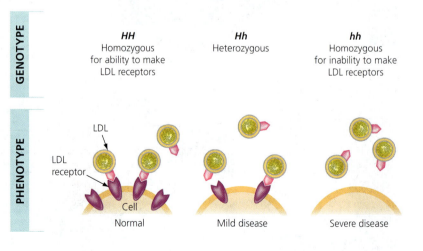

ABO Blood Groups:
An Example of Multiple Alleles and Codominance

So far, we have discussed inheritance patterns involving only two alleles per gene (*H* versus *h*, for example). But most genes can be found in populations in more than two forms, known as multiple alleles. Although each individual carries, at most, two different alleles for a particular gene, in cases of multiple alleles, more than two possible alleles exist in the population.

The **ABO blood groups** in humans involve three alleles of a single gene. Various combinations of these three alleles produce four phenotypes: A person's blood type may be A, B, AB, or O. These letters refer to two carbohydrates, designated A and B, that may be found on the surface of red blood cells **(Figure 9.20)**. A person's red blood cells may be coated with carbohydrate A (giving them type A blood), carbohydrate B (type B), both (type AB), or neither (type O). (In case you are wondering, the "positive" and "negative" notations associated with blood types—referred to as the Rh blood group system—are due to inheritance of a separate, unrelated gene.)

Matching compatible blood groups is critical for safe blood transfusions. If a donor's blood cells have a carbohydrate (A or B) that is foreign to the recipient, then the recipient's immune system produces blood proteins called antibodies that bind to the foreign carbohydrates and cause the donor blood cells to clump together, potentially killing the recipient.

The four blood groups result from various combinations of the three different alleles: I^A (for the ability to make substance A), I^B (for B), and i (for neither A nor B). Each person inherits one of these alleles from each parent. Because there are three alleles, there are six possible genotypes, as listed in Figure 9.21. Both the I^A and I^B alleles are dominant to the i allele. Thus, I^AI^A and I^Ai people have type A blood, and I^BI^B and I^Bi people have type B. Recessive homozygotes (*ii*) have type O blood with neither carbohydrate. Finally, people of genotype I^AI^B make *both* carbohydrates. In other words, the I^A and I^B alleles are **codominant**, meaning that both alleles are expressed in heterozygous individuals (I^AI^B) who have type AB blood. Notice that type O blood reacts with no others, making such a person a universal donor. A person with type AB blood is a universal receiver. Be careful to distinguish codominance (the expression of both alleles) from incomplete dominance (the expression of one intermediate trait). ✓

✅ **CHECKPOINT**

1. Why is a testcross unnecessary to determine whether a snapdragon with red flowers is homozygous or heterozygous?

2. Maria has type O blood, and her sister has type AB blood. What are the genotypes of the girls' parents?

Answers: 1. Only plants homozygous for the dominant allele have red flowers; heterozygotes have pink flowers. 2. One parent is I^Ai, and the other parent is I^Bi.

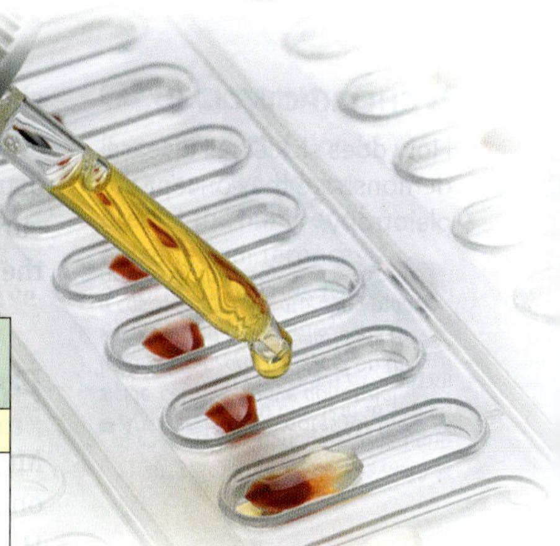

▼ **Figure 9.20 Multiple alleles for the ABO blood groups.** The three versions of the gene responsible for blood type may produce carbohydrate A (allele I^A), carbohydrate B (allele I^B), or neither carbohydrate (allele i). Because each person carries two alleles, six genotypes are possible that result in four different phenotypes. The clumping reaction that occurs between antibodies and foreign blood cells is the basis of blood-typing (shown in the photograph at right) and of the adverse reaction that occurs when someone receives a transfusion of incompatible blood.

Blood Group (Phenotype)	Genotypes	Red Blood Cells	Antibodies Present in Blood	Reactions When Blood from Groups Below Is Mixed with Antibodies from Groups at Left			
				O	A	B	AB
A	I^AI^A or I^Ai	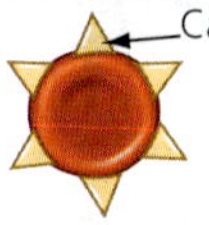 Carbohydrate A	Anti-B	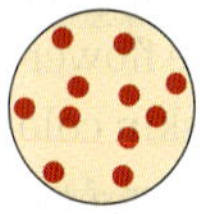		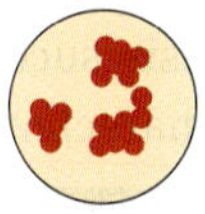	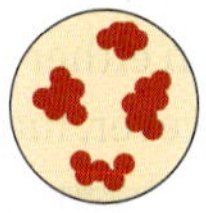
B	I^BI^B or I^Bi	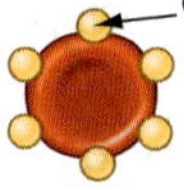 Carbohydrate B	Anti-A		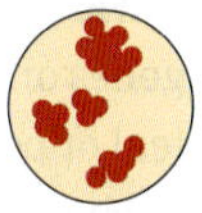	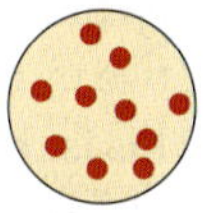	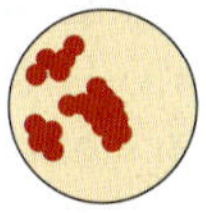
AB	I^AI^B	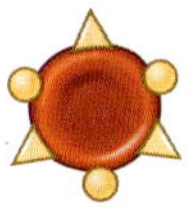	—		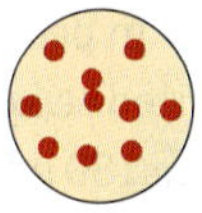	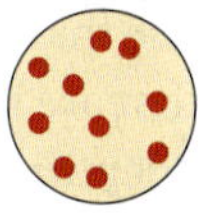	
O	*ii*		Anti-A Anti-B		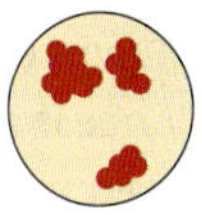	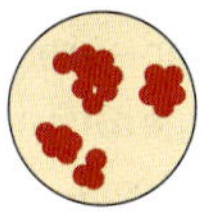	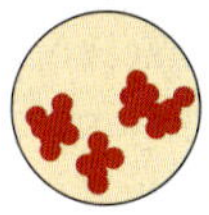

Can DNA and RNA Vaccines Protect Against Viruses?

BACKGROUND

West Nile virus first appeared in the United States in 1999. Most people with the virus do not become ill. But in some cases, the virus causes a potentially fatal swelling of the central nervous system. The virus is spread by mosquitoes. There is no vaccine and no cure for those who become ill.

Medical researchers are hoping to improve our arsenal against West Nile virus by developing DNA and RNA vaccines. Vaccines trigger an immune response to a harmless molecule that mimics some part of the attacking pathogen. Once exposed, the immune system is primed to fight the disease if it detects the real pathogen in the future. A traditional vaccine contains the trigger protein. An RNA or DNA vaccine contains a copy of a viral gene and lets the patient's body make the specific trigger protein.

METHOD

Researchers tested an RNA vaccine containing a gene from West Nile virus using cats and dogs, which, like most mammals, are susceptible to the virus **(Figure 10.29a)**. The cats were given high doses of the vaccine, low doses, or placebos (ineffective treatments that serve as a control). The dogs received only low doses or placebos. All animals received booster vaccines 28 days after their first dose. After that time, the animals were exposed to mosquitoes carrying the virus and tested for the presence of infection.

RESULTS

The results were striking **(Figure 10.29b)**. No animals became infected with the virus after a high dose of vaccine. Clearly, the new vaccine was quite effective, at least among cats and dogs. Although no DNA or RNA vaccines have yet been approved for human use, this type of research suggests that they may became a standard tool in the near future.

TREATMENT	% OF CATS INFECTED	% OF DOGS INFECTED
Placebo	82	93
Low vaccine dose	12.5	0
High vaccine dose	0	(no dose given)

(b) Effects of the RNA vaccine on cats and dogs

▶ **Figure 10.29** Testing an RNA vaccine against West Nile virus.

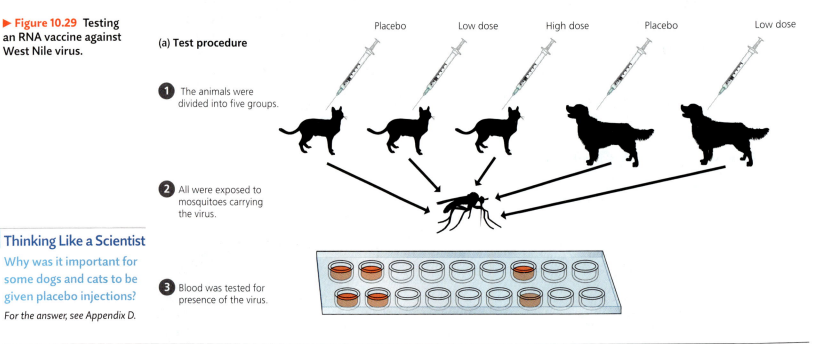

(a) Test procedure

1 The animals were divided into five groups.

2 All were exposed to mosquitoes carrying the virus.

3 Blood was tested for presence of the virus.

Placebo   Low dose   High dose   Placebo   Low dose

Thinking Like a Scientist

Why was it important for some dogs and cats to be given placebo injections?

For the answer, see Appendix D.

HIV, the AIDS Virus

The devastating disease **AIDS** (acquired immunodeficiency syndrome) is caused by **HIV** (human immunodeficiency virus), an RNA virus with some nasty twists. In outward appearance, HIV **(Figure 10.30)** resembles the mumps or flu virus. Its envelope enables HIV to enter and leave a cell much the way the mumps virus does. But HIV has a different mode of reproduction. It is a **retrovirus**, an RNA virus that reproduces by means of a DNA molecule, the reverse of the usual DNA → RNA flow of genetic information. These viruses carry molecules of an enzyme called **reverse transcriptase**, which catalyzes reverse transcription: the synthesis of DNA on an RNA template.

▼ **Figure 10.30** HIV, the AIDS virus.

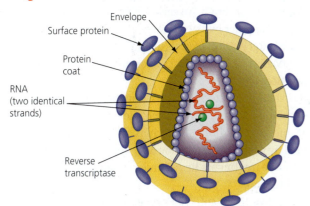

Envelope
Surface protein
Protein coat
RNA (two identical strands)
Reverse transcriptase

Figure 10.31 illustrates what happens after HIV RNA is uncoated in the cytoplasm of a cell. The reverse transcriptase (green) **1** uses the RNA as a template to make a DNA strand and then **2** adds a second, complementary DNA strand. **3** The resulting double-stranded viral DNA then enters the cell nucleus and inserts itself into the chromosomal DNA,

▼ **Figure 10.31** The behavior of HIV nucleic acid in an infected cell.

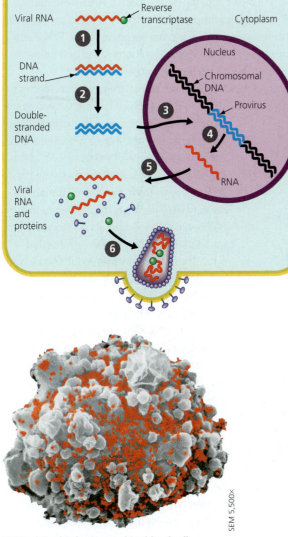

Viral RNA
Reverse transcriptase
Cytoplasm
1
DNA strand
Nucleus
Chromosomal DNA
2
3
Provirus
Double-stranded DNA
4
5
RNA
Viral RNA and proteins
6

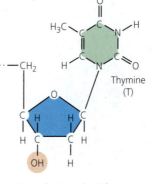

HIV (red dots) infecting a white blood cell

SEM 5,500×

becoming a **provirus**. Occasionally, the provirus is **4** transcribed into RNA **5** and translated into viral proteins. **6** New viruses assembled from these components eventually leave the cell and can then infect other cells. This is the standard reproductive cycle for retroviruses.

HIV infects and eventually kills several kinds of white blood cells that are important in the body's immune system. The loss of such cells causes the body to become susceptible to other infections that it would normally be able to fight off. Such secondary infections cause the syndrome (a collection of symptoms) that eventually kills AIDS patients. Since it was first recognized in 1981, HIV has infected tens of millions of people worldwide, resulting in millions of deaths.

Although there is as yet no cure for AIDS, its progression can be slowed by two categories of anti-HIV drugs. Both types of medicine interfere with the reproduction of the virus. The first type inhibits the action of enzymes called proteases, which help produce the final versions of HIV proteins. The second type, which includes the drug AZT, inhibits the action of the HIV enzyme reverse transcriptase. The key to AZT's effectiveness is its shape. The shape of a molecule of AZT is very similar to the shape of part of the T (thymine) nucleotide (**Figure 10.32**). In fact, AZT's shape is so similar to the T nucleotide that AZT can bind to reverse transcriptase, essentially taking the place of T. But unlike thymine, AZT cannot be incorporated into a growing DNA chain. Thus, AZT "gums up the works," interfering with the synthesis of HIV DNA. Because this synthesis is an essential step in the reproductive cycle of HIV, AZT may block the spread of the virus within the body.

Many HIV-infected people in the United States and other industrialized countries take a "drug cocktail" that contains both reverse transcriptase inhibitors and protease inhibitors, and the combination seems to be much more effective than the individual drugs in keeping the virus at bay and extending patients' lives. In fact, the death rate from HIV infection can be lowered by 80% with proper treatment. However, even in combination, the drugs do not completely rid the body of the virus. Typically, HIV reproduction and the symptoms of AIDS return if a patient discontinues the medications. Because AIDS has no cure yet, prevention (namely, avoiding unprotected sex and staying away from needle sharing) is the only healthy option. ✅

☑ **CHECKPOINT**

Why is HIV called a retrovirus?

■ *Answer: Because it synthesizes DNA from its RNA genome. This is the reverse ("retro") of the usual DNA → RNA information flow.*

▼ **Figure 10.32** AZT and the T nucleotide. The anti-HIV drug AZT (right) has a chemical shape very similar to part of the T (thymine) nucleotide of DNA.

Thymine (T)

Part of a T nucleotide

AZT

191

MAD COW
DISEASE IS
CAUSED BY
ODDLY SHAPED
PROTEINS.

☑ CHECKPOINT

What makes prions so
unusual as pathogens?

■ Answer: Prions, unlike any other infectious agent, have no nucleic acid (DNA or RNA).

Prions

Prions are infectious proteins that cause brain diseases in several animal species. While a virus contains DNA or RNA, a prion consists solely of a misfolded form of a normal brain protein. When the prion gets into a cell containing the normal form of the protein, the prion somehow converts normal protein molecules to misfolded versions. The misfolded proteins then clump together, disrupting brain functions.

Diseases caused by prions include scrapie in sheep; chronic wasting disease in deer and elk; mad cow disease, which infected more than 2 million cattle in the United Kingdom in the 1980s; and Creutzfeldt-Jakob

disease in humans, an incurable and inevitably fatal deterioration of the brain. An early 1900s New Guinea epidemic of kuru, another human disease caused by prions, was halted after anthropologists identified the cause—ritualistic cannibalism of the brain—and convinced locals to stop that practice.

Prions incubate at least 10 years before symptoms develop. This can prevent timely identification of sources of infection. Additionally, prions are not destroyed in food by normal heating. The only hope for developing effective treatments lies in understanding the process of infection.

To close the chapter, let's revisit some other noncellular threats to human health: emerging viruses. ☑

EVOLUTION CONNECTION Deadly Viruses

Emerging Viruses

Viruses that suddenly come to the attention of medical scientists are called **emerging viruses (Figure 10.33)**. We've already explored Zika virus (first recognized in Brazil in 2015) and West Nile virus (which first appeared in North America in 1999). Although each virus had persisted at low levels for many years, each became a much greater threat quite suddenly.

How do viruses give rise to new diseases? First, they can evolve into more dangerous forms. Although viruses are not

alive, they are subject to natural selection, which is accelerated by high mutation rates. Unlike DNA, RNA has no mechanisms to repair copying errors, so RNA viruses can mutate rapidly. Some mutations enable viruses to infect people who had developed resistance to the ancestral strain. This is why we need yearly flu vaccines: Mutations create new influenza virus strains to which people have no immunity.

Second, viral diseases can spread from one host species to another. Scientists estimate that about three-quarters of new human diseases originated in other animals. When humans hunt, live, or raise livestock in new habitats, the risk increases. HIV (which causes AIDS) may have started as a slightly different virus in chimpanzees. Human hunters were probably infected when they butchered infected animals. As the virus mutated in the human hosts, strains that out-competed other varieties for human host cells became increasingly common.

Third, viral diseases from a small, isolated population can spread, leading to an epidemic. AIDS went unnamed and virtually ignored for decades. Several factors, including international travel, intravenous drug use, sexual activity, and delayed effective action allowed it to become a global scourge.

Nobel Prize winner Joshua Lederberg warned: "We live in evolutionary competition with microbes. There is no guarantee that we will be the survivors." If we are to be victorious in the fight against emerging viruses, we must understand molecular biology and evolutionary processes.

▼ **Figure 10.33**
A sample of major emerging virus outbreaks of the past 100 years.

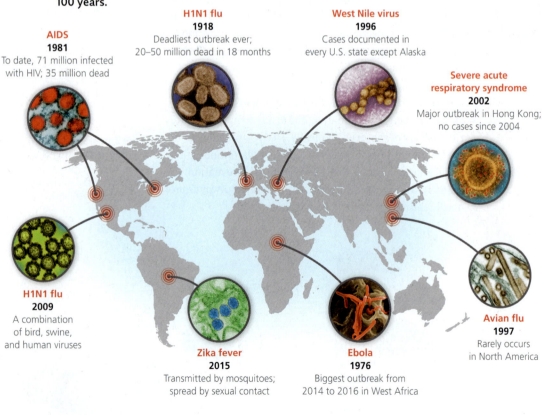

AIDS
1981
To date, 71 million infected with HIV; 35 million dead

H1N1 flu
1918
Deadliest outbreak ever; 20–50 million dead in 18 months

West Nile virus
1996
Cases documented in every U.S. state except Alaska

Severe acute respiratory syndrome
2002
Major outbreak in Hong Kong; no cases since 2004

H1N1 flu
2009
A combination of bird, swine, and human viruses

Zika fever
2015
Transmitted by mosquitoes; spread by sexual contact

Ebola
1976
Biggest outbreak from 2014 to 2016 in West Africa

Avian flu
1997
Rarely occurs in North America

Chapter Review

SUMMARY OF KEY CONCEPTS

DNA: Structure and Replication

DNA and RNA Structure

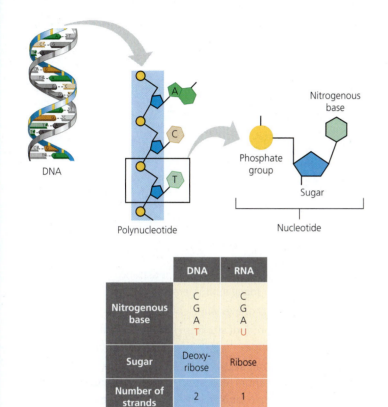

DNA

Polynucleotide

Phosphate group

Sugar

Nitrogenous base

Nucleotide

	DNA	RNA
Nitrogenous base	C G A T	C G A U
Sugar	Deoxy-ribose	Ribose
Number of strands	2	1

Watson and Crick's Discovery of the Double Helix

Watson and Crick worked out the three-dimensional structure of DNA: two polynucleotide strands wrapped around each other in a double helix. Hydrogen bonds between bases hold the strands together. Each base pairs with a complementary partner: A with T, and G with C.

DNA Replication

The structure of DNA, with its comple-mentary base pairing, allows it to function as the molecule of heredity through DNA replication.

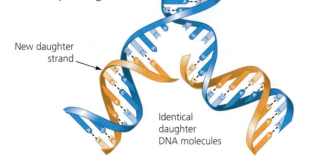

Parental DNA molecule

New daughter strand

Identical daughter DNA molecules

From DNA to RNA to Protein

How an Organism's Genotype Determines Its Phenotype

The information constituting an organism's genotype is carried in the sequence of its DNA bases. The genotype controls phenotype through the expression of proteins.

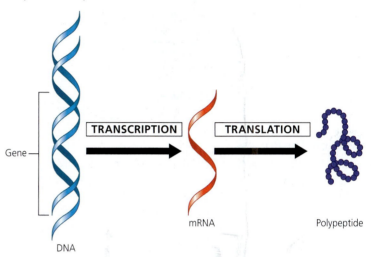

Gene

DNA

TRANSCRIPTION

mRNA

TRANSLATION

Polypeptide

From Nucleotides to Amino Acids: An Overview

The DNA of a gene is transcribed into RNA using the usual base-pairing rules, except that an A in DNA pairs with U in RNA. In the translation of a genetic message, each triplet of nucleotide bases in the RNA, called a codon, specifies one amino acid in the polypeptide.

The Genetic Code

In addition to codons that specify amino acids, the genetic code has one codon that is a start signal and three that are stop signals for translation.

Transcription: From DNA to RNA

In transcription, RNA polymerase binds to the promoter of a gene, opens the DNA double helix there, and catalyzes the synthesis of an RNA molecule using one DNA strand as a template. As the single-stranded RNA transcript peels away from the gene, the DNA strands rejoin.

The Processing of Eukaryotic RNA

The RNA transcribed from a eukaryotic gene is processed before leaving the nucleus to serve as messenger RNA (mRNA). Introns are spliced out, and a cap and tail are added.

Translation: The Players

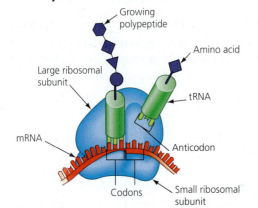

Growing polypeptide

Amino acid

Large ribosomal subunit

tRNA

mRNA

Anticodon

Codons

Small ribosomal subunit

Translation: The Process

In initiation, a ribosome assembles with the mRNA and the initiator tRNA bearing the first amino acid. Beginning at the start codon, the codons of the mRNA are recognized one by one by tRNAs bearing succeeding amino acids. The ribosome bonds the amino acids together. With each addition, the mRNA moves by one codon through the ribosome. When a stop codon is reached, the completed polypeptide is released.

Review: DNA → RNA → Protein

The sequence of codons in DNA, through the sequence of codons in mRNA, spells out the primary structure of a polypeptide.

Mutations

Mutations are changes in the DNA base sequence, caused by errors in DNA replication, recombination, or mutagens. Substituting, deleting, or inserting nucleotides in a gene has varying effects on the polypeptide and organism.

Type of Mutation	Effect
Substitution of one DNA base for another	**Silent** mutations result in no change to amino acids.
	Missense mutations swap one amino acid for another.
	Nonsense mutations change an amino acid codon to a stop codon.
Insertions or **deletions** of DNA nucleotides	**Frameshift** mutations can alter the triplet grouping of codons and greatly change the amino acid sequence.

Viruses and Other Noncellular Infectious Agents

Viruses are infectious particles consisting of genes packaged in protein.

Bacteriophages

When phage DNA enters a lytic cycle inside a bacterium, it is replicated, transcribed, and translated. The new viral DNA and protein molecules then assemble into new phages, which burst from the cell. In the lysogenic cycle, phage DNA inserts into the cell's chromosome and is passed on to generations of daughter cells. Much later, it may initiate phage production.

Plant Viruses

Viruses that infect plants can be a serious agricultural problem. Most have RNA genomes. Viruses enter plants through breaks in the plant's outer layers.

Animal Viruses

Many animal viruses, such as flu viruses, have RNA genomes; others, such as hepatitis viruses, have DNA. Some animal viruses "steal" a bit of cell membrane as a protective envelope. Some, such as the herpesvirus, can remain latent inside cells for long periods.

HIV, the AIDS Virus

HIV is a retrovirus. Inside a cell it uses its RNA as a template for making DNA, which is then inserted into a chromosome.

Prions

Prions are infectious proteins that cause a number of degenerative brain diseases in humans and other animals.

Mastering Biology

For practice quizzes, BioFlix animations, MP3 tutorials, video tutors, and more study tools designed for this textbook, go to Mastering Biology™

SELF-QUIZ

1. A molecule of DNA contains two polymer strands called _____, made by bonding together many monomers called _____.

2. Name the three parts of every nucleotide.

3. List these terms in order of size from largest to smallest: chromosome, codon, gene, nucleotide.

4. A scientist inserts a radioactively labeled DNA molecule into a bacterium. The bacterium replicates this DNA molecule and distributes one daughter molecule (double helix) to each of two daughter cells. How much radioactivity will the DNA in each of the two daughter cells contain? Why?

5. The nucleotide sequence of a DNA codon is GTA. What would be the nucleotide sequence of an mRNA molecule transcribed from this DNA? In the process of protein synthesis, a tRNA pairs with the mRNA codon. What is the nucleotide sequence of the tRNA anticodon that corresponds to this mRNA codon? What amino acid is attached to the tRNA (see Figure 10.10)?

6. Describe the process by which the information in a gene is transcribed and translated into a protein. Correctly use these terms in your description: tRNA, amino acid, start codon, transcription, mRNA, gene, codon, RNA polymerase, ribosome, translation, anticodon, peptide bond, stop codon.

7. Match the following molecules with the cellular process or processes in which they are primarily involved.
   a. ribosomes
   b. tRNA
   c. DNA polymerases
   d. RNA polymerase
   e. mRNA

   1. DNA replication
   2. transcription
   3. translation

8. A geneticist finds that a particular mutation has no effect on the polypeptide encoded by the gene. This mutation probably involves
   a. deletion of one nucleotide.
   b. alteration of the start codon.
   c. insertion of one nucleotide.
   d. substitution of one nucleotide.

9. Scientists have discovered how to put together a bacteriophage with the protein coat of phage A and the DNA of phage B. If this composite phage were allowed to infect a bacterium, the phages produced in the cell would have
   a. the protein of A and the DNA of B.
   b. the protein of B and the DNA of A.
   c. the protein and DNA of A.
   d. the protein and DNA of B.

10. How do some viruses reproduce without ever having DNA?

11. HIV requires an enzyme called _____ to convert its RNA genome to a DNA version. Why is this enzyme a particularly good target for anti-AIDS drugs? (*Hint*: Would you expect such a drug to harm the human host?)

For answers to the Self Quiz, see Appendix D.

IDENTIFYING MAJOR THEMES

For each statement, identify which major theme is evident (the relationship of structure to function, information flow, pathways that transform energy and matter, interactions within biological systems, or evolution) and explain how the statement relates to the theme. If necessary, review the themes (see Chapter 1) and review the examples highlighted in blue in this chapter.

12. Nearly every organism on Earth shares the identical genetic code, indicating that this scheme arose very early in the history of life.

13. The shape of a tRNA molecule, with its anticodon on one end and amino acid attachment site at the other end, hints at how the molecule acts during translation.

14. Genes carry the instructions needed to build an RNA and then a protein.

For answers to Identifying Major Themes, see Appendix D.

THE PROCESS OF SCIENCE

15. A cell containing a single chromosome is placed in a medium containing radioactive phosphate, making any new DNA strands formed by DNA replication radioactive. The cell replicates its DNA and divides. Then the daughter cells (still in the radioactive medium) replicate their DNA and divide, resulting in a total of four cells. Sketch the DNA molecules in all four cells, showing a normal (nonradioactive) DNA strand as a solid line and a radioactive DNA strand as a dashed line.

16. In a classic 1952 experiment, biologists Alfred Hershey and Martha Chase labeled two batches of bacteriophages, one with radioactive sulfur (which only tags protein) and the other with radioactive phosphorus (which only tags DNA). In separate test tubes, they allowed

each batch of phages to bind to nonradioactive bacteria and inject its DNA. After a few minutes, they separated the bacterial cells from the viral parts that remained outside the bacterial cells and measured the radioactivity of both portions. What results do you think they obtained? How would these results help them to determine which viral component—DNA or protein—was the infectious portion?

17. **Interpreting Data** The graph shows the number of cases per week of Zika, Dengue, and Chikungunya virus in Puerto Rico during the period from November 1, 2015 to April 14, 2016. The same mosquitoes spread all three viruses. Did all three diseases show a similar pattern during this time period? Explain.

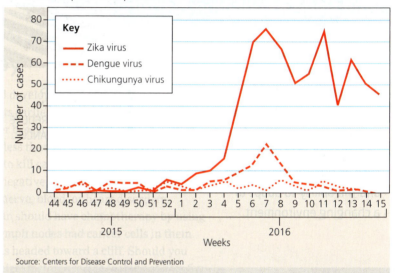

Source: Centers for Disease Control and Prevention

BIOLOGY AND SOCIETY

18. Scientists at the National Institutes of Health (NIH) have worked out thousands of sequences of genes and the proteins they encode, and similar analyses are being carried out at universities and private companies. Knowledge of the nucleotide sequences of genes might be used to treat genetic defects or produce lifesaving medicines. The NIH and some U.S. biotechnology companies have applied for patents on their discoveries. In Britain, the courts have ruled that a naturally occurring gene cannot be patented. Do you think individuals and companies should be able to patent genes and gene products? Before answering, consider the following: What are the purposes of a patent? How might the discoverer of a gene benefit from a patent? How might the public benefit? What negative effects might result from patenting genes?

19. Your college roommate seeks to improve her appearance by visiting a tanning salon. How would you explain the dangers of this to her?

20. Flu vaccines have been shown to be safe, are very reliable at reducing the risk of hospitalization or death from influenza, and are inexpensive. Should children be required to obtain a flu vaccine before going to school? What about hospital workers before reporting to work? Defend your answers to these questions.

Cell Signaling

Cell-to-cell signaling is key to the development and functioning of multicellular organisms. Signal transduction pathways convert molecular messages to cell responses, such as the transcription of particular genes.

Homeotic Genes

Evidence for the evolutionary importance of gene regulation is apparent in homeotic genes, master genes that regulate other genes that in turn control embryonic development.

Visualizing Gene Expression

Researchers can visualize which genes are active in which tissues using a variety of techniques, including the use of probes and DNA microarrays.

Cloning Plants and Animals

The Genetic Potential of Cells

Most differentiated cells retain a complete set of genes, so an orchid plant, for example, can be made to grow from a single orchid cell. Under controlled conditions, animals can also be cloned.

Reproductive Cloning of Animals

Nuclear transplantation is a procedure whereby a donor cell nucleus is inserted into an egg from which the nucleus has been removed. First demonstrated in frogs in the 1950s, reproductive cloning was used in 1996 to clone a sheep from an adult cell and has since been used to create many other cloned animals.

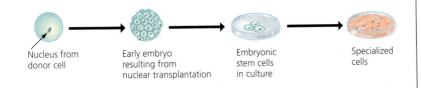

Nucleus from donor cell → Early embryo resulting from nuclear transplantation → Embryo implanted in surrogate mother → Clone of nucleus donor

Therapeutic Cloning and Stem Cells

The purpose of therapeutic cloning is to produce embryonic stem cells for medical uses. Embryonic, umbilical cord, and adult stem cells all show promise for therapeutic uses.

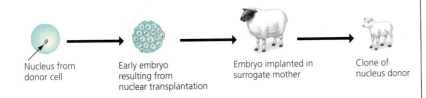

Nucleus from donor cell → Early embryo resulting from nuclear transplantation → Embryonic stem cells in culture → Specialized cells

The Genetic Basis of Cancer

Genes That Cause Cancer

Cancer cells, which divide uncontrollably, can result from mutations in genes whose protein products regulate the cell cycle.

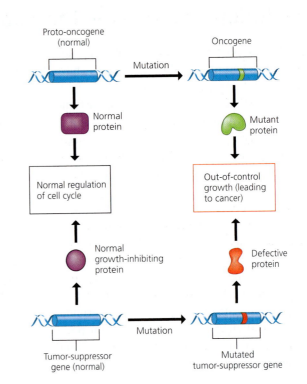

Proto-oncogene (normal) — Mutation → Oncogene

Normal protein / Mutant protein

Normal regulation of cell cycle / Out-of-control growth (leading to cancer)

Normal growth-inhibiting protein / Defective protein

Tumor-suppressor gene (normal) — Mutation → Mutated tumor-suppressor gene

Many proto-oncogenes and tumor-suppressor genes code for proteins active in signal transduction pathways regulating cell division. Mutations of these genes cause malfunction of the pathways. Cancer results from a series of genetic changes in a cell lineage. Researchers have identified many genes that, when mutated, promote the development of cancer.

Cancer Risk and Prevention

Reducing exposure to carcinogens (which induce cancer-causing mutations) and making other healthful lifestyle choices can help reduce cancer risk.

Mastering Biology

For practice quizzes, BioFlix animations, MP3 tutorials, video tutors, and more study tools designed for this textbook, go to Mastering Biology™

SELF-QUIZ

1. Your bone cells, muscle cells, and skin cells look different because
   a. different kinds of genes are present in each kind of cell.
   b. they are present in different organs.
   c. different genes are active in each kind of cell.
   d. different mutations have occurred in each kind of cell.

2. A group of prokaryotic genes with related functions that are regulated as a single unit, along with the control sequences that perform this regulation, is called a(n) _____.

3. The regulation of gene expression must be more complex in multicellular eukaryotes than in prokaryotes because
   a. eukaryotic cells are much larger.
   b. in a multicellular eukaryote, different cells are specialized.
   c. prokaryotes are restricted to stable environments.
   d. eukaryotes have fewer genes, so each gene must do several jobs.

4. A eukaryotic gene was inserted into the DNA of a bacterium. The bacterium then transcribed this gene into mRNA and translated the mRNA into protein. The protein produced was useless and contained many more amino acids than the protein made by the eukaryotic cell. Why?
   a. The mRNA was not spliced as it is in eukaryotes.
   b. Eukaryotes and prokaryotes use different genetic codes.
   c. Repressor proteins interfered with transcription and translation.
   d. Ribosomes were not able to bind to tRNA.

5. How does DNA packing in chromosomes prevent gene expression?

6. What evidence demonstrates that differentiated cells in a plant or animal retain their full genetic potential?

7. The most common procedure for cloning an animal is _____.

8. What is learned from a DNA microarray?

9. Which of the following is a substantial difference between embryonic stem cells and the stem cells found in adult tissues?
   a. In laboratory culture, only adult stem cells are immortal.
   b. In nature, only embryonic stem cells give rise to all the different types of cells in the organism.
   c. Only adult stem cells can be made to differentiate in the laboratory.
   d. Only embryonic stem cells are in every tissue of the adult body.

10. Name three potential sources of stem cells.

11. What is the difference between oncogenes and proto-oncogenes? How can one turn into the other? What function do proto-oncogenes serve?

12. A mutation in one gene may cause a major change in the body of a fruit fly. Yet it takes many genes to produce a wing or leg. How can a change in one gene cause a big change? What are such genes called?

For answers to the Self Quiz, see Appendix D.

IDENTIFYING MAJOR THEMES

For each statement, identify which major theme is evident (the relationship of structure to function, information flow, pathways that transform energy and matter, interactions within biological systems, or evolution) and explain how the statement relates to the theme. If necessary, review the theme descriptions (see Chapter 1) and review the examples highlighted in blue in this chapter.

13. Changing the shape of the *lac* repressor affects how the repressor acts.

14. A cell can produce and secrete chemicals, such as hormones, that affect gene regulation in another cell.

15. Master control genes regulate other genes that determine what body parts will develop in which locations.

For answers to Identifying Major Themes, see Appendix D.

THE PROCESS OF SCIENCE

16. Study the depiction of the *lac* operon in Figure 11.2. Normally, the genes are turned off when lactose is not present. Lactose activates the genes, which code for enzymes that enable the cell to use lactose. Predict how the following mutations would affect the function of the operon in the presence and absence of lactose:
    a. mutation of regulatory gene; repressor will not bind to lactose
    b. mutation of operator; repressor will not bind to operator
    c. mutation of regulatory gene; repressor will not bind to operator

17. The human body has a far greater variety of proteins than genes, highlighting the importance of alternative RNA splicing. Suppose you have samples of two types of adult cells from one person. Design an experiment using microarrays to determine whether different gene expression is due to alternative RNA splicing.

18. Because a cat must have both orange and non-orange alleles to be tortoiseshell (see Figure 11.4), we would expect only female cats, which have two X chromosomes, to be tortoiseshell. Normal male cats (XY) can carry only one of the two alleles. Male tortoiseshell cats are rare and usually sterile. What might be their genotype?

19. Design a DNA microarray experiment that measures the difference in gene expression between normal colon cells and cells from a colon tumor.

20. **Interpreting Data** Review Figure 11.22. We can estimate the deadliness of each type of cancer by dividing the number of deaths by number of cases. (Although someone diagnosed may not die the same year, it's a useful approximation.) If nearly everyone diagnosed with a certain cancer dies, that ratio will be near 1 (100% deadly). If many more people receive diagnoses than die, the ratio will be near 0 (near 0% deadly). Which region of the graph represents the more deadly cancers? The least deadly? Calculate the diagnosis/death rate for different cancers.

BIOLOGY AND SOCIETY

21. A chemical called dioxin is present in Agent Orange, a defoliant used during the Vietnam War. There has been controversy over its effects on soldiers exposed to Agent Orange. Animal tests have suggested that dioxin can cause multiple health problems and at high dosage can be lethal. Researchers have discovered that dioxin enters a cell and binds to a protein that attaches to the cell's DNA. How might this mechanism help explain the variety of effects on different body systems and different animals? How might you determine whether a person became ill as a result of dioxin?

22. There are genetic tests for several types of "inherited cancer." The results cannot usually predict that someone will get cancer. Rather, they indicate only an increased risk of developing cancer. For many cancers, lifestyle changes cannot decrease risk. Therefore, some people consider the tests useless. If your close family had a history of cancer and a test were available, would you get screened? What would you do with this information?

Because growth hormones from other animals are not effective in people, HGH was an early target of genetic engineers. Before genetically engineered HGH became available in 1985, children with an HGH deficiency could only be treated with scarce and expensive supplies of HGH obtained from human cadavers. Another genetically engineered protein helps dissolve blood clots. If administered shortly after a stroke, it reduces the risk of additional strokes and heart attacks.

Although bacteria can produce many human proteins, some proteins can only be made by eukaryotic cells, such as cells from fungi, animals, and plants. Common baker's yeast is currently used to produce proteins used as medicines, including the hepatitis B vaccine, an antimalarial drug, and interferons used to treat cancer and viral infections. In 2015, scientists announced they had transferred 23 genes (from bacteria, plants, and animals) into yeast that allow the recombinant fungi to convert sugar into the painkiller drug hydrocodone. Genetically modified mammalian cells are used to produce erythropoietin (EPO), a hormone used to treat anemia by stimulating production of red blood cells. Researchers have also developed transgenic plant cells that can produce human drugs. The drug factories of the future may be carrots because they are easily grown in culture and are unlikely to be contaminated by human pathogens (such as viruses).

Genetically modified whole animals are also used to produce drugs. **Figure 12.8** shows a transgenic goat that carries a gene for an enzyme called lysozyme. This enzyme, found naturally in breast milk, has antibacterial properties. In another example, the gene for a human blood protein has been inserted into the genome of a goat so that the protein is secreted in the goat's milk. The protein is then purified from the milk. Because

transgenic animals are difficult to produce, researchers may create a single transgenic animal and then breed or clone it. The resulting herd of transgenic animals could serve as a grazing pharmaceutical factory.

DNA technology is also helping medical researchers develop vaccines. A vaccine is a harmless variant or derivative of a disease-causing microorganism—such as a bacterium or virus—that is used to prevent an infectious disease. When a person is inoculated, the vaccine stimulates the immune system to develop lasting defenses against the microorganism. For many viral diseases, the only way to prevent serious harm is to use vaccination to prevent the illness in the first place. The vaccine against hepatitis B, a disabling and sometimes fatal liver disease, is produced by genetically engineered yeast cells that secrete a protein found on the virus's outer surface.

DNA technologies can also identify causes of illnesses. For example, the Centers for Disease Control and Prevention regularly uses DNA technology to identify the precise strain of bacteria that is causing a food poisoning outbreak, allowing officials to implement food safety measures. ✓

Genetically Modified Organisms in Agriculture

Since ancient times, people have selectively bred crops to make them more useful. Today, DNA technology is quickly replacing traditional breeding programs as scientists work to improve the productivity of agriculturally important plants and animals.

GENETICALLY MODIFIED POTATOES COULD SAVE MANY CHILDREN FROM DEATH BY CHOLERA.

In the United States today, nearly all of our corn, soybean, and cotton crops are genetically modified. **Figure 12.9** shows corn that has been genetically engineered to resist attack by an insect called the European corn borer. Growing insect-resistant plants reduces the need for chemical insecticides. In another example, modified strawberry plants produce bacterial proteins that act as a natural antifreeze, protecting the delicate plants from the damages of cold weather. Potatoes and rice have been engineered to produce harmless proteins derived from the cholera bacterium; researchers hope that these modified foods will one day serve as an edible vaccine against cholera, a disease that kills thousands of children in developing nations every year. In India, the insertion of a natural but rare saltwater resistance gene has enabled new varieties of rice to thrive in water three times as salty as seawater, allowing food to be grown in drought-stricken or flooded regions.

▶ **Figure 12.8**
A genetically modified goat.

▼ **Figure 12.9 Genetically modified corn.** The corn plants in this field carry a bacterial gene that helps prevent infestation by the European corn borer (inset).

Scientists are also using genetic engineering to improve the nutritional value of crop plants **(Figure 12.10)**. One example is "golden rice 2," a transgenic variety of rice that carries genes from daffodils and corn. This rice could help prevent vitamin A deficiency and resulting blindness, especially in developing nations that depend on rice as a staple crop. Cassava, a starchy root crop that is a staple for nearly 1 billion people in developing nations, has similarly been modified to produce increased levels of iron and beta-carotene (which is converted to vitamin A in the body). However, controversy surrounds the use of GM foods, as we'll discuss at the end of the chapter.

Genetic engineers are targeting agricultural animals as well as plant crops. Scientists might, for example, identify in one variety of cattle a gene that causes the development of larger muscles (which make up most of the meat we eat) and transfer it to other cattle or even to chickens. Researchers have genetically modified pigs to carry a roundworm gene whose protein converts less healthy fatty acids to omega-3 fatty acids. Meat from the modified pigs contains four to five times as much healthy omega-3 fat as regular pork. In 2015, researchers replaced a gene in dairy cows with one from Angus cattle to produce cattle that lack horns, saving the bulls from painful dehorning. Similar gene-editing techniques produce improved varieties of goats (for meat and cashmere wool), pigs (for agriculture and pets), and dogs. A type of Atlantic salmon has been genetically modified to reach market size in half the normal time (18 months versus 3 years) and to grow twice as large. In late 2015, the FDA approved the sale of this GMO salmon to U.S. consumers, declaring that it is as safe and nutritious as traditional salmon. Although it could be years before the GMO salmon reaches store shelves, this is the first time a transgenic animal product was allowed to be sold as food in the United States. ☑

☑ **CHECKPOINT**

What is a genetically modified organism?

■ *Answer: one that carries DNA introduced through artificial means*

▼ **Figure 12.10 Genetically modified staple crops.** "Golden rice 2," the yellow grains shown here (left) alongside ordinary rice, has been genetically modified to produce high levels of beta-carotene, a molecule that the body converts to vitamin A. Transgenic cassava (right), a starchy root crop that serves as the main food source for nearly a billion people, has been modified to produce extra nutrients.

Chapter Review

SUMMARY OF KEY CONCEPTS

Genetic Engineering

DNA technology, the manipulation of genetic material, is a relatively new branch of biotechnology, the use of organisms to make helpful products. DNA technology often involves the use of recombinant DNA, the combination of nucleotide sequences from two different sources.

Recombinant DNA Techniques

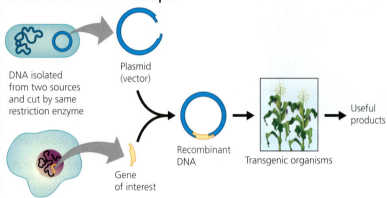

Gene Editing

The CRISPR-Cas9 system can be used to deactivate or edit genes within living cells.

Medical Applications

By transferring a human gene into a bacterium or other easy-to-grow cell, scientists can mass-produce valuable human proteins to be used as drugs or vaccines.

Genetically Modified Organisms in Agriculture

Recombinant DNA techniques have been used to create genetically modified organisms, those that carry artificially introduced genes. Nonhuman cells have been engineered to produce human proteins, genetically modified food crops, and transgenic farm animals. A transgenic organism is one that carries artificially introduced genes, typically from a different species.

Human Gene Therapy

A virus can be modified to include a normal human gene. If this virus is injected into the bone marrow of a person suffering from a genetic disease, the normal human gene may be transcribed and translated, producing a normal human protein that may cure the genetic disease. This technique has been used in gene therapy trials involving a number of inherited diseases. There have been both successes and failures to date, and research continues.

DNA Profiling and Forensic Science

Forensics, the scientific analysis of legal evidence, has been revolutionized by DNA technology. DNA profiling is used to determine whether two DNA samples come from the same individual.

DNA Profiling Techniques

Short tandem repeat (STR) analysis compares DNA fragments using the polymerase chain reaction (PCR) and gel electrophoresis.

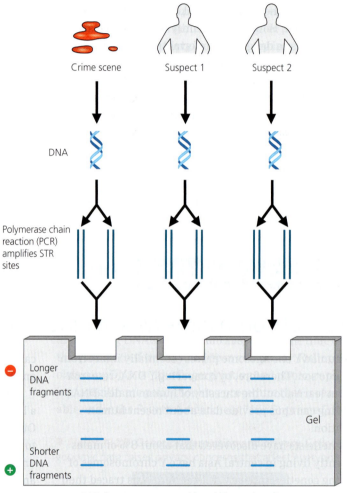

DNA fragments compared by gel electrophoresis
(Bands of shorter fragments move faster toward the positive pole.)

Investigating Murder, Paternity, and Ancient DNA

DNA profiling can be used to establish innocence or guilt of a criminal suspect, identify victims, determine paternity, and contribute to basic research.

Bioinformatics

DNA Sequencing

Automated machines can now sequence many thousands of DNA nucleotides per hour.

Genomics

Advances in DNA sequencing have ushered in the era of genomics, the study of complete genome sets.

Genome-Mapping Techniques

The whole-genome shotgun method involves sequencing DNA fragments from an entire genome and then assembling the sequences.

The Human Genome

The nucleotide sequence of the human genome is providing a wealth of useful data. The 24 different chromosomes of the human genome contain about 3 billion nucleotide pairs and 21,000 genes. The majority of the genome consists of noncoding DNA.

Applied Genomics

Comparing genomes can aid criminal investigations and basic biological research.

Systems Biology

Success in genomics has given rise to proteomics, the systematic study of the full set of proteins found in organisms. Genomics and proteomics both contribute to systems biology, the study of how many parts work together within complex biological systems.

Safety and Ethical Issues

The Controversy over Genetically Modified Foods

The debate about genetically modified crops centers on whether they might harm humans or damage the environment by transferring genes through cross-pollination with other species.

Ethical Questions Raised by Human DNA Technologies

As members of society we must become educated about DNA technologies so that we can intelligently address the ethical questions raised by their use.

Mastering Biology

For practice quizzes, BioFlix animations, MP3 tutorials, video tutors, and more study tools designed for this textbook, go to Mastering Biology™

SELF-QUIZ

1. Suppose you wish to create a large batch of the protein lactase using recombinant DNA. Place the following steps in the order you would have to perform them.
   a. Find the clone with the gene for lactase.
   b. Insert the plasmids into bacteria and grow the bacteria into clones.
   c. Isolate the gene for lactase.
   d. Create recombinant plasmids, including one that carries the gene for lactase.

2. A carrier that moves DNA from one cell to another, such as a plasmid, is called a _____.

3. In making recombinant DNA, what is the benefit of using a restriction enzyme that cuts DNA in a staggered fashion?

4. A paleontologist has recovered a bit of organic material from the 400-year-old preserved skin of an extinct dodo. She would like to compare DNA from the sample with DNA from living birds. The most useful method for initially increasing the amount of dodo DNA available for testing is _____.

5. Why do DNA fragments containing STR sites from different people tend to migrate to different locations during gel electrophoresis?

6. What feature of a DNA fragment causes it to move through a gel during electrophoresis?
   a. the electrical charges of its phosphate groups
   b. its nucleotide sequence
   c. the hydrogen bonds between its base pairs
   d. its double helix shape

7. After a gel electrophoresis procedure is run, the pattern of bars in the gel shows
   a. the order of bases in a particular gene.
   b. the presence of various-sized fragments of DNA.
   c. the order of genes along particular chromosomes.
   d. the exact location of a specific gene in the genome.

8. Name the steps of the whole-genome shotgun method.

9. Put the following steps of human gene therapy in the correct order.
   a. Virus is injected into patient.
   b. Human gene is inserted into a virus.
   c. Normal human gene is isolated and cloned.
   d. Normal human gene is transcribed and translated in the patient.

For answers to the Self Quiz, see Appendix D.

IDENTIFYING MAJOR THEMES

For each statement, identify which major theme is evident (the relationship of structure to function, information flow, pathways that transform energy and matter, interactions within biological systems, or evolution) and explain how the statement relates to the theme. If necessary, review the themes (see Chapter 1) and review the examples highlighted in blue in this chapter.

10. Comparisons of DNA sequences can reveal not just recent paternity, but patterns of ancient lineages, such as when humans diverged from other primates.

11. Bioinformatic tools allow for examinations of all the parts that constitute a living organism (such as proteins and genes), allowing their interrelationships to be studied.

12. Studying whole genomes reveals how genes from one generation can affect the appearance of the next.

For answers to Identifying Major Themes, see Appendix D.

THE PROCESS OF SCIENCE

13. Some scientists once joked that when the DNA sequence of the human genome was complete, "we can all go home" because there would be nothing left for genetic researchers to discover. Why haven't they all "gone home"?

14. **Interpreting Data** When comparing genomes from different species, biologists often calculate the genome density, the number of genes per number of nucleotides in the genome. Refer to Figure 12.20. You can estimate the gene density of each species by dividing the number of genes by the size of the genome (usually expressed in Mb, which is mega base pairs, or 1 million base pairs). Using a spreadsheet, estimate the gene density for every species in the figure. (Don't forget that 1 billion = 1,000 million; for example, humans have 3,000 Mb.) How does the gene density of bacteria compare to humans? Humans and roundworms have nearly the same number of genes, but how do the gene densities of these two species compare? Can you identify any general correlation between gene density and the size or complexity of an organism?

15. Listed below are 4 of the 13 genome sites used to create a standard DNA profile. Each site consists of a number of short tandem repeats: sets of 4 nucleotides repeated in a row within the genome. For each site, the number of repeats found at that site for this individual are listed.

Chromosome number	Genetic site	# of repeats
3	D3S1358	4
5	D5S818	10
7	D7S820	5
8	D8S1179	22

Imagine that you perform a PCR procedure to create a DNA profile for this individual. Which of the following four gels correctly represents the DNA profile of this person?

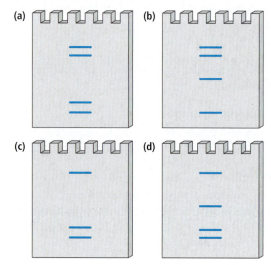

BIOLOGY AND SOCIETY

16. In the not-too-distant future, gene therapy may be used to treat many inherited disorders. What do you think are the most serious ethical issues to face before human gene therapy is used on a large scale? Explain.

17. Today, it is fairly easy to make transgenic plants and animals. What are some safety and ethical issues raised by this use of recombinant DNA technology? What are some dangers of introducing genetically engineered organisms into the environment? What are some reasons for and against leaving such decisions to scientists? Who should decide?

18. In October 2002, the government of the African nation of Zambia announced that it was refusing to distribute 15,000 tons of corn donated by the United States, enough corn to feed 2.5 million Zambians for three weeks. The government rejected the corn because it was likely to contain genetically modified kernels. The government made the decision after its scientific advisers concluded that the studies of the health risks posed by GM crops "are inconclusive." Do you agree with Zambia's decision? Why or why not? Consider that Zambia was facing food shortages, and 35,000 Zambians were expected to starve to death over the next six months. How do the risks posed by GM crops compare with the risk of starvation?

19. From 1977 to 2000, 12 convicts were executed in Illinois. During that same period, 13 death row inmates were exonerated based on DNA evidence. In 2000, the governor of Illinois declared a moratorium on all executions in his state because the death penalty system was "fraught with errors." Do you support the Illinois governor's decision? What rights should death penalty inmates have with regard to DNA testing of old evidence? Who should pay for this additional testing?

Unit 3

Evolution and Diversity

13 How Populations Evolve

CHAPTER THREAD
Evolution in Action

14 How Biological Diversity Evolves

CHAPTER THREAD
Evolution in the Human-Dominated World

15 The Evolution of Microbial Life

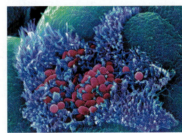

CHAPTER THREAD
Human Microbiota

16 The Evolution of Plants and Fungi

CHAPTER THREAD
Plant-Fungus Interactions

17 The Evolution of Animals

CHAPTER THREAD
Human Evolution

13 How Populations Evolve

CHAPTER CONTENTS

The Diversity of Life 244
Charles Darwin and *The Origin of Species* 246
Evidence of Evolution 248
Natural Selection as the Mechanism for Evolution 252
The Evolution of Populations 254
Mechanisms of Evolution 258

DO YOU LIKE BIG, JUICY TOMATOES? IF IT WEREN'T FOR ARTIFICIAL SELECTION, YOU'D BE EATING TOMATOES THE SIZE OF BLUEBERRIES.

Why Evolution Matters

The abundant diversity of life on Earth, more than a million species, is the product of past evolution. Evolution is also happening right now, somewhere near you—perhaps even within your own body.

THE CHEETAH—THE FASTEST ANIMAL ON EARTH—MAY BE RACING TOWARD EXTINCTION. MANY ENDANGERED SPECIES ARE DOOMED BY THEIR LACK OF GENETIC DIVERSITY.

SCRATCHING YOUR HEAD OVER BUG INFESTATIONS? BECAUSE OF NATURAL SELECTION, HEAD LICE, BED BUGS, AND MOSQUITOES ARE INCREASINGLY HARD TO KILL.

CHAPTER THREAD

Evolution in Action

BIOLOGY AND SOCIETY Mosquitoes and Evolution 243

THE PROCESS OF SCIENCE Did Natural Selection Shape the Beaks of Darwin's Finches? 262

EVOLUTION CONNECTION The Rising Threat of Antibiotic Resistance 265

A malaria ward in Kisii, Kenya.

BIOLOGY AND SOCIETY | Evolution in Action

Mosquitoes and Evolution

What does actor George Clooney have in common with George Washington, Ernest Hemingway, Christopher Columbus, and Mother Teresa? They all survived bouts with malaria, a disease caused by a microscopic parasite that is one of the worst killers in human history. In 1955, the World Health Organization (WHO) launched a campaign to eradicate malaria. Their strategy focused on killing the mosquitoes that carry the parasite from person to person. DDT, a recently developed pesticide in wide use at the time, was deployed in massive spraying operations. But in many locations, early success was followed by rebounding mosquito populations. Although the lethal chemical killed most of the mosquitoes immediately, survivors gave rise to new DDT-resistant populations—an example of evolution in action. Clearly, a single pesticide could not wipe out all the disease-carrying mosquitoes. WHO also learned that the parasite, too, was evolving. Drugs that once cured malaria became less and less effective as resistance evolved in parasite populations. Eradication of the disease is no longer viewed as imminent, but by using a judicious combination of mosquito-control strategies, public health agencies have made progress in the battle against malaria. WHO also monitors the evolution of drug resistance in the parasite's populations around the world.

As you'll learn in this chapter, malaria is not the only disease that has become more difficult to cure as a result of evolution. Dozens of species of bacteria and other microorganisms are increasingly resistant to antibiotics because of evolution. And malaria is not the only disease carried by mosquitoes—Zika virus, dengue (also called breakbone fever, for the intense pain it causes), Chikungunya virus, yellow fever, and West Nile virus are among the others. If you don't live in a region inhabited by the relatively few mosquito species that carry these diseases, you may not be familiar with them. That may soon change, however. As climate change brings rising temperatures, changing patterns of rainfall, and increased flooding to many regions, disease-carrying mosquitoes are expanding their range.

An understanding of evolution informs all of biology, from exploring life's molecules to analyzing ecosystems. And applications of evolutionary biology are transforming medicine, agriculture, biotechnology, and conservation biology. In this chapter, you'll learn how the process of evolution works and read about verifiable, measurable examples of evolution that affect our world.

Charles Darwin and *The Origin of Species*

Darwin's Journey

During his voyage on the *Beagle*, Darwin observed adaptations of organisms that inhabited diverse environments, especially on the Galápagos Islands, off the South American coast. These observations, along with other insights, eventually led Darwin to formulate his theory of evolution.

Darwin's Theory

In his book *On the Origin of Species by Means of Natural Selection*, Darwin made two proposals: (1) Existing species descended from ancestral species and (2) natural selection is the mechanism of evolution.

Evidence of Evolution

Evidence from Fossils

The fossil record shows that organisms have appeared in a historical sequence, and many fossils link ancestral species with those living today.

Evidence from Homologies

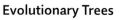

Structural and molecular homologies reveal evolutionary relationships. All species share a common genetic code, suggesting that all forms of life are related through branching evolution from the earliest organisms.

Evolutionary Trees

An evolutionary tree represents a succession of related species, with the most recent at the tips of the branches. Each branch point represents a common ancestor of all species that radiate from it.

Natural Selection as the Mechanism for Evolution

Darwin proposed natural selection as the mechanism that produces adaptive evolutionary change. In a population that varies, individuals best suited for a particular environment are more likely to survive and reproduce than those that are less suited to that environment.

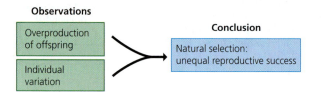

Natural Selection in Action

Natural selection has been observed in many scientific studies, including in the evolution of pesticide-resistant insects.

Key Points about Natural Selection

Individuals do not evolve. Only heritable traits, not those acquired during an individual's lifetime, can be amplified or diminished by natural selection. Natural selection only works on existing variation—new variation does not arise in response to an environmental change. Natural selection does not produce perfect organisms.

The Evolution of Populations

Sources of Genetic Variation

Mutation and sexual reproduction produce genetic variation. Mutation is the ultimate source of genetic variation.

Populations as the Units of Evolution

A population, members of the same species living in the same time and place, is the smallest biological unit that can evolve.

Analyzing Gene Pools

A gene pool consists of all the alleles in all the individuals making up a population. The Hardy-Weinberg formula can be used to calculate the frequencies of genotypes in a gene pool from the frequencies of alleles, and vice versa:

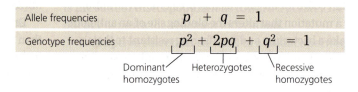

Population Genetics and Health Science

The Hardy-Weinberg formula can be used to estimate the frequency of a harmful allele, which is useful information for public health programs dealing with genetic diseases.

Microevolution as Change in a Gene Pool

Microevolution is generation-to-generation change in allele frequencies in a population.

Mechanisms of Evolution

Natural Selection

Natural selection is the most important mechanism of evolutionary change because it is the only process that promotes adaptation.

Genetic Drift

Genetic drift is a change in the gene pool of a small population due to chance. A bottleneck event (a drastic reduction in population size) and the founder effect (occurring in a new population started by a few individuals) are two situations that can lead to genetic drift.

Gene Flow

A population may gain or lose alleles by gene flow, which is genetic exchange with another population.

Natural Selection: A Closer Look

Of all the causes of evolution, only natural selection promotes evolutionary adaptations. Relative fitness is the contribution an individual makes to the gene pool of the next generation relative to the contributions of other individuals. The outcome of natural selection may be directional, disruptive, or stabilizing. Secondary sexual traits (such as sex-specific plumage or behaviors) can promote sexual selection, a type of natural selection in which mating preferences are determined by inherited traits.

Mastering Biology

For practice quizzes, BioFlix animations, MP3 tutorials, video tutors, and more study tools designed for this textbook, go to Mastering Biology™

SELF-QUIZ

1. Place these levels of classification in order from least inclusive to most inclusive: class, domain, family, genus, kingdom, order, phylum, species.

2. Which of the following is a true statement about Charles Darwin?
   a. He was the first to discover that living things can change, or evolve.
   b. He based his theory on the inheritance of acquired traits.
   c. He proposed natural selection as the mechanism of evolution.
   d. He was the first to realize that Earth is more than 6,000 years old.

3. How did the insights of Lyell and other geologists influence Darwin's thinking about evolution?

4. In a population with two alleles for a particular genetic locus, B and b, the allele frequency of B is 0.7. If this population is in Hardy-Weinberg equilibrium, what is the frequency of heterozygotes? What is the frequency of homozygous dominants? What is the frequency of homozygous recessives?

5. Define fitness from an evolutionary perspective.

6. Which of the following processes is the ultimate source of the genetic variation that serves as raw material for evolution?
   a. sexual reproduction
   b. mutation
   c. genetic drift
   d. natural selection

7. As a mechanism of evolution, natural selection can be most closely equated with
   a. random mating.
   b. genetic drift.
   c. unequal reproductive success.
   d. gene flow.

8. Compare and contrast how the bottleneck effect and the founder effect can lead to genetic drift.

9. In a particular bird species, individuals with average-sized wings are more likely to survive severe storms than other birds in the same population with longer or shorter wings. Of the three general outcomes of natural selection (directional, disruptive, or stabilizing), this example illustrates _____.

For answers to the Self Quiz, see Appendix D.

IDENTIFYING MAJOR THEMES

For each statement below, identify which major theme is evident (the relationship of structure to function, information flow, pathways that transform energy and matter, interactions within biological systems, or evolution) and explain how the statement relates to the theme. If necessary, review the themes (Chapter 1) and review the examples highlighted in blue in this chapter.

10. If two species have homologous genes with sequences that match closely, biologists conclude that these sequences must have been inherited from a relatively recent common ancestor.

11. The body and bill of a diving bird, the blue-footed booby, is streamlined like a torpedo.

12. Darwin hypothesized that as descendants of a remote ancestor spread into various habitats, natural selection resulted in diverse modifications that fit them to their environment.

For answers to Identifying Major Themes, see Appendix D.

THE PROCESS OF SCIENCE

13. **Interpreting Data** A population of snails has recently become established in a new region. The snails are preyed on by birds that break the snails open on rocks, eat the soft bodies, and leave the shells. The snails occur in both striped and unstriped forms. In one area, researchers counted both live snails and broken shells. Their data are summarized here:

	Striped Shells	Unstriped Shells
Number of live snails	264	296
Number of broken snail shells	486	377
Total	750	673

Based on these data, which snail form is subject to more predation by birds? Predict how the frequencies of striped and unstriped individuals might change over time.

14. Imagine that the presence or absence of stripes on the snails from the previous question is determined by a single gene locus, with the dominant allele (S) producing striped snails and the recessive allele (s) producing unstriped snails. Combining the data from both the living snails and broken shells, calculate the following: the frequency of the dominant allele, the frequency of the recessive allele, and the number of heterozygotes in the observed groups.

BIOLOGY AND SOCIETY

15. To what extent are people in a technological society exempt from natural selection? Explain your answer.

16. What plants and animals have you seen near your home or school? What evolutionary adaptations suit them to their environment?

14 How Biological Diversity Evolves

CHAPTER CONTENTS

The Origin of Species 270
Earth History and Macroevolution 279
Mechanisms of Macroevolution 283
Classifying the Diversity of Life 286

SEE A RESEMBLANCE BETWEEN THE NEANDERTHAL AND DARWIN? PIECING TOGETHER EVOLUTIONARY HISTORIES SHOWS WHO'S RELATED TO WHOM.

ENJOY CORN ON THE COB? ALLELES FROM TWO WILD GRASSES COULD BE USED TO PROTECT CORN CROPS FROM FUTURE DISASTERS.

Why Evolution Matters

The diversity of life on Earth is the product of evolutionary processes that have been happening for billions of years. During this time, the overwhelming majority of species that ever lived became extinct. The classification system used by biologists shows how all forms of life on Earth are related.

IF ROCKS COULD SPEAK, WHAT WOULD THEY TELL US? SEVERAL HUNDRED MILLION YEARS OF HISTORY IS "WRITTEN" IN THE LAYERS OF ROCK IN THE GRAND CANYON.

CHAPTER THREAD

Evolution in the Human-Dominated World

BIOLOGY AND SOCIETY Humanity's Footprint 269

THE PROCESS OF SCIENCE Do Human Activities Facilitate Speciation? 276

EVOLUTION CONNECTION Evolution in the Anthropocene 289

BIOLOGY AND SOCIETY Evolution in the Human-Dominated World

Tailings (waste residue) dump from mining oil sands in Alberta, Canada. Oil sands, also called tar sands, contain a sticky, viscous form of petroleum. Mining these deposits leaves behind a large volume of tailings.

Humanity's Footprint

Humanity has had an extraordinary effect on the ecology and geology of Earth. Our indelible footprint includes the transport of organisms far from their natural homes, the prevalence of agriculture and domesticated animals, the existence of manufactured materials such as plastics and concrete, radioactivity from testing nuclear weapons, and climate-altering emissions from burning fossil fuels. Very little of the planet remains untouched by human activities. In 2017, for example, researchers reported that animals living 10 km (6.2 mi) deep in the Pacific Ocean were contaminated with toxic chemicals from industrial waste. The irreversibility of these changes has led some scientists to propose that a new epoch in Earth's history has begun: the Anthropocene (from the Greek *anthropos*, human). The Anthropocene signals a significant shift in the geologic record that includes a high rate of extinction and accelerating change to Earth. As you will learn in this chapter, scientists have divided the 4.6 billion years of Earth's history into a sequence of geologic eras, which are further subdivided into periods and epochs. This geologic time line is formally defined by the International Commission on Stratigraphy using long-established criteria. Naming a new epoch is no small matter, and experts are currently debating the proposal to acknowledge the Anthropocene.

Regardless of whether this new epoch is officially added to the geologic time line, it's a useful idea for nonscientists to consider. We are living in a time that is unique in Earth's 4.6-billion-year history. Humans are not the only organisms that have changed the environment on a global scale. More than 2 billion years ago, oxygen released by photosynthesis in single-celled prokaryotes eventually transformed the biosphere. But we are the only organisms with the ability to recognize our impact and understand its potential consequences for life on Earth. For example, human activities are modifying the global environment to such an extent that many species are disappearing. In the past 400 years— a very short time on a geologic scale—more than 1,000 species are known to have become extinct. Scientists estimate that this is 100 to 1,000 times the extinction rate seen in the past. Human-driven changes in the environment also bring about evolutionary change in populations of organisms, including pesticide-resistant insects and antibiotic-resistant bacteria (as you learned about in Chapter 13). Activities such as agriculture, hunting, and fishing apply selection pressures to specific populations of organisms. New species have arisen in recent geologic times, too, through transportation of plants, animals, and microbes to new environments.

We'll begin this chapter by discussing the birth of new species and then examine how biologists trace the evolution of biological diversity. We'll also take a closer look at how scientists classify living organisms.

The Origin of Species

Natural selection, a microevolutionary mechanism, explains the striking ways in which organisms are suited to their environment. But what accounts for the tremendous diversity of life, the millions of species that have existed during Earth's history? This question intrigued Darwin, who referred to it in his diary as "that mystery of mysteries—the first appearance of new beings on this Earth."

When, as a young man, Darwin visited the Galápagos Islands (see Figure 13.3), he realized that he was visiting a place of origins. Though the volcanic islands were geologically young, they were already home to many plants and animals known nowhere else in the world. Among these unique inhabitants were marine iguanas **(Figure 14.1)**, Galápagos tortoises (see Figure 13.4), and numerous species of small birds called finches, which you will learn more about in this chapter (see Figure 14.12). Surely, Darwin thought, not all of these species could have been among the original colonists. Some of them must have evolved later on, the myriad descendants of the original colonists, modified by natural selection from those original ancestors.

In the century and a half since the publication of Darwin's *On the Origin of Species by Means of Natural Selection*, new discoveries and technological advances—especially in molecular biology—have given scientists a wealth of new information about the evolution of life on Earth. For example, researchers have explained the genetic patterns underlying the homology of vertebrate limbs (see Figure 13.6). Hundreds of thousands more fossil discoveries have been cataloged since Darwin's time, including many of the transitional (intermediate) forms predicted by Darwin. In a fascinating convergence of old and new techniques, researchers have even been able to investigate the genetic material of certain fossils, including our ancient relatives, the Neanderthals (see Figure 14.23). New dating methods have confirmed that Earth is billions of years old, much older than even the most radical geologists of Darwin's time proposed. As you'll learn in this chapter, these dating methods have also enabled researchers to determine the ages of fossils and rocks, providing valuable insight into evolutionary relationships among groups of organisms. In addition, our enhanced understanding of geologic processes, such as the changing positions of continents, explains some of the geographic distributions of organisms and fossils that puzzled Darwin and his contemporaries.

In this chapter, you'll learn how evolution has woven the rich tapestry of life, beginning with **speciation**, the process in which one species splits into two or more species. Other topics include the origin of evolutionary novelty, such as the wings and feathers of birds and the large brains of humans, and the impact of mass extinctions, which clear the way for new adaptive explosions, such as the diversification of mammals following the disappearance of most of the dinosaurs.

▼ **Figure 14.1 A marine iguana (right), an example of the unique species inhabiting the Galápagos.** Darwin noticed that Galápagos marine iguanas—with a flattened tail that aids in swimming—are similar to, but distinct from, land-dwelling iguanas on the islands and on the South American mainland (left).

What Is a Species?

Species is a Latin word meaning "kind" or "appearance." Even as children we learn to distinguish between the kinds of plants and animals—between dogs and cats, for example, or between roses and dandelions—from differences in their appearance. Although the basic idea of species as distinct life-forms seems intuitive, devising a more formal definition is not so easy.

One way of defining a species (and the main definition used in this book) is the **biological species concept**. It defines a **species** as a group of populations whose members have the potential to interbreed with one another in nature and produce fertile offspring (offspring that can reproduce) **(Figure 14.2)**. Geography and culture may conspire to keep a Manhattan businesswoman and a Mongolian dairyman apart. But if the two did meet and mate, they could have viable babies who develop into fertile adults because all humans belong to the same species. In contrast, humans and chimpanzees, despite having a shared evolutionary history, are distinct species because they can't successfully interbreed.

We cannot apply the biological species concept to all situations. For example, basing the definition of species on reproductive compatibility excludes organisms that only reproduce asexually (producing offspring from a single parent), such as most prokaryotes. And because fossils are obviously not currently reproducing sexually, they cannot be evaluated by the biological species concept. In response to such challenges, biologists have developed other ways to define species. For example, most of the species named so far have been classified based on measurable physical traits such as number and type of teeth or flower structures. Another approach defines a species as the smallest group of individuals sharing a common ancestor and forming one branch on the tree of life. Yet another approach proposes defining a species solely on the basis of molecular data, a sort of bar code that identifies each species.

Each species concept is useful, depending on the situation and the questions being asked. The biological species concept, however, is particularly useful when focusing on how species originate—that is, when we ask: What prevents a member of one group from successfully interbreeding with a member of another group? You'll learn about the variety of answers to that question next. ✓

✓ CHECKPOINT

According to the biological species concept, what defines a species?

Answer: the ability of its members to interbreed with one another and produce fertile offspring in a natural setting

▼ **Figure 14.2** The biological species concept is based on reproductive compatibility rather than physical similarity.

Similarity between different species. The eastern meadowlark (left) and the western meadowlark (right) are very similar in appearance, but they are separate species and do not interbreed.

Diversity within one species. Humans, as diverse in appearance as we are, belong to a single species (*Homo sapiens*) and can interbreed.

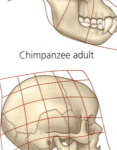

into major structural differences among species. In particular, striking evolutionary transformations can result from genes that alter the rate, timing, and spatial pattern of changes in an organism's form as it develops from a zygote into an adult.

An organism's shape depends in part on the relative growth rates of different body parts during development. For example, in the fetuses of both chimpanzees and humans, the skulls are rounded and the jaws are small, making the face rather flat. As development proceeds, accelerated growth in the jaw produces the elongated skull, sloping forehead, and massive jaws of an adult chimpanzee. In the human evolutionary lineage, mutations slowed the growth of the jaw relative to other parts of the skull. As a result, the skull of an adult human still resembles that of a child—and that of a baby chimpanzee **(Figure 14.19)**.

Our large skull and complex brain are among our most distinctive features. The human brain is proportionately larger than the chimpanzee brain because growth of the organ is switched off much later in human development.

Changes in the rate of developmental events also explain the dramatic differences seen in the homologous limb bones of vertebrates (see Figure 13.6). For instance, increased growth rates produced the extralong "finger" bones in bat wings. Slower growth rates of leg and pelvic bones led to the eventual loss of hind limbs in whales.

Evolutionary changes can also result from alterations in homeotic genes, the master control genes that determine such basic developmental events as where a pair of wings or legs will appear on a fruit fly (see Figure 11.9). A subtle change in the developmental program can have profound effects. Accordingly, changes in the number, nucleotide sequence, and regulation of homeotic genes have led to the huge diversity in body forms.

Next, we see how the process of evolution can produce new, complex structures.

Chimpanzee fetus

Chimpanzee adult

Human fetus

Human adult
(paedomorphic features)

▲ **Figure 14.19 Comparison of human and chimpanzee skull development.** Starting with fetal skulls that are very similar (left), the differential growth rates of the bones making up the skulls produce adult heads with very different proportions. The grid lines will help you relate the fetal skulls to the adult skulls.

The Evolution of Biological Novelty

The two squirrels in Figure 14.6 are different species, but they are very similar animals that live very much the same way. How do we account for the dramatic differences between dissimilar groups—squirrels and birds, for example? Let's see how the Darwinian theory of gradual change can explain the evolution of intricate structures such as eyes or of new (novel) structures such as feathers.

Adaptation of Old Structures for New Functions

The feathered flight of birds is a perfect marriage of structure and function. Consider the evolution of feathers, which are clearly essential to avian aeronautics. In a flight feather, separate filaments called barbs emerge from a central shaft that runs from base to tip. Each barb is linked to the next by tiny hooks that act much like the teeth of a zipper, forming a tightly connected sheet of barbs that is strong but flexible. In flight, the shapes and arrangements of various feathers produce lift, smooth airflow, and help with steering and balance. How did such a beautifully intricate structure evolve? Reptilian features apparent in fossils of *Archaeopteryx*, one of the earliest birds, offered clues in Darwin's time **(Figure 14.20)**, but the

▼ **Figure 14.20 An extinct bird.** Called *Archaeopteryx* ("ancient wing"), this animal lived near tropical lagoons in central Europe about 150 million years ago. Despite its feathers, *Archaeopteryx* has many features in common with reptiles. *Archaeopteryx* is not considered an ancestor of today's birds. Instead, it probably represents an extinct side branch of the bird lineage.

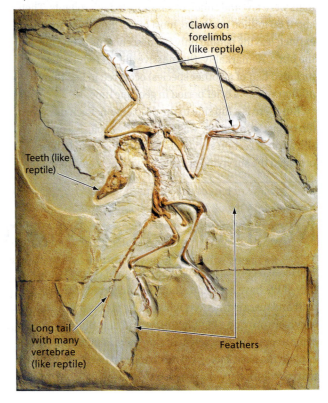

Claws on forelimbs (like reptile)

Teeth (like reptile)

Long tail with many vertebrae (like reptile)

Feathers

definitive answer came in 1996. Birds were not the first feathered animals on Earth—dinosaurs were.

The first feathered dinosaur to be discovered, a 130-million-year-old fossil found in northeastern China, was named *Sinosauropteryx* ("Chinese lizard-wing"). About the size of a turkey, it had short arms and ran on its hind legs, using its long tail for balance. Its unimpressive plumage consisted of a downy covering of hairlike feathers. Since the discovery of *Sinosauropteryx*, thousands of fossils of feathered dinosaurs have been found and classified into more than 30 different species. Although none was unequivocally capable of flying, many of these species had elaborate feathers that would be the envy of any modern bird. But the feathers seen in these fossils could not have been used for flight, nor would their reptilian anatomy have been suited to flying. So if feathers evolved before flight, what was their function? Their first utility may have been for insulation. It is possible that longer, winglike forelimbs and feathers, which increased the surface area of these forelimbs, were co-opted for flight after functioning in some other capacity, such as mating displays, thermoregulation, or camouflage (all functions that feathers also serve today). The first flights may have been only short glides to the ground or from branch to branch in tree-dwelling species. Once flight itself became an advantage, natural selection would have gradually adapted feathers and wings to fit their additional function.

Structures such as feathers that evolve in one context but become co-opted for another function are called **exaptations**. However, exaptation does not mean that a structure evolves in anticipation of future use. Natural selection cannot predict the future; it can only improve an existing structure in the context of its current use. ✅

From Simple to Complex Structures in Gradual Stages

Most complex structures have evolved in small steps from simpler versions having the same basic function—a process of refinement rather than the sudden appearance of complexity. Consider the amazing camera-like eyes of vertebrates and squids. Although these complex eyes evolved independently, the origin of both can be traced from a simple ancestral patch of photoreceptor cells through a series of incremental modifications that benefited their owners at each stage. Indeed, there appears to have been a single evolutionary origin of light-sensitive cells, and all animals with eyes—vertebrates and invertebrates alike—share the same master genes that regulate eye development.

Figure 14.21 illustrates the range of complexity in the structure of eyes among present-day molluscs, a large and diverse phylum of animals. Simple patches of pigmented cells enable limpets, single-shelled molluscs that cling to seaside rocks, to distinguish light from dark. When a shadow falls on them, they hold on more tightly—a behavioral adaptation that reduces the risk of being eaten. Other molluscs have eyecups that have no lenses or other

✅ CHECKPOINT

Explain why the concept of exaptation does not imply that a structure evolves in anticipation of some future environmental change.

■ *Answer: Although an exaptation has new or additional functions in a new environment, it existed earlier because it worked as an adaptation to the old environment.*

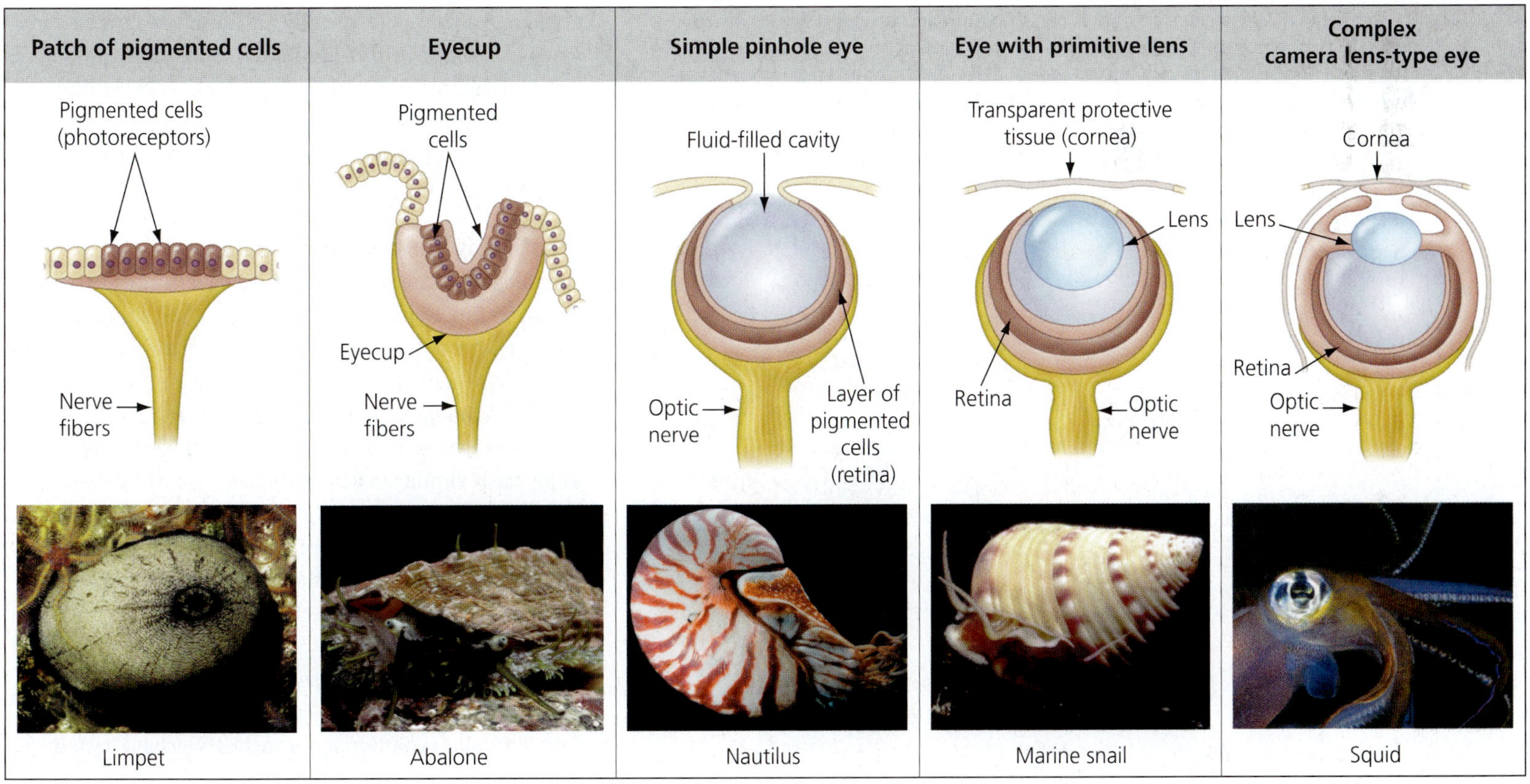

▼ **Figure 14.21** A range of eye complexity among molluscs. The complex eye of the squid evolved in small steps. Even the simplest eye was useful to its owner.

Patch of pigmented cells	Eyecup	Simple pinhole eye	Eye with primitive lens	Complex camera lens-type eye
Limpet	Abalone	Nautilus	Marine snail	Squid

means of focusing images but can indicate light direction. In those molluscs that do have complex eyes, the organs probably evolved in small steps of adaptation. You can see examples of such small steps in Figure 14.21.

Classifying the Diversity of Life

The Linnaean system of taxonomy (see Figure 13.1) is quite a useful method of organizing life's diversity into groups. Ever since Darwin, however, biologists have had a goal beyond simple organization: to have classification reflect evolutionary relationships. In other words, how an organism is named and classified should reflect its place within the evolutionary tree of life. **Systematics**, which includes taxonomy, is a discipline of biology that focuses on classifying organisms and determining their evolutionary relationships.

Classification and Phylogeny

Biologists use **phylogenetic trees** to depict hypotheses about the evolutionary history, or **phylogeny**, of species. These branching diagrams reflect the hierarchical classification of groups nested within more inclusive groups. The tree in **Figure 14.22** shows the classification of some carnivores and their probable evolutionary relationships. Note

ALLELES FROM
TWO WILD GRASSES
COULD BE USED
TO PROTECT CORN
CROPS FROM
FUTURE DISASTERS.

that each branch point represents the divergence of two lineages from a common ancestor. (You may recall Figure 13.8, which is a phylogenetic tree of tetrapods.)

Understanding phylogeny can have practical applications. For example, maize (corn) is an important food crop worldwide; it also provides us with snack favorites such as popcorn, tortilla chips, and corn dog batter. Thousands of years of artificial selection (selective breeding) transformed a scrawny grass with small ears of rock-hard kernels into the maize we know today. In the process, much of the plant's original genetic variation was stripped away. By constructing a phylogeny of maize, researchers have identified two species of wild grasses that may be maize's closest living relatives. The genomes of these plants may harbor alleles that offer disease resistance or other useful traits that could be transferred into cultivated maize by crossbreeding or genetic engineering—insurance against future disease outbreaks or other environmental changes that might threaten corn crops.

Identifying Homologous Characters

Homologous structures in different species may vary in form and function but exhibit fundamental similarities because they evolved from the same structure in a common ancestor. Among the vertebrates, for instance, the whale forelimb is adapted for steering in the water, whereas the bat wing is adapted for flight. Nonetheless, there are many basic similarities in the bones supporting these two structures (see Figure 13.6). Thus, homologous structures are one of the best sources of information for phylogenetic relationships. The greater the number of homologous structures between two species, the more closely the species are related.

There are pitfalls in the search for homology: Not all likeness is inherited from a common ancestor. Species from different evolutionary branches may have certain

▼ **Figure 14.22** **The relationship of classification and phylogeny for some members of the order Carnivora.** The hierarchical classification is reflected in the finer and finer branching of the phylogenetic tree. Each branch point in the tree represents an ancestor common to species to the right of that branch point.

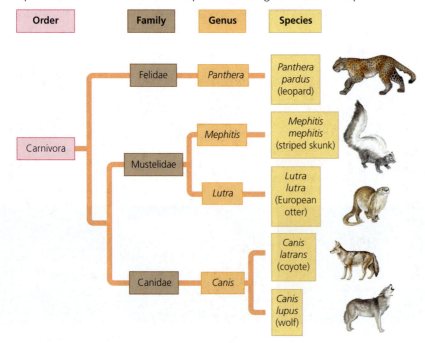

▲ **Figure 14.23** **Artist's reconstruction of Neanderthal.** DNA extracted from Neanderthals, extinct members of the human family, has allowed scientists to study their evolutionary relationship with modern humans.

structures that are superficially similar if natural selection has shaped analogous adaptations. This is called **convergent evolution**. Similarity due to convergence is called **analogy**, not homology. For example, the wings of insects and those of birds are analogous flight equipment: They evolved independently and are built from entirely different structures.

Comparing the embryonic development of two species can often reveal homology that is not apparent in the mature structures (for example, see Figure 13.7). There is another clue to distinguishing homology from analogy: The more complex two similar structures are, the less likely it is they evolved independently. For example, compare the skulls of a human and a chimpanzee (see Figure 14.19). Although each is a fusion of many bones, they match almost perfectly, bone for bone. It is highly improbable that such complex structures matching in so many details could have separate origins. Most likely, the genes required to build these skulls were inherited from a common ancestor.

If homology reflects common ancestry, then comparing the DNA sequences of organisms gets to the heart of their evolutionary relationships. The more recently two species have branched from a common ancestor, the more similar their DNA sequences should be. Scientists have sequenced the genomes of thousands of species. This enormous database has fueled a boom in the study of phylogeny and clarified many evolutionary relationships. In addition, some fossils are preserved in such a way that DNA fragments can be extracted for comparison with living organisms **(Figure 14.23)**. ✓

Inferring Phylogeny from Homologous Characters

Once homologous characters—characters that reflect an evolutionary relationship—have been identified for a group of organisms, how are these characters used to construct phylogenies? The most widely used approach is called cladistics. In **cladistics**, organisms are grouped by common ancestry. A **clade** (from the Greek word for "branch") consists of an ancestral species and all its evolutionary descendants—a distinct branch in the tree of life. Thus, identifying clades makes it possible to construct classification schemes that reflect the branching pattern of evolution.

Cladistics is based on the Darwinian concept of "descent with modification from a common ancestor"—species have some characters in common with their ancestors, but they also differ from them. To identify clades, scientists compare an ingroup with an outgroup **(Figure 14.24)**. The ingroup (for example, the three mammals in Figure 14.24) is the group of species that is actually being analyzed. The outgroup (in Figure 14.24, the iguana, representing reptiles) is a species or group of species known to have diverged before the lineage that contains the groups being studied. By comparing members of the ingroup with each other and with the outgroup, we can determine what characters distinguish the ingroup from the outgroup. All the mammals in the ingroup have hair and mammary glands. These characters were present in the ancestral mammal, but not in the outgroup. Next, gestation, the carrying of offspring in the uterus within the female parent, is absent from the duck-billed platypus (which lays eggs with a shell). From this absence we might infer that the duck-billed platypus represents an early branch point in the mammalian clade. Proceeding in this manner, we can

PIECING TOGETHER EVOLUTIONARY HISTORIES SHOWS WHO'S RELATED TO WHOM.

✅ CHECKPOINT

Our forearms and a bat's wings are derived from the same ancestral prototype; thus, they are _____. In contrast, the wings of a bat and the wings of a bee are derived from totally unrelated structures; thus, they are _____.

■ *Answer: homologous; analogous*

▼ **Figure 14.24** **Simplified example of cladistics.**

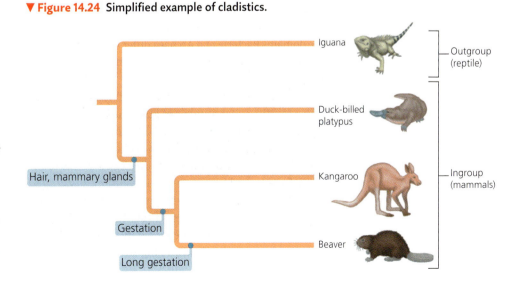

Major Episodes in the History of Life

To put our survey of diversity into perspective and to understand the enormous span of time over which life has evolved, let's look at a brief overview of major events in the history of life on Earth. Our planet's history began 4.6 billion years ago—a period of time that is difficult to grasp. To visualize this immense scale, imagine taking a road trip across North America in which each mile traveled is the equivalent of passing through 1 million years. Our journey will take us from Kamloops, British Columbia, in Canada, to the finish line of the Boston Marathon in Boston, Massachusetts, on a route that covers 4,600 miles **(Figure 15.1)**.

Setting out from Kamloops, we head southwest to Seattle, Washington, and then south toward San Francisco,

California. By the time we reach the California border, nearly 750 million years have passed, and the first rocks have formed on Earth's cooling surface. Our arrival at the Golden Gate Bridge coincides with the appearance of the first cells in the fossil record. After 1,100 million years, life on Earth has begun! Those earliest organisms were all **prokaryotes**, having cells that lack true nuclei. You'll learn more about the origin of these first cells in the next section.

Conditions on the young Earth were very different from conditions today. One difference that was critical to the origin and evolution of life was the lack of atmospheric O_2.

▶ **Figure 15.1 Some major episodes in the history of life.** On this 4,600-mile metaphorical road trip, each mile equals 1 million years in Earth's history.

As we continue driving south on our metaphoric journey, diverse metabolic pathways are evolving among the prokaryotes. However, we won't reach our next milestone for another 800 million years. By that time we have passed through San Diego and turned east across the desert. At Phoenix, Arizona, Earth is 2,700 million years old, and atmospheric O_2 has begun to increase as a result of photosynthesis by autotrophic prokaryotes.

Nine hundred million years later, just past Oklahoma City, we find the first fossils of eukaryotic organisms. **Eukaryotes** are composed of one or more cells that contain nuclei and many other membrane-bound organelles absent in prokaryotic cells. Eukaryotic cells evolved from ancestral host cells that engulfed smaller prokaryotes. The mitochondria of our cells and those of every other eukaryote are descendants of those prokaryotes, as are the chloroplasts of plants and algae. Prokaryotes had been on Earth for 1.7 billion years before eukaryotes evolved. The appearance of the more complex eukaryotes, however, launched a period of tremendous diversification of eukaryotic forms. These new organisms were the protists. Protists are mostly microscopic and unicellular, and as you will learn in this chapter, they are represented today by a great diversity of species.

The next great event in the evolution of life was multicellularity. The oldest fossils that are clearly multicellular are 1.2 billion years old and fall on our metaphorical road trip about midway between St. Louis, Missouri, and Terre Haute, Indiana. The organisms that resulted in these fossils were tiny and not at all complex.

It isn't until 600 million years later (roughly 600 million years ago) that large, diverse, multicellular organisms appeared in the fossil record. At this point, we have traveled 4,000 miles on our road trip—or 4 billion years—and now find ourselves in Erie, near the western edge of Pennsylvania (coincidentally, where one of the authors was born). Less than 15% of our journey remains, and we still have not encountered much diversity of life. But that's about to change.

A great diversification of animals, the so-called Cambrian explosion, marked the beginning of the Paleozoic era, about 541 million years ago (see Table 14.1). By the end of that period, all the major animal body plans, as well as all the major groups, had evolved.

The colonization of land by plants, fungi, and insects also occurred during the Paleozoic. This evolutionary transition began about 500 million years ago, a time that, in our road trip, corresponds to reaching Buffalo, New York.

By the time we get to Albany, New York, we're in the middle of the Mesozoic era, sometimes called the age of dinosaurs. At the end of the Mesozoic, 66 million years ago, we find ourselves about halfway across the state of Massachusetts. As we draw closer to Boston (and our modern time), more and more familiar organisms begin to dominate the landscape—flowering plants, birds, and mammals, including primates.

Modern humans, *Homo sapiens*, appeared roughly 195,000 years ago. At that point in our cross-country tour, we are less than two blocks shy of the finish line of our journey from the origin of Earth to the present day. This metaphorical road trip demonstrates that spans of time that seem lengthy in the context of our own lives are brief moments in the history of life on Earth.

Colonization of land
500 mya

Mid-Mesozoic
180 mya

Homo sapiens
0.195 mya

Boston

Albany

4,600 miles

Buffalo

Large, complex multicellular organisms
600 mya

Erie

4,000 miles

Oldest multicellular fossils
1,200 mya

Terre Haute

St Louis

3,400 miles

ARE WE THERE YET?

The Origin of Plants from Green Algae

▼ **Figure 16.5** Two species of charophytes, the closest algal relatives of plants.

The algal ancestors of plants carpeted moist fringes of lakes or coastal salt marshes more than 500 million years ago. These shallow-water habitats were subject to occasional drying, and natural selection would have favored algae that could survive periodic droughts. Some species accumulated adaptations that enabled them to live permanently above the water line. A modern-day lineage of green algae, the **charophytes (Figure 16.5)**, may resemble

LM 150×

one of these early plant ancestors. Plants and present-day charophytes probably evolved from a common ancestor.

Adaptations making life on dry land possible had accumulated by about 470 million years ago, the age of the oldest known plant fossils. The evolutionary novelties of these first land plants opened the new frontier of a terrestrial habitat. Early plant life would have thrived in the new environment. Bright sunlight was abundant on land, the atmosphere had a wealth of carbon dioxide, and at first there were relatively few pathogens and plant-eating animals. The stage was set for an explosive diversification of plant life.

Plant Diversity

The history of the plant kingdom is a story of adaptation to diverse terrestrial habitats. As we survey the diversity of modern plants, remember that the evolutionary past is the key to the present.

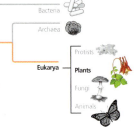

▼ **Figure 16.6** Highlights of plant evolution. This phylogenetic tree highlights the evolution of structures that allowed plants to move onto land; these structures still exist in modern plants. As we survey the diversity of plants, miniature versions of this tree will help you place each plant group in its evolutionary context.

Highlights of Plant Evolution

The fossil record chronicles four major periods of plant evolution, which are also evident in the diversity of modern plants **(Figure 16.6)**. Each stage is marked by the evolution of structures that opened new opportunities on land.

❶ After plants originated from an algal ancestor approximately 470 million years ago, early diversification gave rise to nonvascular plants, including mosses, liverworts, and hornworts. These plants, called **bryophytes**, lack true roots and leaves. Bryophytes also lack lignin, the wall-hardening material that enables other plants to stand tall. Without lignified cell walls, bryophytes have weak upright support. The most familiar bryophytes are **mosses**. A mat of moss actually consists of many plants growing in a tight pack, holding one another up. Structures that protect the gametes and embryos are a terrestrial adaptation that originated in bryophytes.

❷ The second period of plant evolution, which began about 425 million years ago, was the diversification of

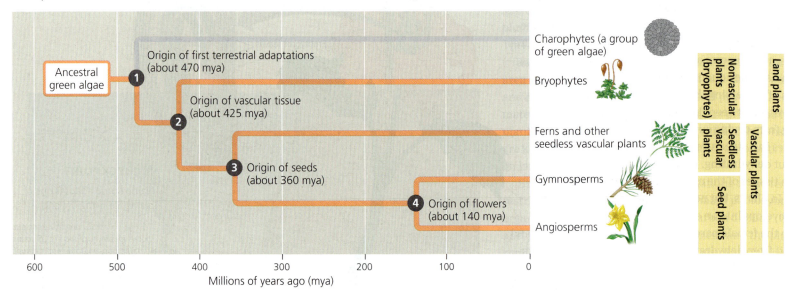

▼ Figure 16.7 The major groups of plants.

PLANT DIVERSITY			
Bryophytes (nonvascular plants)	**Ferns** (seedless vascular plants)	**Gymnosperms** (naked-seed plants)	**Angiosperms** (flowering plants)

plants with vascular tissue. The presence of conducting tissues hardened with lignin allowed vascular plants to grow much taller, rising above the ground to achieve significant height. The earliest vascular plants lacked seeds. Today, this seedless condition is retained by **ferns** and a few other groups of vascular plants.

❸ The third major period of plant evolution began with the origin of the seed about 360 million years ago. Seeds advanced the colonization of land by further protecting plant embryos from drying and other hazards. A **seed** consists of an embryo packaged along with a store of food within a protective covering. The seeds of early seed plants were not enclosed in any specialized chambers. These plants gave rise to the **gymnosperms** ("naked seeds"). Today, the most widespread and diverse gymnosperms are the **conifers**, consisting mainly of cone-bearing trees, such as pines.

❹ The fourth major episode in the evolutionary history of plants was the emergence of flowering plants, or **angiosperms** ("contained seeds"), at least 140 million years ago. The **flower** is a complex reproductive structure that bears seeds within protective chambers called ovaries. This contrasts with the naked seeds of gymnosperms. The great majority of living plants—some 250,000 species—are angiosperms, including all our fruit and vegetable crops, grains and other grasses, and most trees.

With these highlights as our framework, we are now ready to survey the four major groups of modern plants: bryophytes, ferns, gymnosperms, and angiosperms **(Figure 16.7)**. ✅

Bryophytes

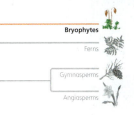

Mosses, which are bryophytes, may sprawl as low mats over acres of land **(Figure 16.8)**. Mosses display two of the key terrestrial adaptations that made the move onto land possible: (1) a waxy cuticle that helps prevent dehydration and (2) the retention of developing embryos within the female plant. However, mosses are not totally liberated from their ancestral aquatic habitat. Mosses need water to reproduce because their sperm need to swim to reach eggs located within the female plant. (A film of rainwater or dew is enough moisture for the sperm to travel.) In addition, because most mosses have no vascular tissue

☑ CHECKPOINT

Name the four major groups of plants. Name an example of each.

■ *Answer: bryophytes (mosses), seedless vascular plants (ferns), gymnosperms (conifers), angiosperms (plants that produce fruits and vegetables)*

▼ Figure 16.8 A peat moss bog in Scotland. Mosses are bryophytes, which are nonvascular plants. Sphagnum mosses, collectively called peat moss, carpet at least 3% of Earth's land surface. They are most commonly found in high northern latitudes. The ability of peat moss to absorb and retain water makes it an excellent addition to garden soil.

319

Gymnosperms

"Coal forests" domi-
nated the North
American and Eurasian
landscapes until near the
end of the Carboniferous period. At
that time, the global climate turned drier and colder,
and the vast swamps began to disappear. This climatic
change provided an opportunity for seed plants, which
can complete their life cycles on dry land and withstand
long, harsh winters. Of the earliest seed plants, the most
successful were the gymnosperms, and several kinds
grew along with the seedless plants in the Carboniferous
swamps. Their descendants include the conifers, or
cone-bearing plants.

Conifers

Perhaps you have had the fun of hiking or skiing through
a forest of conifers, the most common gymnosperms.
Pines, firs, spruces, junipers, cedars, and redwoods
are all conifers. A broad band of coniferous forests covers much of northern Eurasia and North
America and extends southward in mountainous regions (Figure 16.13). Today, about
190 million acres of coniferous forests in the
United States are designated national forests.
Conifers are among the tallest, largest, and
oldest organisms on Earth. Coastal redwoods,
native to the northern California coast, are the
world's tallest trees—up to 110 m, the height of
a 33-story building. Giant sequoias, relatives

of redwoods that grow in the Sierra Nevada mountains of
California, are massive. One, known as the General Sher-
man tree, is about 84 m (275 feet) high and outweighs the
combined weight of a dozen space shuttles. Bristlecone
pines, another species of California conifer, are among the
oldest organisms alive. A recently discovered specimen is
more than 5,000 years old; it was a seedling when people
invented writing.

Nearly all conifers are evergreens, meaning they retain
leaves throughout the year. Even during winter, they
perform a limited amount of photosynthesis on sunny
days. And when spring comes, conifers already have fully
developed leaves that can take advantage of the sunnier
days. The needle-shaped leaves of pines and firs are also
adapted to survive dry seasons. A thick cuticle covers the
leaf, and the stomata are located in pits, further reducing
water loss.

Coniferous forests are highly productive; you probably
use products harvested from them every day. For example, conifers provide much of our lumber for building and
wood pulp for paper production. What we call wood is
actually an accumulation of vascular tissue with lignin,
which gives the tree structural support.

Terrestrial Adaptations of Seed Plants

Compared with ferns, most gymnosperms have three
additional adaptations that make survival in diverse
terrestrial habitats possible: (1) further reduction of
the gametophyte, (2) pollen, and (3) seeds.

The first adaptation is an even greater development of
the diploid sporophyte compared with the haploid gametophyte generation (Figure 16.14). A pine tree or other

▼ **Figure 16.13 A
coniferous forest in Tetlin
National Wildlife Refuge,
Alaska.** Coniferous forests
are widespread in northern
North America and Eurasia;
conifers also grow in the
Southern Hemisphere,
though they are less
numerous there.

▶ **Figure 16.14
Three variations
on alternation
of generations in
plants.**

Key

- Haploid (*n*)
- Diploid (2*n*)

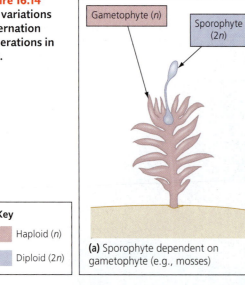

(a) Sporophyte dependent on
gametophyte (e.g., mosses)

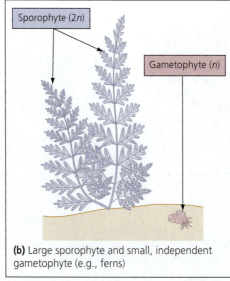

(b) Large sporophyte and small, independent
gametophyte (e.g., ferns)

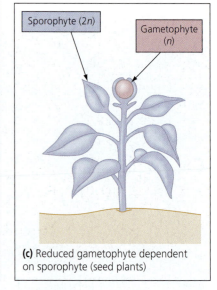

(c) Reduced gametophyte dependent
on sporophyte (seed plants)

conifer is a sporophyte with tiny gametophytes living in its cones (Figure 16.15). In contrast to what is seen in bryophytes and ferns, gymnosperm gametophytes are totally dependent on and protected by the tissues of the parent sporophyte.

A second adaptation of seed plants to dry land came with the evolution of pollen. A **pollen grain** is actually the much-reduced male gametophyte; it houses cells that will develop into sperm. In the case of conifers, **pollination**, the delivery of pollen from the male parts of a plant to the female parts of a plant, occurs via wind. This mechanism for sperm transfer contrasts with the swimming sperm of mosses and ferns. In seed plants, the use of tough, airborne pollen that carries sperm-producing cells to the egg is a terrestrial adaptation that led to even greater success and diversity of plants on land.

The third important terrestrial adaptation of seed plants is the seed itself. A seed consists of a plant embryo packaged along with a food supply within a protective coat. Seeds develop from **ovules**, structures that contain the female gametophytes (Figure 16.16). In conifers, the ovules are located on the scales of female cones. Once released from the parent plant, the seed can remain

dormant for days, months, or even years. Under favorable conditions, the seed can then **germinate**, or sprout: Its embryo emerges through the seed coat as a seedling. Some seeds drop close to their parents, and others are carried far by the wind or animals. ✓

✅ **CHECKPOINT**

Contrast the modes of sperm delivery in ferns versus conifers.

■ *Answer: The flagellated sperm of ferns must swim through water to reach eggs. In contrast, the airborne pollen of conifers delivers sperm-producing cells to eggs in ovules without the need to go through water.*

▼ **Figure 16.16** From ovule to seed.

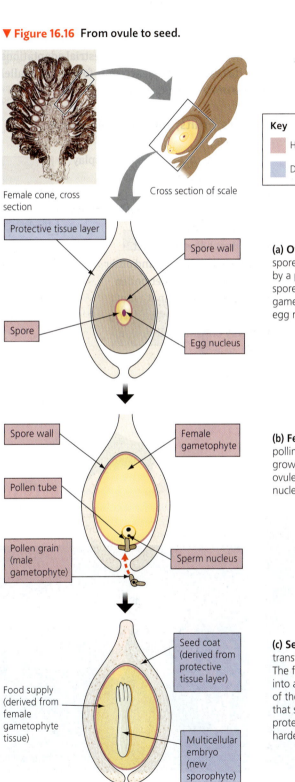

Female cone, cross section

Cross section of scale

Key

Haploid (*n*)

Diploid (2*n*)

Protective tissue layer

Spore wall

Spore

Egg nucleus

(a) Ovule. The sporophyte produces spores within a tissue surrounded by a protective tissue layer. The spore develops into a female gametophyte, which produces an egg nucleus.

Spore wall

Female gametophyte

Pollen tube

Pollen grain (male gametophyte)

Sperm nucleus

(b) Fertilized ovule. After pollination, the pollen grain grows a tiny tube that enters the ovule, where it releases a sperm nucleus that fertilizes the egg.

Seed coat (derived from protective tissue layer)

Food supply (derived from female gametophyte tissue)

Multicellular embryo (new sporophyte)

(c) Seed. Fertilization triggers the transformation of ovule to seed. The fertilized egg (zygote) develops into a multicellular embryo; the rest of the gametophyte forms a tissue that stockpiles food; and the protective tissue layer of the ovule hardens to become the seed coat.

▼ **Figure 16.15** **A pine tree, the sporophyte, bearing two types of cones containing gametophytes.** The leaf-like structures, or scales, of the female cone bears a structure called an ovule containing a female gametophyte. Male cones release clouds of millions of pollen grains, the male gametophytes. Some of these pollen grains land on female cones on trees of the same species. The sperm can fertilize eggs in the ovules of the female cones. The ovules eventually develop into seeds.

Scale

Ovule-producing cones; the scales contain female gametophytes

Pollen-producing cones; they produce male gametophytes

Ponderosa pine

323

Gymnosperms

A drier and colder global climate near the end of the Carboniferous period favored the evolution of the first seed plants. The
most successful were the gymnosperms, represented by conifers.
Needle-shaped leaves with thick cuticles and sunken stomata are
adaptations to dry conditions. Conifers and most other gymnosperms have three
additional terrestrial adaptations: (1) further reduction of the haploid gametophyte and greater development of the diploid sporophyte; (2) sperm-bearing
pollen, which doesn't require water for transport; and (3) seeds, which consist of
a plant embryo packaged along with a food supply inside a protective coat.

Angiosperms

Angiosperms supply nearly all our food and much of our fiber
for textiles. The evolution of the flower and more efficient water
transport help account for the success of the angiosperms. The
dominant stage is a sporophyte with gametophytes in its flowers.
The female gametophyte is located within an ovule, which in turn resides
within a chamber of the ovary. Fertilization of an egg in the female gametophyte produces a zygote, which develops into an embryo. The whole
ovule develops into a seed. The seed's enclosure within an ovary is what
distinguishes angiosperms from gymnosperms, which have naked seeds. A
fruit is the ripened ovary of a flower. Fruits protect and help disperse seeds.
Angiosperms are a major food source for animals, and animals aid plants
in pollination and seed dispersal. Agriculture constitutes a unique kind of
evolutionary relationship among plants, people, and other animals.

Plant Diversity as a Nonrenewable Resource

Deforestation to meet the demand of human activities for space and natural
resources is causing the extinction of plant species at an unprecedented
rate. The problem is especially critical for tropical forests.

Fungi

Characteristics of Fungi

Fungi are unicellular or multicellular eukaryotes; they are heterotrophs that
digest their food externally and absorb the nutrients from the environment.
They are more closely related to animals than to plants. A fungus usually
consists of a mass of threadlike hyphae, forming a mycelium. The cell walls
of fungi are mainly composed of chitin. Although most fungi are nonmo-
tile, a mycelium can grow very quickly, extending the tips of its hyphae into
new territory. Mushrooms are reproductive structures that extend from the
underground mycelium. Fungi reproduce and disperse by releasing spores
that are produced either sexually or asexually.

The Ecological Impact of Fungi

Fungi and bacteria are the principal decomposers of ecosystems. Many
molds destroy fruit, wood, and human-made materials. About 500 species
of fungi are known to be parasites of people and other animals.

Commercial Uses of Fungi

A variety of fungi are eaten or used in food production. In addition, some
fungi are medically valuable.

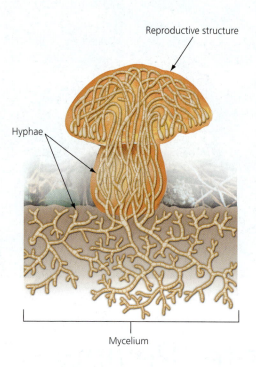
Reproductive structure
Hyphae
Mycelium

Mastering Biology

For practice quizzes, BioFlix animations, MP3 tutorials, video tutors, and more
study tools designed for this textbook, go to Mastering Biology™

SELF-QUIZ

1. Which of the following structures is common to all four major plant
groups? vascular tissue, flowers, seeds, cuticle, pollen

2. Angiosperms are distinguished from all other plants because only
angiosperms have reproductive structures called _____.

3. Complete the following analogies:
   a. Gametophyte is to haploid as _____ is to diploid.
   b. _____ are to conifers as flowers are to _____.
   c. Ovule is to seed as ovary is to _____.

4. Under a microscope, a piece of a mushroom would look most like
   a. jelly.   c. grains of sand.
   b. a tangle of string. d. a sponge.

5. During the Carboniferous period, the dominant plants, which later
formed the great coal beds, were mainly
   a. mosses and other bryophytes.
   b. ferns and other seedless vascular plants.
   c. charophytes and other green algae.
   d. conifers and other gymnosperms.

6. You discover a new species of plant. Under the microscope, you find
that it produces flagellated sperm. A genetic analysis shows that its
dominant generation has diploid cells. What kind of plant do you have?

7. How does the evergreen nature of pines and other conifers adapt the plants for living where the growing season is very short?

8. Which of the following terms includes all others in the list? angiosperm, fern, vascular plant, gymnosperm, seed plant

9. Plant diversity is greatest in
   a. tropical forests.
   b. the temperate forests of Europe.
   c. deserts.
   d. the oceans.

10. What is a fruit?

11. Mycorrhizae are symbiotic associations between _____ and _____.

12. Contrast the heterotrophic nutrition of a fungus with your own heterotrophic nutrition.

For answers to the Self-Quiz, see Appendix D.

IDENTIFYING MAJOR THEMES

For each statement below, identify which major theme is evident (the relationship of structure to function, information flow, pathways that transform energy and matter, interactions within biological systems, or evolution) and explain how the statement relates to the theme. If necessary, review themes (Chapter 1) and review the examples highlighted in blue in this chapter.

13. Gametophytes produce gametes that unite to form zygotes, which develop into new sporophytes. And sporophytes produce spores that give rise to new gametophytes, transmitting DNA through an alternation of generations.

14. Roots typically have many fine branches that thread among the grains of soil, providing a large surface area that maximizes contact with mineral-bearing water in the soil.

15. Vascular tissue was an adaptation that allowed ferns to colonize a greater variety of habitats than mosses.

For answers to Identifying Major Themes, see Appendix D.

THE PROCESS OF SCIENCE

16. In April 1986, an accident at a nuclear power plant in Chernobyl, Ukraine, scattered radioactive fallout for hundreds of miles. In assessing the biological effects of the radiation, researchers found mosses to be especially valuable as organisms for monitoring the damage. Radiation damages organisms by causing mutations. Explain why it is faster to observe the genetic effects of radiation on mosses than on other types of plants. Imagine that you are conducting tests shortly after a nuclear accident. Using potted moss plants as your experimental organisms, design an experiment to test the hypothesis that the frequency of mutations decreases with the organism's distance from the source of radiation.

17. **Interpreting Data** Airborne pollen of wind-pollinated plants such as pines, oaks, weeds, and grasses causes seasonal allergy symptoms in many people. As global warming lengthens the growing season for plants, scientists predict longer periods of misery for allergy sufferers. However, global warming does not affect all regions equally (see Figure 18.44). The table below shows the length of the average season in nine locations for ragweed pollen, an allergen that affects millions of people. Calculate the change in length of the pollen season from 1995 to 2009 for each location and graph this information against latitude. Is there a latitudinal trend in the length of ragweed season? You may want to record the data on a map to help you visualize the geographic locations at which samples were taken.

Average Length of Ragweed Pollen Season in Nine Locations (averages obtained from at least 15 years of data)				
Location	Latitude (°N)	Length of Pollen Season in 1995 (days)	Length of Pollen Season in 2009 (days)	Change in Length of Pollen Season (days)
Oklahoma City, OK	35.47	88	89	
Rogers, AR	36.33	64	69	
Papillion, WI	41.15	69	80	
Madison, WI	43.00	64	76	
La Crosse, WI	43.80	58	71	
Minneapolis, MN	45.00	62	78	
Fargo, ND	46.88	36	52	
Winnipeg, MB, Canada	50.07	57	82	
Saskatoon, SK, Canada	52.07	44	71	

Data from: L. Ziska et al., Recent Warming by Latitude Associated with Increased Length of Ragweed Pollen Season in Central North America. *Proceedings of the National Academy of Sciences* 108: 4248–4251 (2011).

BIOLOGY AND SOCIETY

18. Why are tropical forests being destroyed so rapidly? What kinds of social, technological, and economic factors are responsible? Most forests in more industrialized Northern Hemisphere countries have already been cut. Do the more industrialized nations have a right to pressure the less industrialized nations in the Southern Hemisphere to slow or stop the destruction of their forests? Defend your answer. What kinds of benefits, incentives, or programs might slow the destruction of tropical forests?

19. Many prescription drugs are derived from natural plant products. Numerous other plant substances, including caffeine and nicotine, have effects in the human body as well. There is also a wide array of plant products, in the form of pills, powders, or teas, marketed as herbal medicines. Some people prefer taking these "natural" products to pharmaceuticals. Others use herbal supplements to boost energy, promote weight loss, strengthen the immune system, relieve stress, and more. The U.S. Federal Drug Administration, which approves pharmaceuticals, is also responsible for regulating herbal remedies. What does the label "FDA-approved" on an herbal remedy mean? How does that compare to FDA approval of a drug? The FDA website is a good place to start your research. (Note that the FDA classifies herbal remedies as dietary supplements.)

17 The Evolution of Animals

CHAPTER CONTENTS

The Origins of Animal Diversity 338
Major Invertebrate Phyla 341
Vertebrate Evolution and Diversity 354
The Human Ancestry 361

Why Animal Diversity Matters

We are members of the animal kingdom, a group that encompasses a vast diversity of species ranging in size from microscopic to gargantuan. The evolutionary history of animals spans more than 600 million years, including several million years of human evolution.

IS THAT AN ANIMAL OR A PLANT IN YOUR BATHTUB? A LOOFAH IS THE SKELETON OF A PLANT RELATED TO CUCUMBERS. A SEA SPONGE IS THE SKELETON OF AN ANIMAL.

BEWARE OF RAW FISH! FISH TAPEWORM, FOUND IN SPECIES THAT LIVE OR BREED IN FRESH WATER, IS THE LARGEST PARASITE THAT INFECTS HUMANS.

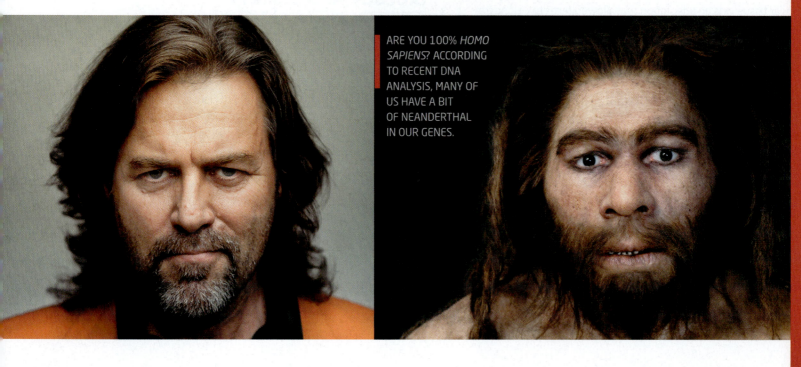

ARE YOU 100% *HOMO SAPIENS?* ACCORDING TO RECENT DNA ANALYSIS, MANY OF US HAVE A BIT OF NEANDERTHAL IN OUR GENES.

CHAPTER THREAD

Human Evolution

BIOLOGY AND SOCIETY   Evolving Adaptability 337

THE PROCESS OF SCIENCE   What Can Lice Tell Us About Ancient Humans? 366

EVOLUTION CONNECTION   Are We Still Evolving? 367

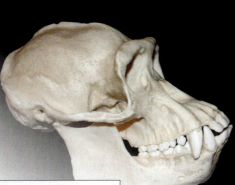

BIOLOGY AND SOCIETY | Human Evolution

Comparison of human and chimpanzee skulls, showing a large difference in brain size.

Evolving Adaptability

Humans (*Homo sapiens*) have managed to colonize almost every major habitat on Earth. What makes humans such successful animals? Much of our success is due to brain power. Species larger than us, such as elephants and whales, do have larger brains, but our brains are much larger than you might predict based on our body size. For example, the ratio of brain volume to body mass in humans is roughly 2.5 times the brain volume to body mass ratio in chimpanzees, our closest primate relatives. Brain size is not the only difference. Our brain cells make more connections with each other, and those connections can be adjusted as needed, making our brains exceptionally flexible. In addition, the configuration of the human brain is distinctive. The part of our brain that deals with problem solving, language, logic, and understanding other people is particularly well-developed.

Although body size has remained roughly the same for about the last 1.5 million years of human evolution, brain size has increased by about 40%. There is evidence that our ancestors' brain size increased as the environment became less predictable. In a changing world, individuals who can plan, imagine, evaluate, and change their behavior have an advantage over individuals who rely more on instinct. Our large, flexible brains have allowed us to create a global society that can share information, culture, and technology. Other species can fly, breathe underwater, or produce millions of offspring in their lifetimes, but none can match our ability to learn and change our behavior.

Humans are just one of the 1.3 million species of animals that have been named and described by biologists. This amazing diversity arose through hundreds of millions of years of evolution as natural selection shaped animal adaptations to Earth's many environments. In this chapter, we'll look at the 9 most abundant and widespread of the roughly 35 phyla (major groups) in the kingdom Animalia. We'll give special attention to the major milestones in animal evolution and conclude by reconnecting with the fascinating subject of human evolution.

The Origins of Animal Diversity

Animal life began in Precambrian seas with the evolution of multicellular creatures that ate other organisms. We are among their descendants.

What Is an Animal?

Animals are eukaryotic, multicellular, heterotrophic organisms that obtain nutrients by eating. This mode of nutrition contrasts animals with plants and other organisms that construct organic molecules through photosynthesis. It also contrasts with fungi, which obtain nutrients by absorption after digesting the food outside their bodies (see Figure 16.23). Most animals digest food within their bodies after ingesting other organisms, dead or alive, whole or by the piece **(Figure 17.1)**.

In addition, animal cells lack the cell walls that provide strong support in the bodies of plants and fungi. And most animals have muscle cells for movement and nerve cells that control the muscles. The most complex animals can use their muscular and nervous systems for many functions other than eating. Some species even use massive networks of nerve cells called brains to interpret complex sensory information and control the body's activities.

Most animals are diploid and reproduce sexually; eggs and sperm are the only haploid cells. The life cycle of a sea star **(Figure 17.2)** includes basic stages found in most animal life cycles. **1** Male and female adult animals make haploid gametes by meiosis, and **2** an egg and a sperm

fuse, producing a zygote. **3** The zygote divides by mitosis, forming **4** an early embryonic stage called a **blastula**, which is usually a hollow ball of cells. **5** In most animals, one side of the blastula folds inward, forming a stage called a **gastrula**. **6** The gastrula develops into a saclike embryo with inner, outer, and middle cell layers and an opening at one end. After the gastrula stage, many animals develop directly into adults. Others, such as the sea star, develop into a **larva** **7**, an immature individual that looks different from the adult animal. (A tadpole, a larval frog, is another example.) **8** The larva undergoes a major change of body form, called **metamorphosis**, in becoming an adult capable of reproducing sexually. ☑

☑ CHECKPOINT
What mode of nutrition distinguishes animals from fungi, both of which are heterotrophs?

■ Answer: ingestion (eating)

▼ **Figure 17.1 Nutrition by ingestion, the animal way of life.** Few animals ingest a piece of food as large as the gazelle being eaten by this rock python. The snake will spend two weeks or more digesting its meal.

▼ **Figure 17.2 The life cycle of a sea star as an example of animal development.**

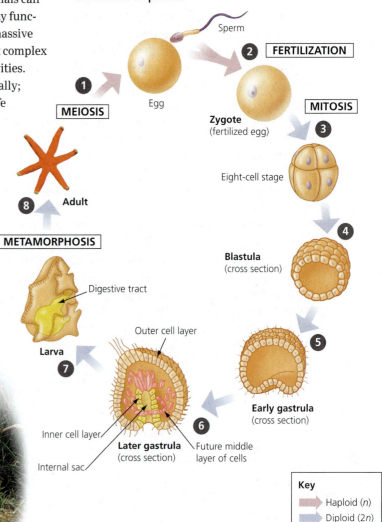

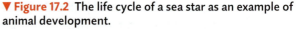

Early Animals and the Cambrian Explosion

Scientists hypothesize that animals evolved from a colonial flagellated protist (Figure 17.3). Although molecular data point to a much earlier origin, the oldest animal fossils that have been found are about 560 million years old. Animal evolution must have been under way already for some time prior to that—the fossils reveal a variety of shapes, and sizes range from 1 cm to 1 m in length (Figure 17.4).

Animal diversification appears to have accelerated rapidly from 535 to 525 million years ago, during the Cambrian period. Because so many animal body plans and new phyla appear in the fossils from such an evolutionarily short time span, biologists call this episode the Cambrian explosion. The most celebrated source of Cambrian fossils is located in the mountains of British Columbia, Canada. The Burgess Shale, as it is known, provided a cornucopia of perfectly preserved animal fossils. In contrast to the Precambrian animals, many Cambrian animals had hard parts such as shells, and many are clearly related to existing animal groups. For example, scientists have classified more than a third of the species found in the Burgess Shale as arthropods, the group that includes present-day crabs, shrimps, and insects (Figure 17.5). Other fossils are more difficult to place. Some are downright weird, like the spiky creature near the center of the drawing, known as *Hallucigenia*, and *Opabinia*, the five-eyed predator grasping a worm with the long, flexible appendage that protrudes in front of its mouth.

What ignited the Cambrian explosion? Scientists have proposed several hypotheses, including increasingly complex predator-prey relationships and an increase in atmospheric oxygen. **But whatever the cause of the rapid diversification, it is likely that the set of "master control" genes—the genetic framework of information flow for building complex bodies—was already in place.** Much of the diversity in body form among the animal phyla is associated with variations in where and when these genes are expressed within developing embryos.

In the last half billion years, animal evolution has to a large degree merely generated variations of the animal forms that originated in the Cambrian seas. Continuing research will help test hypotheses about the Cambrian explosion. But even as the explosion becomes less mysterious, it will seem no less wondrous. ✅

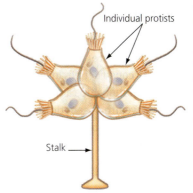

◄ **Figure 17.3 Hypothetical common ancestor of animals.** As you'll learn shortly, the individual cells of this colonial flagellated protist resemble the feeding cells of sponges.

Individual protists

Stalk

▶ **Figure 17.4 Fossils of Precambrian animals.** All of the oldest animal fossils are impressions of soft-bodied animals. Most of them do not appear to be related to any living group of animals.

Sea pen, possibly related to present-day colonial Cnidarians

Impression of upper surface of *Tribrachidium heraldicum*, which had a hemispheric shape and three-part symmetry unlike any living animal (up to 5 cm across)

▼ **Figure 17.5 A Cambrian seascape.** This drawing is based on fossils from the Burgess Shale. The flat-bodied animals are extinct arthropods called trilobites. A photo of a fossil trilobite is shown at the right.

✅ CHECKPOINT

Why is animal evolution during the early Cambrian referred to as an "explosion"?

Answer: because a great diversity of animals evolved in a relatively short time span

Cnidarians

Cnidarians (phylum Cnidaria) are characterized by the presence of body tissues—as are all the remaining animals we will discuss—as well as by radial symmetry and tentacles with stinging cells. Cnidarians include sea anemones, hydras, corals, and jellies (sometimes called jellyfish, although they are not fish). Most of the 10,000 cnidarian species are marine.

The basic body plan of a cnidarian is a sac with a central digestive compartment, the **gastrovascular cavity**. A single opening to this cavity functions as both mouth and anus. This basic body plan has two variations: the stationary **polyp** and the floating **medusa (Figure 17.10)**. Polyps adhere to larger objects and extend their tentacles,

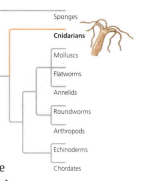

waiting for prey. Examples of the polyp body plan are corals, sea anemones, and hydras. A medusa (plural, *medusae*) is a flattened, mouth-down version of the polyp. It moves freely by a combination of passive drifting and contractions of its bell-shaped body. The largest jellies are medusae with tentacles 60–70 m long (more than half the length of a football field) dangling from an umbrella-like body up to 2 m in diameter. There are some species of cnidarians that live only as polyps, others only as medusae, and still others that pass through both a medusa stage and a polyp stage in their life cycle.

Cnidarians are carnivores that use tentacles arranged in a ring around the mouth to capture prey and push the food into the gastrovascular cavity, where digestion begins. The undigested remains are eliminated through the mouth/anus. The tentacles are armed with batteries of cnidocytes ("stinging cells") that function in defense and in the capture of prey **(Figure 17.11)**. The phylum Cnidaria is named for these stinging cells. ✅

▼ **Figure 17.10 Polyp and medusa forms of cnidarians.** Note that cnidarians have two tissue layers, distinguished in the diagrams by blue and yellow. The gastrovascular cavity has only one opening, which functions as both mouth and anus.

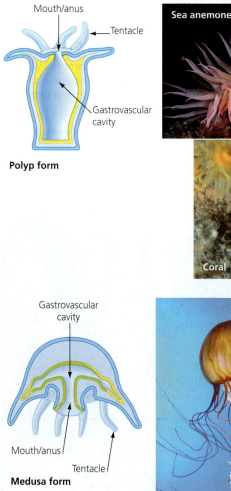

Polyp form

Medusa form

▼ **Figure 17.11 Cnidocyte action.** When a trigger on a tentacle is stimulated by touch, a fine thread shoots out from a capsule. Some cnidocyte threads entangle prey, while others puncture the prey and inject a poison.

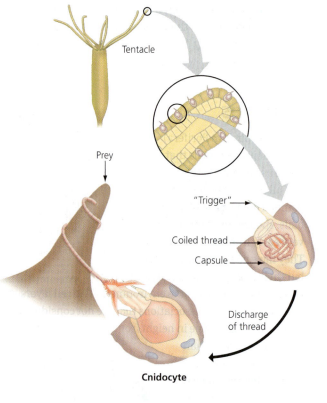

Molluscs

Snails and slugs, oysters and clams, and octopuses and squids are all **molluscs** (phylum Mollusca). Molluscs are soft-bodied animals, but most are protected by a hard shell. Many molluscs feed by extending a file-like organ called a **radula** to scrape up food. For example, the radula of some aquatic snails slides back and forth like a backhoe, scraping and scooping algae off rocks. You can observe a radula in action by watching a snail graze on the glass wall of an aquarium. In cone snails, a group of predatory marine molluscs, the radula is modified to inject venom into prey. The sting of some cone snails is painful, or even fatal, to people.

There are 100,000 known species of molluscs, with most being marine animals. All molluscs have a similar body plan (**Figure 17.12**). The body has three main parts: a muscular foot, usually used for movement; a visceral mass containing most of the internal organs; and a fold of tissue called the mantle. The **mantle** drapes over the visceral mass and secretes the shell if one is present. The three major groups of molluscs are gastropods, bivalves, and cephalopods (**Figure 17.13**).

Most **gastropods**, including snails, are protected by a single spiraled shell into which the animal can retreat when threatened. Slugs and sea slugs lack shells. Many gastropods have a distinct head with eyes at the tips of tentacles (think of a garden snail). Marine, freshwater, and terrestrial gastropods make up about three-quarters of the living mollusc species.

The **bivalves**, including clams, oysters, mussels, and scallops, have shells divided into two halves hinged together. None of the bivalves have a radula. There are both marine and freshwater species, with most being sedentary, using their muscular foot for digging and anchoring in sand or mud.

Cephalopods are all marine animals and generally differ from gastropods and sedentary bivalves in that their bodies are fast and agile. A few have large, heavy shells, but in most the shell is small and internal (as in squids) or missing (as in octopuses). Cephalopods have large brains and sophisticated sense organs, which contribute to their success as mobile predators. They use beak-like jaws and a radula to crush or rip apart prey. The mouth is at the base of the foot, which is drawn out into several long tentacles for catching and holding prey. The colossal squid, discovered in the ocean depths near Antarctica, is the largest living invertebrate. Scientists estimate that this massive cephalopod grows to an average length of 13 m—as long as a school bus. ✓

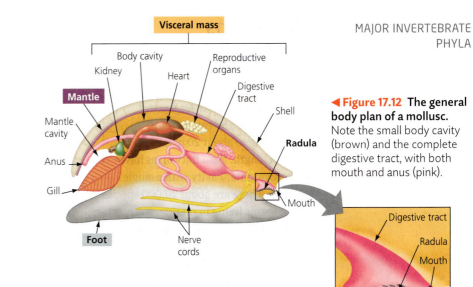
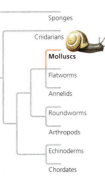

◄ **Figure 17.12** The general body plan of a mollusc. Note the small body cavity (brown) and the complete digestive tract, with both mouth and anus (pink).

☑ CHECKPOINT

Classify these molluscs: A garden snail is an example of a _____; a clam is an example of a _____; a squid is an example of a _____.

■ Answer: gastropod, bivalve, cephalopod

▼ **Figure 17.13** Mollusc diversity.

MAJOR GROUPS OF MOLLUSCS		
Gastropods	**Bivalves** (hinged shell)	**Cephalopods** (large brain and tentacles)
Snail (spiraled shell) **Sea slug** (no shell)	**Scallop.** This scallop has many eyes (small round structures) peering out between the two halves of the hinged shell.	**Octopus.** Octopuses live on the seafloor, where they search for crabs and other food. They have no shell. The brain of an octopus is larger and more complex, proportionate to body size, than that of any other invertebrate. **Nautilus.** The shell of the nautilus is a coiled series of chambers. The animal inhabits only the outermost chamber; the other chambers contain gas and fluid that enable the nautilus to regulate its buoyancy.

others are free-swimming. Their heads are well-equipped with sensory organs for moving about in search of small invertebrates to eat. Segmental appendages aid locomotion in many species.

Sedentarians, which include earthworms, many tube-dwellers, and leeches, tend to be less mobile than errantians. Tube-dwellers, such as the one shown in Figure 17.16, build tubes by secreting calcium carbonate or by mixing mucus with bits of sand and broken shells. The circlet of feathery tentacles seen at the mouth of each tube extends from the head of the worm inside. The tentacles are coated with mucus that traps suspended food particles. The tentacles also function in gas exchange. Leeches are notorious for the bloodsucking

Medicinal leech

habits of some species. However, most species are free-living carnivores that eat small invertebrates such as snails and insects. A few terrestrial species inhabit damp vegetation in the tropics, but the majority of leeches live in fresh water. Leeches that feed on blood have razor-like jaws with hundreds of tiny teeth that cut through the skin. They secrete saliva containing an anesthetic and an anticoagulant into the wound. The anesthetic makes the bite virtually painless, and the anticoagulant prevents clotting as the leech drains excess blood from the wound. A European freshwater species called *Hirudo medicinalis* is occasionally used to drain blood from bruised tissues and to help relieve swelling in fingers or toes that have been sewn back on after accidents. ✅

Roundworms

Roundworms (also called **nematodes**, members of the phylum Nematoda) get their common name from their cylindrical body, which is usually tapered at both ends **(Figure 17.17)**. Roundworms are among the most numerous and widespread of all animals. About 25,000 species of roundworms are known, and perhaps ten times that number actually exist. Roundworms range in length from about 1 mm to 1 m. They are found in most aquatic habitats, in wet soil,

Sponges
Cnidarians
Molluscs
Flatworms
Annelids
Roundworms
Arthropods
Echinoderms
Chordates

and as parasites in the body fluids and tissues of plants and animals.

Free-living roundworms are important decomposers. They live virtually everywhere there is decaying organic matter, and their numbers are huge. Ninety thousand nematodes have been found in a single rotting apple. Recently, researchers even found nematodes living 2 miles underground, where they survive by grazing on microbes. Other species of nematodes thrive as parasites in plants and animals. Some are major agricultural pests that attack the roots of plants. At least 50 parasitic roundworm species infect people; they include pinworms, hookworms, and the parasite that causes trichinosis. ✅

▼ **Figure 17.17** Roundworm diversity.

(a) A free-living roundworm. This species has the classic roundworm shape: cylindrical with tapered ends. The ridges indicate muscles that run the length of the body.

(b) Parasitic roundworms in pork. The potentially fatal disease trichinosis is caused by eating undercooked pork infected with *Trichinella* roundworms. The worms (shown here in pork tissue) burrow into a person's intestine and then invade muscle tissue.

(c) Head of hookworm. Hookworms sink their hooks into the wall of the host's small intestine and feed on blood. Although the worms are small (less than 1 cm), a severe infestation can cause serious anemia.

Arthropods

Arthropods (phylum Arthropoda) are named for their jointed appendages. Crustaceans (such as crabs and lobsters), arachnids (such as spiders and scorpions), and insects (such as grasshoppers and moths) are examples of arthropods **(Figure 17.18)**. Zoologists estimate that the total arthropod population numbers about a billion billion (10^{18}) individuals. Researchers have identified more than a million arthropod species, mostly insects. In fact, two out of every three species of life that have been scientifically described are arthropods. And arthropods are represented in nearly all habitats of the biosphere. In species diversity, distribution, and sheer numbers, arthropods must be regarded as the most successful animal phylum.

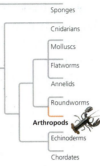

◀ **Figure 17.18** Arthropod diversity.

MAJOR GROUPS OF ARTHROPODS

Arachnids

Crustaceans

Millipedes and Centipedes

Insects

General Characteristics of Arthropods

Arthropods are segmented animals. In contrast with the repeating similar segments of annelids, however, arthropod segments and their appendages have become specialized for a great variety of functions. This evolutionary flexibility contributed to the great diversification of arthropods. Specialization of segments (or of fused groups of segments) provides for an efficient division of labor among body regions. For example, the appendages of different segments may be adapted for walking, feeding, sensory reception, swimming, or defense **(Figure 17.19)**.

The body of an arthropod is completely covered by an **exoskeleton**, an external skeleton. This coat is constructed from layers of protein and a tough polysaccharide called chitin. The exoskeleton can be a thick, hard armor over some parts of the body (such as the head), yet be paper-thin and flexible in other locations (such as the joints). The exoskeleton protects the animal and provides points of attachment for the muscles that move the appendages. There are, of course, advantages to wearing hard parts on the outside. Our own skeleton is interior to most of our soft tissues, an arrangement that doesn't provide much protection from injury. But our skeleton does offer the advantage of being able to grow along with the rest of our body. In contrast, a growing arthropod must occasionally shed its old exoskeleton and secrete a larger one. This process, called molting, leaves the animal temporarily vulnerable to predators and other dangers. The following sections explore the major groups of arthropods. ✓

✅ **CHECKPOINT**

Seafood lovers look forward to soft-shell crab season, when almost the entire animal can be eaten without cracking a hard shell to get at the meat. An individual crab is only a "soft-shell" for a matter of hours; harvesters therefore capture crabs and keep them in holding tanks until they are ready to be cooked. What process makes soft-shell crabs possible?

■ Answer: Crabs have a thick exoskeleton that must be shed (molted) to allow the crab's body to grow. Soft-shell crabs have molted but have not yet secreted a new hard shell.

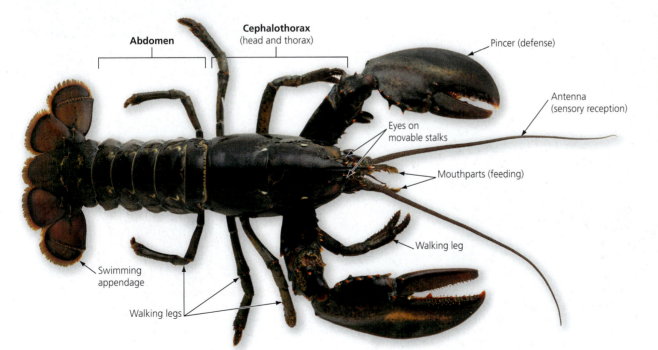

Abdomen

Cephalothorax (head and thorax)

Pincer (defense)

Antenna (sensory reception)

Eyes on movable stalks

Mouthparts (feeding)

Walking leg

Swimming appendage

Walking legs

◀ **Figure 17.19** Anatomy of a lobster, a crustacean. The whole body, including the appendages, is covered by an exoskeleton. The body is segmented, but this characteristic is obvious only in the abdomen.

Arachnids

Arachnids include scorpions, spiders, ticks, and mites **(Figure 17.20)**. Most arachnids live on land. Members of this arthropod group usually have four pairs of walking legs and a specialized pair of feeding appendages.

In spiders, these feeding appendages are fang-like and equipped with poison glands. As a spider uses these appendages to immobilize and dismantle its prey, it spills digestive juices onto the torn tissues and sucks up its liquid meal.

▼ **Figure 17.20** Arachnid characteristics and diversity.

Pair of feeding appendages

Leg (four pairs)

Pair of silk-spinning appendages

Scorpion. Scorpions have a pair of large pincers that function in defense and food capture. The tip of the tail bears a poisonous stinger. Scorpions sting people only when prodded or stepped on.

Dust mite. This microscopic house dust mite is a ubiquitous scavenger in our homes. Dust mites are harmless except to people who are allergic to the mites' feces.

Spider. Like most spiders, including the tarantula in the large photo above, this black widow spins a web of liquid silk, which solidifies as it comes out of specialized glands. A black widow's venom can kill small prey but is rarely fatal to humans.

Wood tick. Wood ticks and other species carry bacteria that cause Rocky Mountain spotted fever. Lyme disease is carried by several different species of ticks.

Crustaceans

Crustaceans, which are nearly all aquatic, include delectable species of crabs, lobsters, crayfish, and shrimps **(Figure 17.21)**. Barnacles, which anchor themselves to rocks, boat hulls, and even whales, are also crustaceans. One group of crustaceans, the isopods, is represented on land by pill bugs. All of these animals exhibit the arthropod characteristic of multiple pairs of specialized appendages.

▼ **Figure 17.21** Crustacean characteristics and diversity.

Two feeding appendages

Antennae

Leg (three or more pairs)

Crab. Ghost crabs are common along shorelines throughout the world. They scurry along the surf's edge and then quickly bury themselves in sand.

Shrimp. Naturally found in Pacific waters from Africa to Asia, giant prawns are widely cultivated as food.

Pill bug. Commonly found in moist locations with decaying leaves, such as under logs, pill bugs get their name from their tendency to roll up into a tight ball when they sense danger.

Crayfish. The red swamp crayfish, cultivated worldwide for food, is native to the southeastern United States. When released outside its native range, it competes aggressively with other species, endangering the local ecosystem.

Barnacles. Barnacles are stationary crustaceans with exoskeletons hardened into shells by calcium carbonate (lime). The jointed appendages projecting from the shell capture small plankton.

Millipedes and Centipedes

Millipedes and **centipedes** are terrestrial arthropods that have similar segments over most of the body. Although they superficially resemble annelids, their jointed legs reveal they are arthropods **(Figure 17.22)**.

Millipedes eat decaying plant matter. They have two pairs of short legs per body segment. Centipedes are carnivores, with a pair of poison claws used in defense and to paralyze prey, such as cockroaches and flies. Each of a centipede's body segments bears a single pair of legs.

▶ **Figure 17.22** Millipedes and centipedes.

Two pairs of legs per segment

One pair of legs per segment

Millipede. Like most millipedes, this one has an elongated body with two pairs of legs per trunk segment.

Centipede. Centipedes can be found in dirt and leaf litter. Their venomous claws can harm cockroaches and spiders but not people.

Insect Anatomy

Like the grasshopper in **Figure 17.23**, most **insects** have a three-part body: head, thorax, and abdomen. The head usually bears a pair of sensory antennae and a pair of eyes. The mouthparts of insects are adapted for particular kinds of eating—for example, for biting and chewing plant material in grasshoppers, for lapping up fluids in houseflies, and for piercing skin and sucking blood in mosquitoes. Most adult insects have three pairs of legs and one or two pairs of wings, all extending from the thorax.

Flight is obviously one key to the great success of insects. An animal that can fly can escape many predators, more readily find food and mates, and disperse to new habitats much faster than an animal that must crawl on the ground. Because their wings are extensions of the exoskeleton and not true appendages, insects can fly without sacrificing legs. By contrast, the flying vertebrates—birds and bats—have one of their two pairs of legs modified for wings, which explains why these vertebrates are generally not very swift on the ground.

▼ **Figure 17.23** Anatomy of a grasshopper.

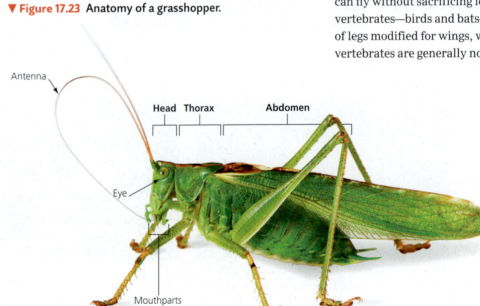

Antenna

Head   Thorax   Abdomen

Eye

Mouthparts

Insect Diversity

In species diversity, insects outnumber all other forms of life combined **(Figure 17.24)**. They live in almost every terrestrial habitat and in fresh water, and flying insects fill the air. Insects are rare in the seas, where crustaceans are the dominant arthropods. The oldest insect fossils date back to about 400 million years ago. Later, the evolution of flight sparked an explosion in insect variety.

▼ **Figure 17.24** Insect diversity.

Robber fly. This predatory insect injects prey with enzymes that liquefy the tissues and then sucks in the resulting fluid.

Rhinoceros beetle. Only males have a "horn," which is used for fighting other males and for digging.

Rainbow shield bugs. The smaller insects are juveniles of the same species.

Greater arid-land katydid. Often called the red-eyed devil, this scary-looking predator displays its wings and spiny legs when threatened.

Praying mantis. There are more then 2,000 species of praying mantises throughout the world.

Weevil. Light reflected from varying thicknesses of chitin produces the brilliant colors of this weevil.

Buckeye butterfly. The eyespots on the wings of this butterfly may startle predators.

Blue dasher dragonfly. The huge multifaceted eyes of dragonflies wrap around the head for nearly 360° vision. Each of the four wings works independently, giving dragonflies excellent maneuverability as they pursue prey.

Many insects undergo metamorphosis in their development. In the case of grasshoppers and some other insect groups, the young resemble adults but are smaller and have different body proportions. The animal goes through a series of molts, each time looking more like an adult, until it reaches full size. In other cases, insects have distinctive larval stages specialized for eating and growing that are known by such names as maggots (fly larvae), grubs (beetle larvae), or caterpillars (larvae of moths and butterflies). The larval stage looks entirely different from the adult stage, which is specialized for dispersal and reproduction. Metamorphosis from the larva to the adult occurs during a pupal stage **(Figure 17.25)**.

Forensic entomologists (forensic scientists who study insects) use their knowledge of insect life cycles to help solve criminal cases. For example, blowfly maggots feed on decaying flesh. Female blowflies, which can smell a dead body from up to a mile away, typically arrive and lay their eggs in a fresh corpse within minutes. By knowing the length of each stage in the life cycle of a blowfly, an entomologist can determine how much time has passed since death occurred.

Animals so numerous, diverse, and widespread as insects are bound to affect the lives of all other terrestrial organisms, including people, in many ways. **Thus, insects provide many examples of interactions within biological systems.** The bees, flies, and other insects that pollinate our crops and orchards exemplify a beneficial interaction. Other interactions are harmful to people. For example, insects are carriers of the microbes that cause many human diseases, such as malaria and West Nile disease. Insects also compete with people for food by eating our field crops. In an effort to minimize their losses, farmers in the United States spend billions of dollars each year on pesticides, spraying crops with massive doses of insecticide poisons. But try as they might, not even humans have significantly challenged the preeminence of insects and their arthropod kin. Rather, the evolution of pesticide resistance has caused humans to change their pest-control methods (see Figure 13.11). ✅

✅ CHECKPOINT

Which major arthropod group is mainly aquatic? Which is the most numerous?

◼ *Answer: crustaceans; insects*

▼ **Figure 17.25** **Metamorphosis of a monarch butterfly.**

The **larva (caterpillar)** spends its time eating and growing, molting as it grows.

After several molts, the larva becomes a **pupa** encased in a cocoon.

Within the pupa, the larval organs break down and adult organs develop from cells that were dormant in the larva.

Finally, the **adult** emerges from the cocoon.

The butterfly flies off and reproduces, nourished mainly by calories stored when it was a caterpillar.

Echinoderms

The **echinoderms** (phylum Echinodermata) are named for their spiny surfaces (*echin* is Greek for "spiny"). Among the echinoderms are sea stars, sea urchins, sea cucumbers, and sand dollars **(Figure 17.26)**.

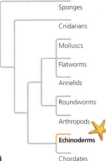

Echinoderms include about 7,000 species, all of them marine. Most move slowly, if at all. Echinoderms lack body segments, and most have radial symmetry as adults. Both the external and the internal parts of a sea star, for instance, radiate from the center like the spokes of a wheel. In contrast to the adult, the larval stage of echinoderms is bilaterally symmetrical. This supports other evidence that echinoderms are not closely related to other radial animals, such as cnidarians, that never show bilateral symmetry. Most echinoderms have an **endoskeleton** (interior skeleton) constructed from hard plates just beneath the skin.

Bumps and spines of this endoskeleton account for the animal's rough or prickly surface. Unique to echinoderms is the **water vascular system**, a network of water-filled canals that circulate water throughout the echinoderm's body, facilitating gas exchange (the entry of O_2 and the removal of CO_2) and waste disposal. The water vascular system also branches into extensions called tube feet. A sea star or sea urchin pulls itself slowly over the seafloor using its suction-cup-like tube feet. Sea stars also use their tube feet to grip prey during feeding.

Looking at sea stars and other adult echinoderms, you may think they have little in common with humans and other vertebrates. But as shown by the phylogenetic tree at the left, echinoderms share an evolutionary branch with chordates, the phylum that includes vertebrates. Analysis of embryonic development can differentiate the echinoderms and chordates from the evolutionary branch that includes molluscs, flatworms, annelids, roundworms, and arthropods. With this context in mind, we're now ready to make the transition in our discussion from invertebrates to vertebrates. ✓

✓ CHECKPOINT

Contrast the skeleton of an echinoderm with that of an arthropod.

■ *Answer: An echinoderm has an endoskeleton; an arthropod has an exoskeleton.*

▼ **Figure 17.26 Echinoderm diversity.**

Sea star. When a sea star encounters an oyster or clam, it grips the mollusc's shell with its tube feet (see inset) and positions its mouth next to the narrow opening between the two halves of the prey's shell. The sea star then pushes its stomach out through its mouth and the crack in the mollusc's shell.

Sea urchin. In contrast to sea stars, sea urchins are spherical and have no arms. If you look closely, you can see the long tube feet projecting among the spines. Unlike sea stars, which are mostly carnivorous, sea urchins mainly graze on seaweed and other algae.

Tube feet

Sea cucumber. On casual inspection, this sea cucumber does not look much like other echinoderms. However, a closer look would reveal many echinoderm traits, including five rows of tube feet.

Sand dollar. Live sand dollars have a skin of short, movable spines covering a rigid skeleton. A set of five pores (arranged in a star pattern) allows seawater to be drawn into the sand dollar's body.

Vertebrate Evolution and Diversity

Most of us are curious about our family ancestry. Biologists are also interested in the larger question of tracing human ancestry within the animal kingdom. In this section, we trace the evolution of the vertebrates, the group that includes humans and our closest relatives. All vertebrates have endoskeletons, a characteristic shared with most echinoderms. However, vertebrate endoskeletons are unique in having a skull and a backbone, a series of bones called vertebrae (singular, *vertebra*), for which the group is named **(Figure 17.27)**. Our first step in tracing the vertebrate lineage is to determine where vertebrates fit in the animal kingdom.

Characteristics of Chordates

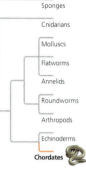

The last phylum in our survey of the animal kingdom is the phylum Chordata. **Chordates** share four key features that appear in the embryo and sometimes in the adult **(Figure 17.28)**. These four chordate characteristics are (1) a **dorsal, hollow nerve cord**; (2) a **notochord**, which is a flexible, longitudinal rod located between the digestive tract and the nerve cord; (3) **pharyngeal slits**, which are grooves in the pharynx, the region of the digestive tube just behind the mouth; and (4) a **post-anal tail**, which is a tail to the rear of the anus. Although these chordate characteristics are often difficult to recognize in the adult animal, they are always present in chordate embryos. For example, the notochord, for which our phylum is named, persists in adult humans only in the form of the cartilage disks that function as cushions between the vertebrae. Back injuries described as "ruptured disks" or "slipped disks" refer to these notochord remnants.

Body segmentation is another chordate characteristic. Chordate segmentation is apparent in the backbone of vertebrates (see Figure 17.27) and is also evident in the segmental muscles of all chordates (see the chevron-shaped—<<<<—muscles in the lancelet in Figure 17.29). Segmental musculature is not so obvious in adult humans unless one is motivated to sculpt "washboard abs."

▼ **Figure 17.27** **A vertebrate endoskeleton.** This snake skeleton, like those of all vertebrates, has a skull and a backbone consisting of vertebrae.

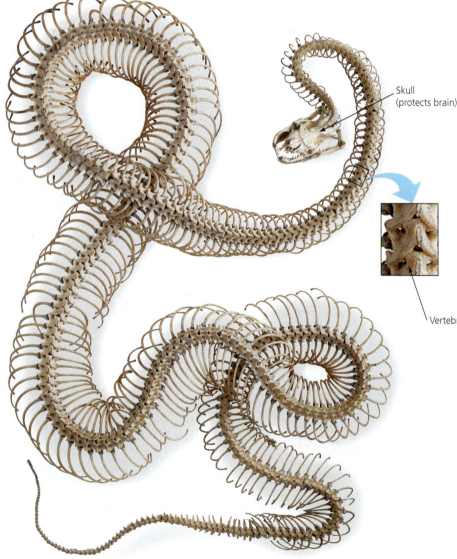

Skull
(protects brain)

Vertebra

▼ **Figure 17.28** Chordate characteristics.

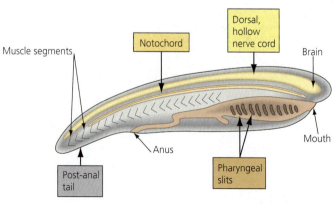

Muscle segments

Notochord

Dorsal, hollow nerve cord

Brain

Mouth

Pharyngeal slits

Anus

Post-anal tail

Two groups of chordates, **tunicates** and **lancelets** (Figure 17.29), are invertebrates. All other chordates are **vertebrates**, which retain the basic chordate characteristics but have additional features that are unique—including, of course, the backbone. **Figure 17.30** is an overview of chordate and vertebrate evolution that will provide a context for our survey. ✓

▼ **Figure 17.29** Invertebrate chordates.

Mouth

Tail

Lancelet. This marine invertebrate owes its name to its bladelike shape. Only a few centimeters long, lancelets wiggle backward into the gravel, leaving their mouth exposed, and filter tiny food particles from the seawater.

Tunicates. The tunicate, or sea squirt, is a stationary animal that filters food from the water. These pastel sea squirts get their nickname from their coloration and the fact that they can quickly expel water to startle intruders.

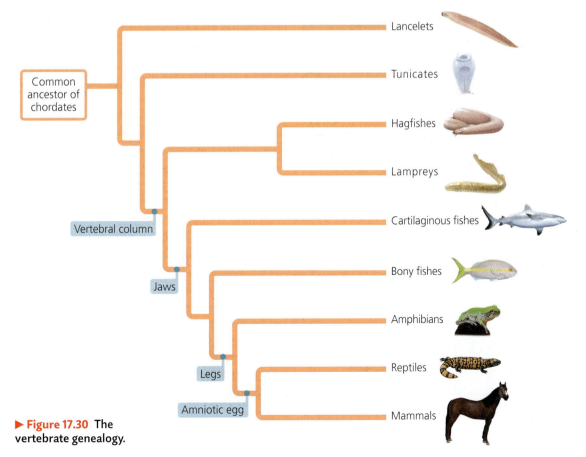

Lancelets

Tunicates

Common ancestor of chordates

Hagfishes

Lampreys

Vertebral column

Cartilaginous fishes

Jaws

Bony fishes

Amphibians

Legs

Reptiles

Amniotic egg

Mammals

▶ **Figure 17.30** The vertebrate genealogy.

✓ **CHECKPOINT**

During our early embryonic development, what four features do we share with invertebrate chordates such as lancelets?

■ *Answers: (1) dorsal, hollow nerve cord; (2) notochord; (3) pharyngeal slits; (4) post-anal tail*

Fishes

The first vertebrates were aquatic and probably evolved during the early Cambrian period about 542 million years ago. In contrast with most living vertebrates, they lacked jaws, hinged bone structures that work the mouth.

Two types of jawless fishes survive today: hagfishes and lampreys. Present-day hagfishes scavenge dead or dying animals on the cold, dark seafloor **(Figure 17.31a)**. When threatened, a hagfish exudes an enormous amount of slime from special glands on the sides of its body. Recently, some hagfishes have become endangered because their skin is used to make "eel-skin" belts, purses, and boots. Most species of lampreys are parasites that use their jawless mouths as suckers to attach to the sides of large fish **(Figure 17.31b)**. The rasping tongue then penetrates the skin, allowing the lamprey to feed on its victim's blood and tissues.

We know from the fossil record that the first jawed vertebrates were fishes that evolved about 440 million years ago. They had two pairs of fins, making them agile swimmers. Some early fishes were active predators up to 10 m (33 ft) in length that could chase prey and bite off chunks of flesh. Even today, most fishes are carnivores.

Cartilaginous fishes, such as sharks and rays, have a flexible skeleton made of cartilage **(Figure 17.31c)**. Most sharks are adept predators because they are fast swimmers with streamlined bodies, acute senses, and powerful jaws. A shark does not have keen eyesight, but its sense of smell is very sharp. In addition, special electrosensors on the head can detect minute electrical fields produced by muscle contractions in nearby animals. Sharks also have a **lateral line system**, a row of sensory organs running along each side of the body. Sensitive to changes in water pressure, the lateral line system enables a shark to detect minor vibrations caused by animals swimming in its neighborhood. There are about 1,000 living species of cartilaginous fishes, nearly all of them marine.

The skeletons of **bony fishes** are reinforced by calcium **(Figure 17.31d)**. Bony fishes have a lateral line system, a keen sense of smell, and excellent eyesight. On each side of the head, a protective flap called the **operculum** (plural, *opercula*) covers a chamber housing the gills, feathery external organs that extract oxygen from water. Movement of the operculum allows the fish to breathe without swimming. By contrast, sharks lack opercula and must swim to pass water over their gills. The need to move water over the gills is why a shark must keep moving to stay alive. Also unlike sharks, bony fishes have a **swim bladder**, a gas-filled sac that enables the fish to control its buoyancy. Thus, many bony fishes can conserve energy by remaining almost motionless, in contrast to sharks, which sink if they stop swimming.

Most bony fishes, including familiar species such as tuna, trout, and goldfish, are **ray-finned fishes**. Their fins are webs of skin supported by thin, flexible skeletal rays, the feature for which the group was named. There are approximately 27,000 species of ray-finned fishes, the greatest number of species of any vertebrate group.

A second evolutionary branch includes the **lobe-finned fishes**. In contrast to the ray-finned fishes, their fins are muscular and are supported by stout bones that are homologous to amphibian limb bones. Early lobe-fins lived in coastal wetlands and may have used their fins to "walk" underwater. Today, three lineages of lobe-fins survive. The coelacanth is a deep-sea dweller once thought to be extinct. The lungfishes are represented by several Southern Hemisphere species that inhabit stagnant waters and gulp air into lungs connected to the pharynx. The third lineage of lobe-fins adapted to life on land and gave rise to amphibians, the first terrestrial vertebrates. ✅

▼ **Figure 17.31** Fish diversity.

(a) Hagfish

(b) Lamprey (inset: mouth)

(c) Shark, a cartilaginous fish

Lateral line   Operculum

(d) Bony fish

✅ **CHECKPOINT**

A shark has a _____ skeleton, whereas a tuna has a _____ skeleton.

■ *Answer: cartilaginous, bony*

Lancelets
Tunicates
Hagfishes
Lampreys
Cartilaginous fishes
Bony fishes
Amphibians
Reptiles
Mammals

Amphibians

In Greek, the word *amphibios* means "living a double life." Most **amphibians**, which include frogs and salamanders, exhibit a mixture of aquatic and terrestrial adaptations **(Figure 17.32a)**. Most species are tied to water because their eggs, lacking shells, dry out quickly in the air. A frog may spend much of its time on land, but it lays its eggs in water **(Figure 17.32b)**. An egg develops into a larva called a tadpole, a legless, aquatic algae-eater with gills, a lateral line system resembling that of fishes, and a long-finned tail **(Figure 17.32c)**. The tadpole undergoes a radical metamorphosis when changing into a frog **(Figure 17.32d)**. When a

Lancelets
Tunicates
Hagfishes
Lampreys
Cartilaginous fishes
Bony fishes
Amphibians
Reptiles
Mammals

young frog crawls onto shore and begins life as a terrestrial insect-eater, it has four legs, air-breathing lungs instead of gills, external eardrums, and no lateral line system. But even as adults, amphibians are most abundant in damp habitats, such as swamps and rain forests. This is partly because amphibians depend on their moist skin to supplement lung function in exchanging gases with the environment. Thus, even those frogs that are adapted to relatively dry habitats spend much of their time in humid burrows or under piles of damp leaves. With their thin, permeable skin and susceptibility to dehydration, amphibians are especially vulnerable to environmental threats such as water pollution and climate change.

Amphibians were the first vertebrates to colonize land. They descended from fishes that had lungs and fins with muscles and skeletal supports strong enough to enable some movement, however clumsy, on land **(Figure 17.33)**. **The fossil record chronicles the evolution of four-limbed amphibians from fish-like ancestors.** Terrestrial vertebrates—amphibians, reptiles, and mammals—are collectively called **tetrapods**, which means "four feet." ✓

✓ CHECKPOINT

Amphibians were the first _____, four-footed terrestrial vertebrates.

Answer: tetrapods

▼ **Figure 17.32** Amphibian diversity.

Malayan horned frog

Texas barred tiger salamander

(a) Frogs and salamanders: the two major groups of amphibians

(b) Frog eggs

(c) Tadpole

(d) Adult tree frog

▼ **Figure 17.33** The origin of tetrapods.

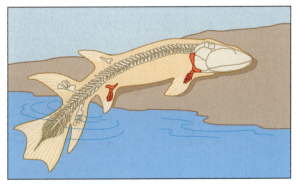

Lobe-finned fish. Fossils of some lobe-finned fishes have skeletal supports extending into their fins.

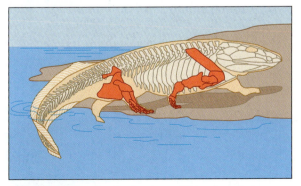

Early amphibian. Fossils of early amphibians have limb skeletons that probably functioned in helping them move on land.

357

Reptiles

Reptiles (including birds) and mammals are **amniotes**. The evolution of amniotes from an amphibian ancestor included many adaptations for living on land. The adaptation that gives the group its name is the **amniotic egg**, a fluid-filled egg with a waterproof shell that encloses the developing embryo **(Figure 17.34)**. The amniotic egg functions as a self-contained "pond" that enables amniotes to complete their life cycle on land.

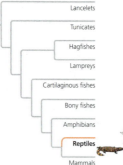

Lancelets
Tunicates
Hagfishes
Lampreys
Cartilaginous fishes
Bony fishes
Amphibians
Reptiles
Mammals

The **reptiles** include snakes, lizards, turtles, crocodiles, alligators, and birds, along with a number of extinct groups, including most of the dinosaurs. The European grass snake in Figure 17.34 displays two reptilian adaptations to living on land: scaled waterproof skin, preventing dehydration in dry air, and amniotic eggs with shells, providing a watery, nutritious internal environment where the embryo can develop. These adaptations allowed reptiles to break their ancestral ties to aquatic habitats. Reptiles cannot breathe through their dry skin and so obtain most of their oxygen through their lungs.

Nonbird Reptiles

Nonbird reptiles are sometimes referred to as "cold-blooded" animals because they do not use their metabolism extensively to control body temperature. Reptiles do regulate body temperature, but largely through behavioral adaptations. For example, many lizards regulate their internal temperature by basking in the sun when the air is cool and seeking shade when the air is too warm. Because lizards and other nonbird reptiles absorb external heat rather than generating much of their own, they are said to be **ectotherms**, a term more accurate than "cold-blooded." By heating directly with solar energy rather than through the metabolic breakdown of food, a nonbird reptile can survive on less than 10% of the calories required by a mammal of equivalent size.

As successful as reptiles are today, they were far more widespread, numerous, and diverse during the Mesozoic era, which is sometimes known as the "age of reptiles." Reptiles diversified extensively during that era, producing a dynasty that lasted until about 66 million years ago. Dinosaurs, the most diverse reptile group, included the largest animals ever to inhabit land. Some were gentle giants that lumbered about while browsing vegetation. Others were voracious carnivores that chased their larger prey on two legs.

The age of reptiles began to fade about 70 million years ago. Around that time, the global climate became cooler and more variable. This was a period of mass extinctions that claimed all the dinosaurs by about 66 million years ago, except for one lineage (see Table 14.1). That lone surviving lineage is represented today by the reptilian group we know as birds. ☑

☑ CHECKPOINT

What is an amniotic egg?

■ *Answer: a shelled egg surrounding fluid that contains an embryo*

▶ **Figure 17.34** Reptile diversity.

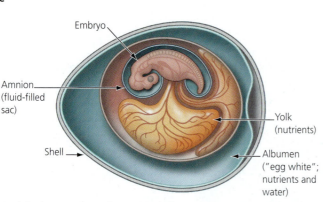

Embryo

Amnion (fluid-filled sac)

Shell

Yolk (nutrients)

Albumen ("egg white"; nutrients and water)

Aminiotic egg. The embryo and its life-support system are enclosed in a waterproof shell.

Snake. The shells of nonbird reptile eggs are leathery and flexible. This European grass snake is nonvenomous. When threatened, it may hiss and strike. If the bluff fails, it goes limp, pretending to be dead.

Lizard. The Gila monster, a desert-dweller of the Southwest, is the only venomous lizard native to the United States. Although large (up to 2 feet long), it moves too slowly to pose any danger to people.

Birds

You may have noticed that the reptilian egg resembles the more familiar chicken egg. Birds have scaly skin on their legs and feet; even feathers—their signature feature—are modified scales. Genetic and fossil evidence shows that **birds** are indeed reptiles, having evolved from a lineage of small, two-legged dinosaurs called theropods. Today, birds look quite different from reptiles because of their distinctive flight equipment—almost all of the 10,000 living bird species are airborne. The few flightless species, including the ostrich and the penguin, evolved from flying ancestors.

Almost every element of bird anatomy is adapted in some way that enhances flight. The bones have a honeycombed structure that makes them strong but light. (The wings of airplanes have the same basic construction.) For example, a huge seagoing species called the frigate bird has a wingspan of more than 2 m (6.6 ft), but its whole skeleton weighs only about 113 g (a mere 4 ounces, less than an iPhone). Another adaptation that reduces the weight of birds is the absence of some internal organs found in other vertebrates. Female birds, for instance, have only one ovary instead of a pair. Also, today's birds are toothless, an adaptation that trims the weight of the head, preventing uncontrolled nosedives. Birds do not chew food in the mouth but grind it in the gizzard, a muscular chamber of the digestive tract near the stomach.

Flying requires a great expenditure of energy and an active metabolism. Unlike other reptiles, birds are **endotherms**, meaning they use their own metabolic heat to maintain a warm, constant body temperature.

A bird's most obvious flight equipment is its wings. Bird wings are airfoils that illustrate the same principles of aerodynamics as the wings of an airplane **(Figure 17.35)**. A bird's flight motors are its powerful breast muscles, which are anchored to a keel-like breastbone. It is mainly these flight muscles that we call "white meat" on chicken and turkey breasts. Some birds, such as eagles and hawks, have wings adapted for soaring on air currents and flap their wings only occasionally. Other birds, including hummingbirds, excel at maneuvering but must flap continuously to stay aloft. Feathers are made of the same protein that forms the scales of reptiles. Feathers may have functioned first as insulation, helping birds retain body heat, or for courtship displays. Only later were they adapted as flight gear. ☑

☑ CHECKPOINT

Birds differ from other reptiles in their main source of body heat, with birds being _____ and other reptiles being _____.

■ Answer: endotherms; ectotherms

▶ **Figure 17.35 The aerodynamics of a bald eagle in flight.** Both birds and airplanes owe their "lift" to changes in air pressure caused by the shape of their wings.

Lower air pressure

Higher air pressure

Airfoil

Crocodile. Found throughout central and southern Africa, the Nile crocodile can grow up to 20 feet long and weigh nearly 2,000 pounds.

Birds. These red-crowned cranes, native to China, are performing an elaborate courtship dance.

Dinosaur. This *Herrerasaurus* skeleton is from a carnivorous theropod discovered in Argentina.

opposable thumb; that is, they can touch the tips of all four fingers with their thumb.

Our closest anthropoid relatives are the nonhuman apes: gibbons, orangutans, gorillas, and chimpanzees. They live only in tropical regions of the Old World. Except for some gibbons, apes are larger than monkeys, with relatively long arms and short legs, and most have

no tail. Although all apes are capable of living in trees, only gibbons and orangutans are primarily arboreal. Gorillas and chimpanzees are highly social. Apes have larger brains proportionate to body size than monkeys, and their behavior is more adaptable. And, of course, the apes include humans. **Figure 17.38** shows examples of primates.

▼ **Figure 17.38** Primate diversity.

Red ruffed lemur

Tarsier

Black spider monkey
(New World monkey)

Orangutan (ape)

Patas monkey (Old World monkey)

Gibbon (ape)

Gorilla (ape)

Chimpanzee (ape)

Human

The Emergence of Humankind

Humanity is one very young twig on the tree of life. In the continuum of life spanning 3.5 billion years, the fossil record and molecular systematics indicate that humans and chimpanzees have shared a common African ancestry for all but the last 6–7 million years (see Figure 17.37). Put another way, if we compressed the history of life to a year, the human branch has existed for only 18 hours.

Some Common Misconceptions

Certain misconceptions about human evolution persist in the minds of many, long after these myths have been debunked by the fossil evidence. One of these myths is expressed in the question "If chimpanzees were our ancestors, then why do they still exist?" In fact, scientists do not think that humans evolved from chimpanzees. Rather, as illustrated in Figure 17.37, the lineages that led to present-day humans and chimpanzees diverged from a common ancestor several million years ago. Each branch then evolved separately. As an analogy, consider a large family reunion attended by the descendants of a man who was born in 1830. Although the attendees share this common ancestor who lived several generations ago, they may be only distantly related to each other, as sixth or seventh cousins. Similarly, chimpanzees are not our parent species, but more like our very distant phylogenetic cousins, related through a common ancestor that lived hundreds of thousands of generations in the past.

Another myth envisions human evolution as a ladder with a series of steps leading directly from an ancestral anthropoid to *Homo sapiens*. This is often illustrated as a parade of fossil **hominins** (members of the human family) becoming progressively more modern as they march across the page. In fact, as **Figure 17.39** shows, there were times in hominin history when several human species coexisted. Scientists have identified about 20 species of fossil hominins, demonstrating that human phylogeny is more like a multibranched bush than a ladder, with our species being the tip of the only twig that still lives.

Although evidence from hundreds of thousands of hominin fossils debunks these and other myths about human evolution, many fascinating questions about our ancestry remain. The discovery of each new hominin fossil brings scientists a little closer to solving the puzzle of how we became human. In the following pages, you'll learn about some of the important clues that have been discovered so far.

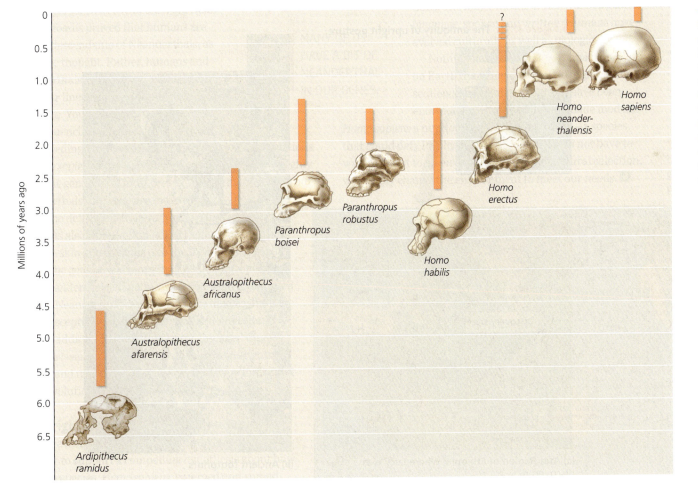

◄ **Figure 17.39** A **timeline of human evolution.** The orange bars indicate the time span during which each species lived. Notice that there have been times when two or more hominin species coexisted. The skulls are all drawn to the same scale so you can compare brain sizes.

Millions of years ago

Ardipithecus ramidus

Australopithecus afarensis

Australopithecus africanus

Paranthropus boisei

Paranthropus robustus

Homo habilis

Homo erectus

Homo neander- thalensis

Homo sapiens

What Can Lice Tell Us About Ancient Humans?

BACKGROUND

Imagine standing naked in a snowstorm. You probably wouldn't last long. When did humans first start wearing clothes? Archeological evidence provides a very rough time frame. Ancient members of genus *Homo* in Europe had tools for scraping animal hides about 780,000 years ago. Animal hides could have been used for clothing but would have had other uses as well. Needles, which indicate the construction of more complex clothing, were in use by 50,000 years ago. To zero in on a narrower date range for the origin of clothing, researchers turned to some of our most unwelcome guests: lice.

Lice are tiny blood-sucking parasites **(Figure 17.42a)** that attach their eggs to fur or similar fibers. Our distant ancestors, which were fur-covered like present-day primates, were parasitized by one species of louse. (Pubic lice, or "crabs," have a different species evolutionary history and are members of a different genus.) How did lice persist in modern humans when, with the exception of a few patches of hair, we are furless?

Around 3 million years ago, environmental changes required hominins to exert more energy in search of food and water. Muscle activity generates a great deal of heat; evaporation of sweat from the skin cools the body. However, fur prevents sweat from evaporating efficiently. By about 1.2 million years ago, hominins had lost most of their body hair, stranding lice on our scalps—the "head lice" that plague present-day humans. We can also be parasitized by clothing lice, which belong to the same species as head lice but have adaptations for grasping cloth fibers rather than hair or fur. Researchers hypothesized that analyzing the evolutionary divergence of lice into two distinct types would allow them to estimate when our ancestors first wore clothes.

METHOD

As you may recall, populations that don't interbreed because they occupy different habitats may accumulate genetic differences (see Figure 14.7). The longer two populations have been separated, the greater the divergence of their genes. Thus, the number of genetic differences between populations can be used to estimate how long the populations have been separated, a method known as a molecular clock **(Figure 17.42b)**. Imagine that a population of people all sang one song. As groups formed new isolated communities, differences would creep into their lyrics. Knowing the average number of word changes per year would let you estimate how long the populations had been singing separately. To construct a molecular clock for the divergence of head lice and clothing lice, researchers compared four DNA sequences for which the mutation rate—the average number of changes—was known.

RESULTS

The molecular clock shows that clothing offered a new habitat for the lice between 83,000 and 170,000 years ago, when populations of head lice and clothing lice began to diverge **(Figure 17.42c)**. Thus, modern humans originated the use of clothing, probably in Africa. Humans spreading north (see Figure 17.41) would have encountered the exceptionally cold climate that resulted from a series of ice ages, making clothing particularly useful.

▼ **Figure 17.42** Using lice to date the origin of clothing.

Head lice: *Pediculus humanus capitis*

Body lice: *Pediculus humanus humanus*

(a) Human head lice (top) and clothing lice.

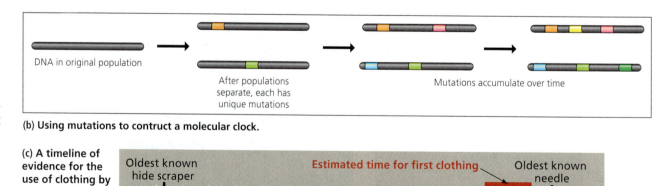

DNA in original population

After populations separate, each has unique mutations

Mutations accumulate over time

(b) Using mutations to construct a molecular clock.

(c) A timeline of evidence for the use of clothing by hominins.

Oldest known hide scraper

Estimated time for first clothing

Oldest known needle

| 800 | 700 | 600 | 500 | 400 | 300 | 200 | 100 | Present day |

Thousands of years ago

Thinking Like a Scientist

Why is a molecular clock more likely to be accurate when it is based on more genes?

For answers, see Appendix D.

Are We Still Evolving?

Imagine that you could take a time machine 100,000 years into the past and bring back a *Homo sapiens* man. If you dressed him in jeans and a T-shirt and took him for a stroll around campus, chances are that no one would look twice. Did we stop evolving after becoming *Homo sapiens*? In some ways, yes. The human body has not changed much in the past 100,000 years. And by the time *Homo sapiens* began to travel out of Africa, all of the complex characteristics that define our humanity, including our big-brained intelligence and our capacity for language and symbolic thought, had already evolved.

But as humans wandered far from their site of origin and settled in diverse environments, populations encountered different selective forces. Some traits of people today reflect evolutionary responses of ancient ancestors to their physical and cultural environment. For example, the high frequency of sickle hemoglobin (see Figure 9.21) in certain populations is an adaptation that protects against the deadly disease malaria. In other malarial regions, a group of inherited blood disorders called thalassemia serves the same adaptive function.

Diet has had a significant effect on human evolution, too. For example, the ability to digest lactose as adults (see the Evolution Connection section of Chapter 3) evolved in populations that kept dairy herds. Reliance by early farmers on starchy crops such as rice or tubers also left a genetic trace: extra copies of the gene that encodes the starch-digesting enzyme amylase.

One of the most striking differences among people is skin color **(Figure 17.43)**. The loss of skin pigmentation in humans who migrated north from Africa is thought to be an adaptation to low levels of ultraviolet (UV) radiation in northern latitudes. Dark pigment blocks the UV radiation necessary for synthesizing vitamin D—essential for proper bone development—in the skin. Recent research has turned up numerous other examples of adaptations that enabled us to colonize Earth's varied environments. For instance, indigenous people in the Andes Mountains in South America live at altitudes up to 12,000 feet (2.3 miles), where the air has 40% less oxygen than at sea level **(Figure 17.44)**. Researchers have identified genes that have undergone evolutionary changes in response to this challenging environment. High-altitude adaptations have also been discovered in people who are native to the Tibetan Plateau in Asia. Despite evolutionary tweaks such as these, however, we remain a single species.

▼ **Figure 17.44** Quechuans, indigenous people adapted to living at high altitude.

▼ **Figure 17.43** People with different adaptations to UV radiation.

Chapter Review

SUMMARY OF KEY CONCEPTS

The Origins of Animal Diversity

What Is an Animal?

Animals are eukaryotic, multicellular, heterotrophic organisms that obtain nutrients by ingestion. Most animals reproduce sexually and develop from a zygote to a blastula and then to a gastrula. After the gastrula stage, some develop directly into adults, whereas others pass through a larval stage.

Early Animals and the Cambrian Explosion

Animals probably evolved from a colonial flagellated protist. Precambrian animals were soft-bodied. Animals with hard parts appeared during the Cambrian period. Between 535 and 525 million years ago, animal diversity increased rapidly.

Animal Phylogeny

Major branches of animal evolution are defined by two key evolutionary differences: the presence or absence of tissues and radial versus bilateral body symmetry. A tissue-lined body cavity evolved in a number of later branches.

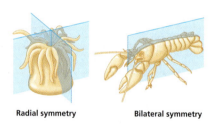

Radial symmetry   **Bilateral symmetry**

Major Invertebrate Phyla

This tree shows the eight major invertebrate phyla, as well as chordates, which include a few invertebrates.

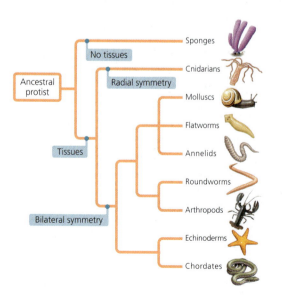

Sponges

Sponges (phylum Porifera) are stationary animals with porous bodies but no tissues. Specialized cells draw water through pores in the sides of the body and trap food particles.

Cnidarians

Cnidarians (phylum Cnidaria) have radial symmetry, a gastrovascular cavity with a single opening, and tentacles with stinging cnidocytes. The body is either a stationary polyp or a floating medusa.

Molluscs

Molluscs (phylum Mollusca) are soft-bodied animals often protected by a hard shell. The body has three main parts: a muscular foot, a visceral mass, and a fold of tissue called the mantle.

MOLLUSCS		
Gastropods	**Bivalves**	**Cephalopods**

Flatworms

Flatworms (phylum Platyhelminthes) are the simplest bilateral animals. They may be free-living (such as planarians) or parasitic (such as tapeworms).

Annelids

Annelids (phylum Annelida) are segmented worms with complete digestive tracts. They may be free-living or parasitic.

Roundworms

Roundworms, also called nematodes (phylum Nematoda), are unsegmented and cylindrical with tapered ends. They may be free-living or parasitic.

Arthropods

Arthropods (phylum Arthropoda) are segmented animals with an exoskeleton and specialized, jointed appendages.

ARTHROPODS			
Arachnids	**Crustaceans**	**Millipedes and Centipedes**	**Insects**

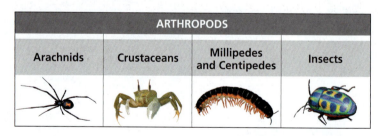

Echinoderms

Echinoderms (phylum Echinodermata) are stationary or slow-moving marine animals that lack body segments and possess a unique water vascular system. Bilaterally symmetric larvae usually change to radially symmetric adults. Echinoderms have a bumpy endoskeleton.

Vertebrate Evolution and Diversity

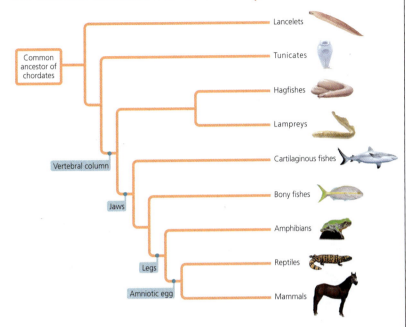

Characteristics of Chordates

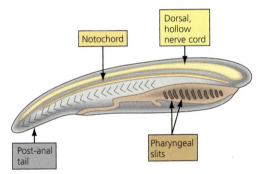

Tunicates and lancelets are invertebrate chordates. The vast majority of chordates are vertebrates, possessing a skull and backbone.

Fishes

Hagfishes and lampreys are jawless vertebrates. Cartilaginous fishes, such as sharks, are mostly predators with powerful jaws and a flexible skeleton made of cartilage. Bony fishes have a stiff skeleton reinforced by calcium. Bony fishes are further classified into ray-finned fishes and lobe-finned fishes (including lungfishes).

Amphibians

Amphibians are tetrapod vertebrates that usually deposit their eggs (lacking shells) in water. Aquatic larvae typically undergo a radical metamorphosis into the adult stage. Their moist skin requires that amphibians spend much of their adult life in humid environments.

Reptiles

Reptiles are amniotes, vertebrates that develop in a fluid-filled egg enclosed by a shell. Reptiles include terrestrial ectotherms with lungs and waterproof skin covered by scales. Scales and amniotic eggs enhanced reproduction on land. Birds are endothermic reptiles with wings, feathers, and other adaptations for flight.

Mammals

Mammals are endothermic vertebrates with mammary glands and hair. There are three major groups of mammals: Monotremes lay eggs; marsupials use a placenta but give birth to tiny embryonic offspring that usually complete development while attached to nipples inside the mother's pouch; and eutherians, or placental mammals, use their placenta in a longer-lasting association between the mother and her developing young.

MAMMALS		
Monotremes	**Marsupials**	**Eutherians**

The Human Ancestry

The Evolution of Primates

The first primates were small, arboreal mammals that evolved from insect-eating mammals about 65 million years ago. Anthropoids consist of New World monkeys (with prehensile tails), Old World monkeys (without prehensile tails), apes, and humans.

The Emergence of Humankind

Chimpanzees and humans evolved from a common ancestor about 6–7 million years ago. Species of the genus *Australopithecus*, which lived at least 4 million years ago, walked upright and had a small brain. Enlargement of the human brain in *Homo habilis* came later, about 2.4 million years ago. *Homo erectus* was the first species to extend humanity's range from its birthplace in Africa to other continents. *Homo erectus* gave rise to regionally diverse descendants, including the Neanderthals (*Homo neanderthalensis*). Current data indicate a relatively recent dispersal of modern Africans that gave rise to today's human diversity.

Mastering Biology

For practice quizzes, BioFlix animations, MP3 tutorials, video tutors, and more study tools designed for this textbook, go to Mastering Biology™

SELF-QUIZ

1. Bilateral symmetry in the animal kingdom is best correlated with
   a. an ability to sense equally in all directions.
   b. the presence of a skeleton.
   c. motility and active predation and escape.
   d. development of a body cavity.

2. Identify which of the following categories includes all others in the list: arthropod, arachnid, insect, butterfly, crustacean, millipede.

3. The oldest group of tetrapods is the _____.

4. Reptiles are much more extensively adapted to life on land than amphibians because reptiles
   a. have a complete digestive tract.
   b. lay eggs that are enclosed in shells.
   c. are endothermic.
   d. go through a larval stage.

5. What is the name of the phylum to which humans belong? For what anatomical structure is the phylum named? Where in your body is a derivative of this anatomical structure found?

6. Which of the following types of animals is not included in the human ancestry? (*Hint*: See Figure 17.30.)
   a. a bird
   b. a bony fish
   c. an amphibian
   d. a primate

7. Put the following list of species in order, from the oldest to the most recent: *Homo erectus*, *Australopithecus* species, *Homo habilis*, *Homo sapiens*.

8. Match each of the following animals to its phylum:
   a. human   1. Echinodermata
   b. leech   2. Arthropoda
   c. sea star   3. Cnidaria
   d. lobster   4. Chordata
   e. sea anemone   5. Annelida

For answers to the Self-Quiz, see Appendix D.

IDENTIFYING MAJOR THEMES

For each statement below, identify which major theme is evident (the relationship of structure to function, information flow, pathways that transform energy and matter, interactions within biological systems, or evolution) and explain how the statement relates to the theme. If necessary, review the themes (Chapter 1) and review the examples highlighted in blue in this chapter.

9. It is likely that the set of "master control" genes that allow building complex bodies was in place before rapid diversification.

10. The head of a tapeworm is equipped with suckers and hooks that lock the worm to the intestinal lining of the host.

11. Bees and flies pollinate our crops and orchards. Other insects are carriers of the microbes that cause many human diseases. Insects also compete with people for food by eating our crops.

For answers to Identifying Major Themes, see Appendix D.

THE PROCESS OF SCIENCE

12. Imagine that you are a marine biologist. As part of your exploration, you dredge up an unknown animal from the seafloor. Describe some of the characteristics you should look at to determine the phylum to which the creature should be assigned.

13. Vegan and vegetarian diets are increasingly popular. While vegans consume no animals or animal products, many vegetarians are less strict. Talk to acquaintances who describe themselves as vegetarians, or who follow a meat-free diet, and determine which taxonomic groups they avoid eating (see Figures 17.6 and 17.30). Try to generalize about their diet. What do they consider "meat"? For example, do they avoid eating vertebrates but eat some invertebrates? Do they avoid only birds and mammals? What about fish? Do they eat dairy products or eggs?

14. What adaptations inherited from our primate ancestors enable humans to make and use tools?

15. **Interpreting Data** Average brain size, relative to body mass, gives a rough indication of the intelligence of a species. Graph the data below for anthropoids. *Homo naledi*, which was discovered in 2015, had a mean brain of 510 cm³; its mean body mass was 37 kg. What do these data imply about the intelligence of *H. naledi* compared to other hominids?

Anthropoid Species	Mean Brain Volume (cm³)	Mean Body Mass (kg)
Australopithecus afarensis	440	37
Chimpanzee	405	46
Gorilla	500	105
Homo erectus	940	58
Homo habilis	610	34
Homo neanderthalensis	1480	65
Homo sapiens	1330	64
Paranthropus boisei	490	41

BIOLOGY AND SOCIETY

16. Coral reefs harbor a greater diversity of animals than any other environment in the sea. Australia's Great Barrier Reef has been protected as a marine reserve and is a mecca for scientists and nature enthusiasts. Elsewhere, such as in Indonesia and the Philippines, coral reefs are in danger. Many reefs have been depleted of fish, and runoff from the shore has covered coral with sediment. Nearly all the changes in the reefs can be traced back to human activities. What kinds of activities do you think might be contributing to the decline of the reefs? What are some reasons to be concerned about this decline? Do you think the situation is likely to improve or worsen in the future? Why? What might the local people do to halt the decline? Should the more industrialized countries help? Why or why not?

17. The size of the human brain has not changed in the last 100,000 years, but human culture has changed a great deal. As a result of our culture, we change the environment at a rate far greater than the rate at which many species, including our own, can evolve. What evidence of rapid environmental change do you see around you? What aspects of human culture are responsible for these changes? Do you see any evidence of a decrease in the rate of human-caused environmental changes?

Ecology

18 An Introduction to Ecology and the Biosphere

CHAPTER THREAD
Climate Change

19 Population Ecology

CHAPTER THREAD
Biological Invasions

20 Communities and Ecosystems

CHAPTER THREAD
Importance of Biodiversity

18 An Introduction to Ecology and the Biosphere

CHAPTER CONTENTS

An Overview of Ecology 374
Living in Earth's Diverse Environments 376
Biomes 380
Climate Change 394

Why Ecology Matters

From the icy polar regions to the lush tropics, from the vibrant coral reefs to the deepest ocean, Earth's biosphere is a treasure trove of life that is still being explored. The study of ecology helps us understand the interactions of organisms with these diverse environments and offers insight into our own impact on the natural world.

WHAT HAPPENS TO ALL THE TRASH WE DISCARD? LANDFILLS AREN'T THE ONLY PLACE WHERE GARBAGE ACCUMULATES.

VAMPIRE ALERT! YOUR HOME MAY BE INFESTED WITH "VAMPIRE" DEVICES THAT CONSUME ELECTRICITY EVEN WHILE YOU SLEEP.

POLLUTION DOESN'T JUST VANISH INTO THIN AIR. AIRBORNE POLLUTANTS FROM CAR EXHAUST CAN COMBINE WITH WATER AND RETURN TO EARTH AS ACID PRECIPITATION FAR AWAY.

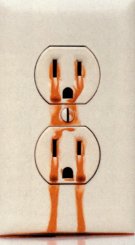

BIOLOGY AND SOCIETY   Penguins, Polar Bears, and People in Peril 373

THE PROCESS OF SCIENCE   How Does Climate Change Affect Species Distribution? 397

EVOLUTION CONNECTION   Climate Change as an Agent of Natural Selection 399

BIOLOGY AND SOCIETY   Climate Change

A polar bear with her two cubs on Arctic ice. Climate change is shrinking the permanent sea ice that polar bears use as hunting platforms.

Penguins, Polar Bears, and People in Peril

Ninety-seven percent of climate scientists agree: The global climate is changing due to human activities. The change is driven by a rapid rise in temperatures—the ten hottest years on record have occurred since 1998. The current rate of warming is *ten times* faster than the average rate during the warm-up that followed the last ice age. So what do we know about climate change now, and what can we expect for the future?

The northernmost regions of the Northern Hemisphere and the Antarctic Peninsula have heated up the most. In parts of Alaska, for example, winter temperatures have risen by 5–6°F. The permanent Arctic sea ice is shrinking; each summer brings thinner ice and more open water. Polar bears, which stalk their prey on ice and need to store up body fat for the warmer months when there is no ice, are showing signs of starvation as their winter hunting grounds melt away. At the other end of the planet, diminishing sea ice near the Antarctic Peninsula limits the access of Adélie penguins to their food supply, and spring blizzards of unprecedented frequency and severity are taking a heavy toll on their eggs and chicks. But these charismatic animals are just a distress signal that should alert us to our own danger. We are already feeling the effects of climate change in more frequent and larger wildfires, increased coastal flooding, deadly heat waves, and altered precipitation patterns that bring drought to some regions and torrential downpours to others.

What does climate change mean for the future of life on Earth? Any predictions that scientists make now about future impacts of global climate change are based on incomplete information. Much remains to be discovered about species diversity and about the complex interactions of organisms (living things) with each other and with their environments. There is overwhelming evidence that human enterprises are responsible for the changes that are occurring. How we respond to this crisis will determine whether circumstances improve or worsen. And the process begins with understanding the basic concepts of ecology, which we start to explore in this chapter.

An Overview of Ecology

In your study of biology so far, you have learned about the diversity of life on Earth and about the molecular and cellular structures and processes that make life tick. **Ecology**, the scientific study of the interactions between organisms and their environments, offers a different perspective on life—biology from the skin out, so to speak.

Humans have always had an interest in other organisms and their environments. As hunters and gatherers, prehistoric people had to learn where and when game and edible plants could be found in greatest abundance. Naturalists, from Aristotle to Darwin and beyond, made the process of observing and describing organisms in their natural habitats an end in itself rather than simply a means of survival. We can still gain valuable insight from watching nature and recording its structure and processes **(Figure 18.1)**. As you might expect, hypothesis-driven science performed in natural environments is fundamental to ecology. But ecologists also test hypotheses using laboratory experiments, where conditions can be simplified and controlled. And some ecologists take a theoretical approach, devising mathematical and computer models, which enable them to simulate large-scale experiments that are impossible to conduct in the field.

▼ **Figure 18.1** **Discovery science in a rain forest canopy.** A biologist collects insects in a rain forest on the eastern slope of the Andes mountain range in Argentina.

Ecology and Environmentalism

Technological innovations have enabled people to colonize just about every environment on Earth. Even so, our survival depends on Earth's resources, which have been profoundly altered by human activities **(Figure 18.2)**. Climate change is just one of the many environmental issues that have stirred public concern in recent decades. Some of our industrial and agricultural practices have contaminated the air, soil, and water. Our relentless quest for land and other resources has endangered a lengthy list of plant and animal species and has even driven some to extinction.

The science of ecology can provide the understanding needed to solve environmental problems. But these problems cannot be solved by ecologists alone, because they require making decisions based on values and ethics. On a personal level, each of us makes daily choices that affect our ecological impact. And legislators and corporations, motivated by environmentally aware voters and consumers, must address questions that have wider implications: How should the use of land and water be regulated? Should we try to save all species or just certain ones? What alternatives to environmentally destructive practices can be developed? How can we balance environmental impacts with economic needs?

▼ **Figure 18.2** **Human impact on the environment.** Cleanup crews burn oil collected from the surface of the Gulf of Mexico after the Deepwater Horizon disaster in 2010.

A Hierarchy of Interactions

Many different factors can potentially affect an organism's interactions with its environment. **Biotic factors**—all of the organisms in the area—make up the living component of the environment. Other organisms may compete with an individual for food and other resources, prey upon it, or change its physical and chemical environment. **Abiotic factors** make up the environment's nonliving component and include chemical and physical factors, such as temperature, light, water, minerals, and air. An organism's **habitat**, the specific environment it lives in, includes the biotic and abiotic factors of its surroundings.

When we study the interactions between organisms and their environments, it is convenient to divide ecology into four increasingly comprehensive levels: organismal ecology, population ecology, community ecology, and ecosystem ecology.

An **organism** is an individual living thing. **Organismal ecology** is concerned with the evolutionary adaptations that enable organisms to meet the challenges posed by their abiotic environments. The distribution of organisms is limited by the abiotic conditions they can tolerate. For example, amphibians such as the salamander in **Figure 18.3a** cannot live in cold climates because they gain most of their body warmth by absorbing heat from their surroundings. Temperature and precipitation shifts due to global climate change have already affected the distributions of some salamander species, and many more will feel the impact in the coming decades.

The next level of organization in ecology is the **population**, a group of individuals of the same species living in a particular geographic area. **Population ecology** concentrates mainly on factors that affect population density and growth **(Figure 18.3b)**. Biologists who study endangered species are especially interested in this level of ecology.

A **community** consists of all the organisms that inhabit a particular area; it is an assemblage of populations of different species. Questions in **community ecology** focus on how interactions between species, such as predation and competition, affect community structure and organization **(Figure 18.3c)**.

An **ecosystem** includes all the abiotic factors in addition to the community of species in a certain area. For example, a savanna ecosystem includes not only the organisms, such as diverse plants and animals, but also the soil, water sources, sunlight, and other abiotic factors of the environment. Thus, an ecosystem encompasses a wide variety of interactions within and between biological systems. In **ecosystem ecology**, questions concern energy flow and the cycling of chemicals among the various biotic and abiotic factors **(Figure 18.3d)**.

▼ **Figure 18.3** Examples of questions at different levels of ecology.

(a) Organismal ecology. What range of temperatures can a red salamander tolerate?

(b) Population ecology. What factors affect the survival of emperor penguin chicks?

(c) Community ecology. How do predators such as this beech marten affect the diversity of rodents in a community?

(d) Ecosystem ecology. What processes recycle vital chemical elements such as nitrogen within a savanna ecosystem in Africa?

The **biosphere** is the global ecosystem—the sum of all the planet's ecosystems, or all of life and where it lives. The most complex level in ecology, the biosphere includes the atmosphere to an altitude of several kilometers, the land down to water-bearing rocks about 1,500 m (almost a mile) deep, lakes and streams, caves, and the oceans to a depth of several kilometers. But despite its grand scale, organisms within the biosphere are linked; events in one part may have far-reaching effects. ✅

✓ CHECKPOINT

1. What does the ecosystem level of classification have in common with the community level of classification?

2. What does the ecosystem level include that the community level does not?

■ *Answers:* 1. *all the biotic factors of the area.* 2. *the abiotic factors of the area.*

375

The Evolutionary Adaptations of Organisms

The ability of organisms to live in Earth's diverse environments demonstrates the close relationship between the fields of ecology and evolutionary biology. Charles Darwin was an ecologist, although he predated the word *ecology*. It was the geographic distribution of organisms and their exquisite adaptations to specific environments that provided Darwin with evidence for evolution. Evolutionary adaptation via natural selection results from the interactions of organisms with their environments, which brings us back to our definition of ecology. Thus, events that occur in the short term, during the course of an individual's lifetime, may translate into effects over the longer scale of evolutionary time. For instance, because the availability of water affects a plant's growth and ultimately its reproductive success, precipitation has an impact on the gene pool of a plant population. **After a period of lower-than-average rainfall, drought-resistant individuals may be more prevalent in a plant population—an example of evolution.** Organisms also evolve in response to biotic interactions, such as predation and competition. ✅

✅ **CHECKPOINT**

How are the fields of ecology and evolution linked?

■ *Answer: The process of evolutionary adaptation via natural selection results from the interactions of organisms with their environments (ecology).*

▼ **Figure 18.9** A mourning dove demonstrating its physiological response to cold weather.

Adjusting to Environmental Variability

The abiotic factors in a habitat may vary from year to year, seasonally, or over the course of a day. An individual's abilities to adjust to environmental changes that occur during its lifetime are themselves adaptations refined by natural selection. For instance, if you see a bird on a cold day, it may look unusually fluffy **(Figure 18.9)**. Small muscles in the skin raise the bird's feathers, a physiological response that traps insulating pockets of air. Some species of birds adjust to seasonal cold by growing heavier feathers. And some bird species respond to the onset of cold weather by migrating to warmer regions—a behavioral response. Note that these responses occur during the lifetime of an individual, so they are not examples of evolution, which is change in a population over time.

Physiological Responses

Like birds, mammals can adjust to a cold day by contracting skin muscles—in this case attached to hairs—to create a temporary layer of insulation. (Our own muscles do this, too, but we just get "goose bumps" instead of a furry insulation.) The blood vessels in the skin also constrict, which slows the loss of body heat. In both cases, the adjustment occurs in just seconds.

A gradual, though still reversible, physiological adjustment that occurs in response to an environmental change is called **acclimation**. For example, suppose you moved from Boston, which is essentially at sea level, to the mile-high city of Denver, where there is less oxygen. One physiological response to your new environment would be a gradual increase in the number of your red blood cells, which transport O_2 from your lungs to other parts of your body. Acclimation can take days or weeks. This is why high-altitude climbers, such as those attempting to scale Mount Everest, need extended stays at a high-elevation base camp before proceeding to the summit.

The ability to acclimate is generally related to the range of environmental conditions a species naturally experiences. Species that live in very warm climates, for example, usually cannot acclimate to extreme cold. Among vertebrates, birds and mammals can generally tolerate the greatest temperature extremes because, as endotherms, they use their metabolism to regulate internal temperature. In contrast, ectothermic reptiles can tolerate only a more limited range of temperatures **(Figure 18.10)**.

▼ **Figure 18.10** The number of lizard species in different regions of the contiguous United States. Notice that there are fewer and fewer lizard species in more northern regions. This reflects lizards' ectothermic physiology, which depends on environmental heat for keeping the body warm enough for the animal to be active.

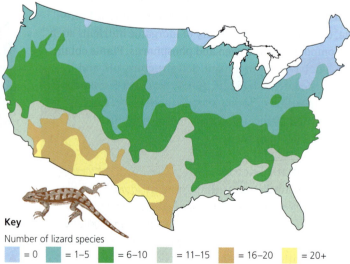

Key

Number of lizard species

| = 0 | = 1–5 | = 6–10 | = 11–15 | = 16–20 | = 20+ |

Anatomical Responses

Many organisms respond to environmental challenge with some type of change in body shape or structure. When the change is reversible, the response is an example of acclimation. Many mammals, for example, grow a heavier coat of fur before the winter cold sets in and shed it when summer comes. In some animals, fur or feather color changes seasonally as well, camouflaging the animal against winter snow and summer vegetation (Figure 18.11).

Other anatomical changes are irreversible over the lifetime of an individual. Environmental variation can affect growth and development so much that there may be remarkable differences in body shape within a population. You can see an example in Figure 18.12, which shows the "flagging" that wind causes in certain trees. In general, plants are more anatomically changeable than animals. Rooted and unable to move to a better location, plants rely entirely on their anatomical and physiological responses to survive environmental fluctuations.

▲ **Figure 18.11** The arctic fox in winter and summer coats.

Behavioral Responses

In contrast to plants, most animals can respond to an unfavorable change in the environment by moving to a new location. Such movement may be fairly localized. For example, many desert ectotherms, including reptiles, maintain a reasonably constant body temperature by shuttling between sun and shade. Some animals are capable of migrating great distances in response to such environmental cues as the changing seasons. Many migratory birds overwinter in Central and South America, returning to northern latitudes to breed during summer. And we humans, with our large brains and available technology, have an especially rich range of behavioral responses available to us (Figure 18.13). ✅

✅ CHECKPOINT

What is acclimation?

■ *Answer: a gradual, reversible change in anatomy or physiology in response to an environmental change*

▼ **Figure 18.12** **Wind as an abiotic factor that shapes trees.** The mechanical disturbance of the prevailing wind hinders limb growth on the windward side of this fir tree near the timberline in the Rocky Mountains, while limbs on the other side grow normally. This anatomical response is an evolutionary adaptation that reduces the number of limbs that are broken during strong winds.

▼ **Figure 18.13** **Behavioral responses have expanded the geographic range of humans.** Dressing for the weather is a thermoregulatory behavior unique to people.

Biomes

The abiotic factors you learned about in the previous section are largely responsible for the distribution of life on Earth. (You'll learn about the role of biotic factors in species distribution in Chapter 20.) Using various combinations of these factors, ecologists have categorized Earth's environments into biomes. A **biome** is a major terrestrial or aquatic life zone, characterized by vegetation type in terrestrial biomes and the physical environment in aquatic biomes. In this section, we'll briefly survey the aquatic biomes, followed by the terrestrial biomes.

Aquatic biomes, which occupy roughly 75% of Earth's surface, are determined by their salinity and other physical factors. Freshwater biomes (lakes, streams and rivers, and wetlands) typically have salt concentrations of less than 1%. The salt concentrations of marine biomes (oceans, intertidal zones, and coral reefs) are generally around 3%.

☑ CHECKPOINT

Why does sewage cause heavy algal growth in lakes?

■ *Answer: Sewage adds mineral nutrients that stimulate growth of the algae.*

Freshwater Biomes

Freshwater biomes cover less than 1% of Earth, and they contain a mere 0.01% of its water. But they harbor a disproportionate share of biodiversity—an estimated 6% of all described species. Moreover, we depend on freshwater biomes for drinking water, crop irrigation, sanitation, and industry.

Freshwater biomes fall into two broad groups: standing water, which includes lakes and ponds, and flowing water, such as rivers and streams. The difference in water movement results in profound differences in ecosystem structure.

Lakes and Ponds

Standing bodies of water range from small ponds only a few square meters in area to large lakes, such as North America's Great Lakes, that are thousands of square kilometers **(Figure 18.14)**.

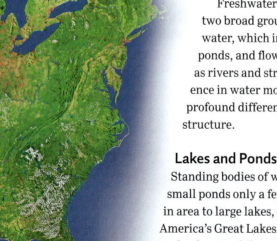

◄ **Figure 18.14** A satellite view of the Great Lakes.

In lakes and large ponds, communities of plants, algae, and animals are distributed according to the depth of the water and the distance from shore **(Figure 18.15)**. Shallow water near shore and the upper layer of water away from shore make up the **photic zone**, so named because light is available for photosynthesis. Microscopic algae and cyanobacteria grow in the photic zone, joined by rooted plants and floating plants such as water lilies in the photic area near shore. If a lake or pond is deep enough or murky enough, it has an **aphotic zone**, where light levels are too low to support photosynthesis.

The **benthic realm** is at the bottom of all aquatic biomes. Made up of sand and organic and inorganic sediments, the benthic realm is occupied by communities of organisms that may include algae, aquatic plants, worms, insect larvae, molluscs, and microorganisms. Dead material that "rains" down from the productive surface waters of the photic zone is a major source of food for animals of the benthic realm.

The mineral nutrients nitrogen and phosphorus typically regulate the growth of **phytoplankton**, the collective name for microscopic algae and cyanobacteria that drift near the surfaces of aquatic biomes. Many lakes and ponds are affected by large inputs of nitrogen and phosphorus from sewage and runoff from fertilized lawns and farms. These nutrients often produce heavy growth of algae, which reduces light penetration. When the algae die and decompose, a pond or lake can suffer serious oxygen depletion, killing fish that are adapted to high-oxygen conditions. ☑

▼ **Figure 18.15** Zones in a lake.

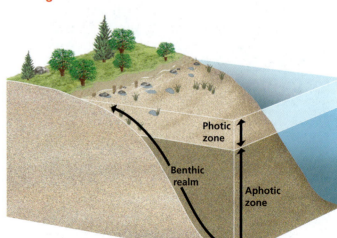

▲ **Figure 18.16** **A stream in the Appalachian Mountains.**

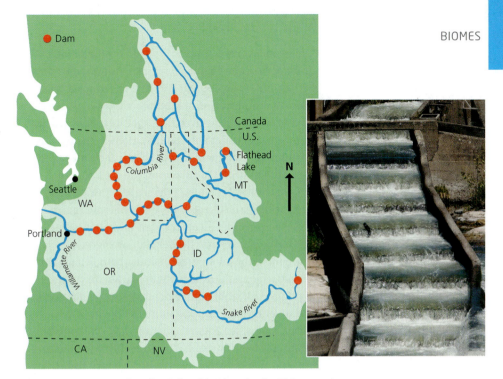

▲ **Figure 18.17** **Damming the Columbia River basin.** This map shows only the largest of the 250 dams that have altered freshwater ecosystems throughout the Pacific Northwest. These great concrete obstacles make it difficult for salmon to swim upriver to their breeding streams, though many dams now have "fish ladders" that provide detours (inset).

Rivers and Streams

Rivers and streams, which are bodies of flowing water, generally support communities of organisms quite different from those of lakes and ponds **(Figure 18.16)**. A river or stream changes greatly between its source (perhaps a spring or snowmelt in the mountains) and the point at which it empties into a lake or the ocean. Near a source, the water is usually cold, low in nutrients, and clear. The channel is often narrow, with a swift current that does not allow much silt to accumulate on the bottom. The current also inhibits the growth of phytoplankton; most of the organisms found here are supported by the photosynthesis of algae attached to rocks or by organic material (such as leaves) carried into the stream from the surrounding land. The most abundant benthic animals are usually insects that eat algae, leaves, or one another. Trout are often the predominant fishes, locating their food, including insects, mainly by sight in the clear water.

Downstream, a river or stream typically widens and slows. There the water is usually warmer and may be murkier because of sediments and phytoplankton suspended in it. Worms and insects that burrow into mud are often abundant, as are waterfowl, frogs, and catfish and other fishes that find food more by scent and taste than by sight.

People have altered rivers by constructing dams to control flooding, to provide reservoirs of drinking water, or to generate hydroelectric power. In many cases, dams have completely changed the downstream ecosystems, altering the rate and volume of water flow and affecting fish and invertebrate populations **(Figure 18.17)**. Many streams and rivers have also been affected by pollution from human activities.

Wetlands

A **wetland** is a transitional biome between an aquatic ecosystem and a terrestrial one. Freshwater wetlands include swamps, bogs, and marshes **(Figure 18.18)**. Covered with water either permanently or periodically, wetlands support the growth of aquatic plants and are rich in species diversity. Migrating waterfowl and many other birds depend on wetland "pit stops" for food and shelter during their journeys. In addition, wetlands provide water storage areas that reduce flooding. Wetlands also improve water quality by trapping pollutants such as metals and organic compounds in their sediments.

▼ **Figure 18.18** **A wetland near Kent, Ohio.**

Marine Biomes

Gazing out over a vast ocean, you might think that it is the most uniform environment on Earth. But marine habitats can be as different as night and day. The deepest ocean, where hydrothermal vents are located, is perpetually dark. In contrast, the vivid coral reefs nearer the surface are utterly dependent on sunlight. Habitats near shore are different from those in mid-ocean, and the seafloor hosts different communities than the open waters.

As in freshwater biomes, the seafloor is known as the benthic realm **(Figure 18.19)**. The **pelagic realm** of the oceans includes all open water. In shallow areas, such as the continental shelves (the submerged parts of continents), the photic zone includes both pelagic and benthic regions. In these sunlit areas, photosynthesis by phytoplankton and multicellular algae provides energy for a diverse community of animals. Sponges, burrowing worms, clams, sea anemones, crabs, and echinoderms inhabit the benthic realm. **Zooplankton** (free-floating animals, including many microscopic ones), fishes, marine mammals, and many other types of animals are abundant in the pelagic photic zone.

The **coral reef** biome occurs in the photic zone of warm tropical waters in scattered locations around the globe. A coral reef is built up slowly by successive generations of coral animals—a diverse group of cnidarians that secrete a hard external skeleton—and by multicellular algae encrusted with limestone **(Figure 18.20)**. Unicellular algae live within the coral's cells, providing the coral with food. The physical structure and productivity of coral reefs support a huge variety of invertebrates and fishes.

The photic zone extends down a maximum of 200 m (about 656 feet) in the ocean. Although there is not enough light for photosynthesis between 200 and 1,000 m (down a little more than half a mile), some light does reach these depths of the aphotic zone. This dimly lit world, sometimes called the twilight zone, is dominated by a fascinating variety of small fishes and crustaceans. Food sinking from the photic zone provides some sustenance for these animals. In addition, many of them migrate to the surface at night to feed. Some fishes in the twilight zone have enlarged eyes, enabling them to see in the very dim light, and light-emitting organs that attract mates and prey.

Below 1,000 m—a depth greater than the height of two Empire State Buildings—the ocean is completely and permanently dark. Adaptation to this environment has produced many bizarre-looking creatures. Most of the benthic organisms here are deposit feeders, animals that consume dead organic material in the sediments on the seafloor. Crustaceans, annelid worms, sea anemones, and echinoderms such as sea cucumbers, sea stars, and sea urchins are common. Food is scarce, however. The

▼ **Figure 18.19** **Ocean life.** (Zone depths and organisms not drawn to scale.)

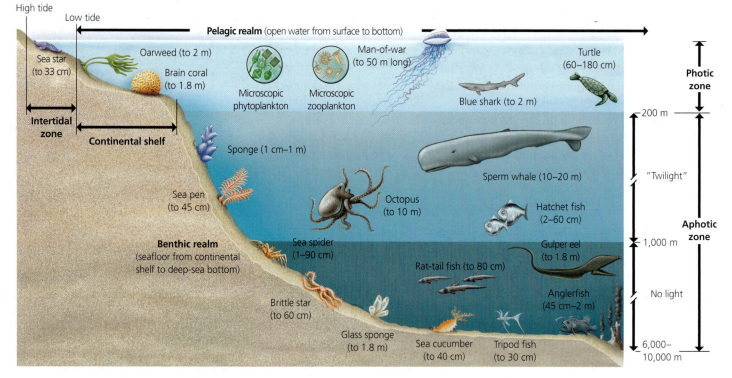

▲ Figure 18.20 A coral reef in the Red Sea off the coast of Egypt.

▼ Figure 18.21 Organisms clinging to the rocks of an intertidal zone on the Pacific coast of Washington.

density of animals is low except at hydrothermal vents, the prokaryote-powered ecosystems mentioned earlier (see Figure 18.6).

The marine environment also includes distinctive biomes, such as the intertidal zone and estuaries, where the ocean interfaces with land or with fresh water. In the **intertidal zone**, where the ocean meets land, the shore is pounded by waves during high tide and exposed to the sun and drying winds during low tide. The rocky intertidal zone is home to many sedentary organisms, such as sea stars, barnacles, and mussels, which attach to rocks and thus are prevented from being washed away (Figure 18.21). On sandy beaches, suspension-feeding worms, clams, and predatory crustaceans bury themselves in the ground.

LANDFILLS AREN'T THE ONLY PLACE WHERE GARBAGE ACCUMULATES.

Figure 18.22 shows an **estuary**, a transition area between a river and the ocean. The saltiness of estuaries ranges from nearly that of fresh water to that of the ocean. With their waters enriched by nutrients from rivers, estuaries, like freshwater wetlands, are among the most productive areas on Earth. Oysters, crabs, and many fishes live or reproduce in estuaries. Estuaries are also crucial nesting and feeding areas for waterfowl. Mudflats and salt marshes are extensive coastal wetlands that often border estuaries.

For centuries, people viewed the ocean as a limitless resource, harvesting its bounty with increasingly effective and indiscriminate technologies and using it as a dumping ground for wastes. The negative effects of these practices are now becoming clear. Populations of commercial fish species are declining. Trillions of small bits of plastic debris float beneath the surface of vast swaths of ocean, concentrated by converging currents in several regions dubbed "garbage patches." Many marine habitats are polluted by nutrients or toxic chemicals; it will be

years before the full extent of the damage from the massive Deepwater Horizon oil spill in the Gulf of Mexico in 2010 (see Figure 18.2) is known. Because of their nearness to land, estuaries are especially vulnerable. Many have been completely replaced by development on landfill; other threats include pollution and alteration of freshwater inflow. Coral reefs are imperiled by ocean acidification and rising sea surface temperatures due to global warming.

Meanwhile, our knowledge of marine biomes is woefully incomplete. A recent decade-long scientific census of marine life announced the discovery of more than 6,000 new species. ✅

☑ **CHECKPOINT**

What are phytoplankton? Why are they essential to other oceanic life?

■ *Answer: Phytoplankton are photosynthetic algae and bacteria. They are food for animals in the photic zone; those animals in turn may become food for animals in the aphotic zone.*

▼ Figure 18.22 Waterfowl in an estuary on the southeast coast of England.

How Climate Affects Terrestrial Biome Distribution

Terrestrial biomes are determined primarily by climate, especially temperature and rainfall. Before we survey these biomes, let's look at the broad patterns of global climate that help explain their locations.

Earth's global climate patterns are largely the result of the input of solar energy—which warms the atmosphere, land, and water—and of the planet's movement in space. Because of Earth's curvature, the intensity of sunlight varies according to latitude **(Figure 18.23)**. The equator receives the greatest intensity of solar radiation and thus has the highest temperatures, which in turn evaporate water from Earth's surface. As this warm, moist air rises, it cools, diminishing its ability to hold moisture. The water vapor condenses into clouds and eventually falls as rain **(Figure 18.24)**. This process largely explains why rain forests are concentrated in the **tropics**—the region from the Tropic of Cancer to the Tropic of Capricorn.

After losing moisture over equatorial zones, dry high-altitude air masses spread away from the equator until they cool and descend at latitudes of about 30° north and south. Many of the world's great deserts—the Sahara in North Africa and the Arabian on the Arabian Peninsula, for example—are centered at these latitudes because of the dry air they receive.

Latitudes between the tropics and the Arctic Circle and the Antarctic Circle are called **temperate zones**. Generally, these regions have milder climates than the tropics or the polar regions. Notice in Figure 18.24 that some of the descending dry air moves into the latitudes above 30°. At first these air masses pick up moisture, but they tend to drop it as they cool at higher latitudes. This is why the north and south temperate zones tend to be relatively wet. Coniferous forests dominate the landscape at the wet but cool latitudes around 60° north.

Proximity to large bodies of water and the presence of landforms such as mountain ranges also affect climate. Oceans and large lakes moderate climate by absorbing heat when the air is warm and releasing heat to cold air. Mountains affect climate in two major ways. First, air temperature drops as elevation increases. As a result, driving up a tall mountain offers a quick tour of several biomes. **Figure 18.25** shows the scenery you might encounter on a journey from the scorching lowlands of the Sonoran Desert to a cool coniferous forest at an elevation of 11,000 feet above sea level.

▼ **Figure 18.23** Uneven heating of Earth.

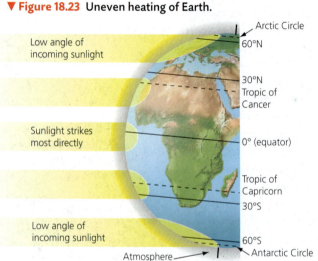

▼ **Figure 18.24** How uneven heating of Earth produces various climates.

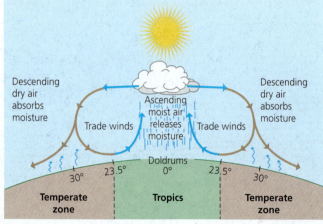

▶ **Figure 18.25** The effect of altitude on vegetation. The zones shown are typical of the Sonoran Desert region in southwestern North America.

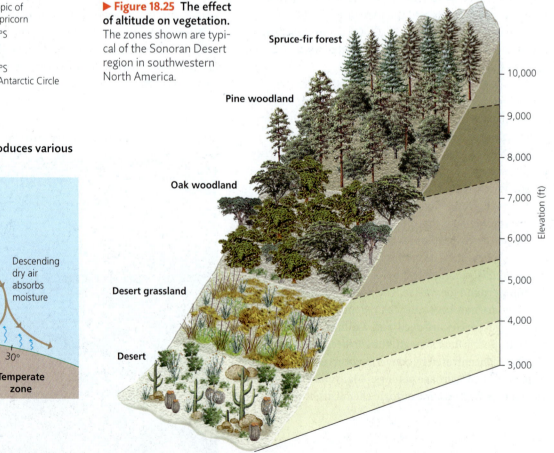

Second, mountains can block the flow of moist air from a coast, causing radically different climates on opposite sides of a mountain range. In the example shown in **Figure 18.26**, moist air moves in off the Pacific Ocean and encounters the Coast Range in California. Air flows upward, cools at higher altitudes, and drops a large amount of rainfall. The world's tallest trees, the coastal redwoods, thrive here. Precipitation increases again farther inland as the air moves up and over higher mountains (the Sierra Nevada). By the time it reaches the eastern side of the Sierra, the air contains little moisture; as this dry air descends, it absorbs moisture. As a result, there is little precipitation on the eastern side of the mountains. This effect, called a rain shadow, is responsible for the desert that covers much of central Nevada. ✅

▼ **Figure 18.26** How mountains affect rainfall.

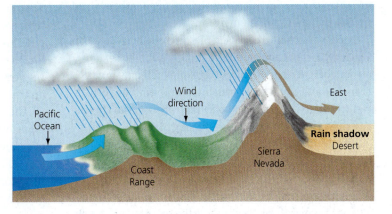

✅ **CHECKPOINT**
Why is there so much rainfall in the tropics?

■ *Answer: Air at the equator rises as it is warmed by direct sunlight. As the air rises, it cools. This causes cloud formation and rainfall because cool air holds less moisture than warm air.*

Terrestrial Biomes

Terrestrial ecosystems are grouped into biomes primarily on the basis of their vegetation type **(Figure 18.27)**. By providing food, shelter, and nesting sites for animals, as well as much of the organic material for the decomposers that recycle mineral nutrients, plants build the foundation for the communities of organisms typical of each biome. The geographic distribution of plants, and thus of biomes, largely depends on climate, with temperature and rainfall often the key factors determining the kind of biome that exists in a particular region. If the climate in two geographically separate areas is similar, the same type of biome may occur in both. Coniferous forests, for instance, extend in a broad band across North America, Europe, and Asia.

Each biome is characterized by a type of biological community rather than an assemblage of particular species. For example, the groups of species living in the deserts of southwestern North America and in the Sahara Desert of Africa are different, but both groups are adapted to desert conditions. Organisms in widely separated

▼ **Figure 18.27** **A map of the major terrestrial biomes.** Although this map has sharp boundaries, biomes actually grade into one another. We'll use smaller versions of this map, highlighted by color coding, during our closer look at the terrestrial biomes in the next several pages.

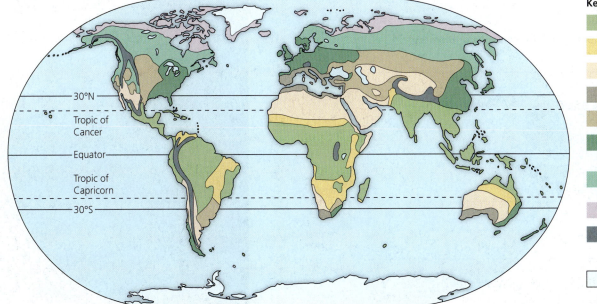

Key

- Tropical forest
- Savanna
- Desert
- Chaparral
- Temperate grassland
- Temperate broadleaf forest
- Northern coniferous forest
- Tundra
- High mountains (coniferous forest and alpine tundra)
- Polar ice

biomes may look alike because of convergent evolution, the appearance of similar traits in independently evolved species living in similar environments.

Local variation within each biome gives the vegetation a patchy, rather than a uniform, appearance. For example, in northern coniferous forests, snowfall may break branches and small trees, causing openings where broadleaf trees such as aspen and birch can grow. Local storms and fires also create openings in many biomes.

The graph in **Figure 18.28** shows the ranges of precipitation and temperature that characterize terrestrial biomes. The *x*-axis shows the range of annual average precipitation, and the *y*-axis displays the range of annual average temperature. By studying the plots on this graph, we can compare these abiotic factors in different biomes. For example, although the range of precipitation in temperate broadleaf forests is similar to that of northern coniferous forests, the lower range of temperatures in northern coniferous forests reveals a significant difference in the abiotic environments of these two biomes. Grasslands are typically drier than forests, and deserts are drier still.

Today, concern about global warming is generating intense interest in the effect of climate on vegetation patterns. Using powerful new tools such as satellite imagery, scientists are documenting shifts in latitude of biome borders, decreases in snow and ice coverage, and changes in the length of the growing season. At the same time, many natural biomes have been fragmented and altered by human activity. We'll discuss both of these issues after we survey the major terrestrial biomes, beginning near the equator and generally approaching the poles.

☑ CHECKPOINT

Why are climbing plants common in tropical rain forests?

■ Answer: Climbing is a plant adaptation for reaching sunlight in a closed canopy, where little sunlight reaches the forest floor.

Tropical Forest

Tropical forests occur in equatorial areas where the temperature is warm year-round. The type of vegetation is determined primarily by rainfall. In tropical rain forests, like the one shown in **Figure 18.29**, rain falls throughout the year, totaling 200 to 400 cm (6.6 to 13 *feet*!) of rain annually. Rainfall is less plentiful in other tropical forests.

The layered structure of tropical rain forests provides many different habitats. Treetops form a closed canopy over one or two layers of smaller trees and a shrub understory. Few plants grow in the deep shade of the forest floor. Many trees are covered by woody vines growing toward the light. Other plants, such as orchids, gain access to sunlight by growing on the branches or trunks of tall trees. Scattered trees reach full sunlight by towering above the canopy. Many of the animals also dwell in trees, where food is abundant. Monkeys, birds, insects, snakes, bats, and frogs find food and shelter many meters above the ground.

Tropical dry forests predominate in lowland areas that have a prolonged dry season or scarce rainfall at any time. The plants found there are a mixture of thorny shrubs and trees and succulents. In regions with distinct wet and dry seasons, deciduous trees that conserve water by shedding their leaves during the dry season are common. ☑

▼ **Figure 18.29** Tropical rain forest in Borneo.

▼ **Figure 18.28** A climate graph for some major biomes in North America.

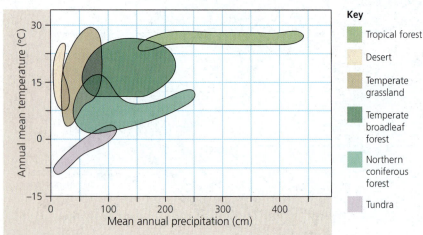

Key

Tropical forest

Desert

Temperate grassland

Temperate broadleaf forest

Northern coniferous forest

Tundra

Savanna

Savannas, such as the one shown in **Figure 18.30**, are dominated by grasses and scattered trees. The temperature is warm year-round. Rainfall averages 30 to 50 cm (roughly 12 to 20 inches) per year, with dramatic seasonal variation.

Fire, caused by lightning or human activity, is an important abiotic factor in the savanna. The grasses survive burning because the growing points of their shoots are below ground. Other plants have seeds that sprout rapidly after a fire. Poor soil and lack of moisture, along with fire and grazing animals, prevent the establishment of most trees in the first place. The luxuriant growth of grasses and small broadleaf plants during the rainy season provides a rich food source for plant-eating animals.

Many of the world's large grazing mammals and their predators inhabit savannas. African savannas are home to zebras and many species of antelope, as well as to lions and cheetahs. Several species of kangaroo are the dominant grazers of Australian savannas. Oddly, though, the large grazers are not the dominant plant-eaters in savannas. That distinction belongs to insects, especially ants and termites. Other animals include burrowers such as mice, moles, gophers, and ground squirrels.

Desert

Deserts are the driest of all biomes, characterized by low and unpredictable rainfall—less than 30 cm (about 12 inches) per year. Some deserts are very hot, with daytime soil surface temperatures above 60°C (140°F) and large daily temperature fluctuations. Other deserts, such as those west of the Rocky Mountains and the Gobi Desert, spanning northern China and southern Mongolia, are relatively cold. Air temperatures in cold deserts may fall below −30°C (−22°F).

Desert vegetation typically includes water-storing plants, such as cacti, and deeply rooted shrubs. Various snakes, lizards, and seed-eating rodents are common inhabitants. Arthropods such as scorpions and insects also thrive in the desert. Evolutionary adaptations of desert plants and animals include a remarkable array of mechanisms that conserve water. For example, the "pleated" stem of saguaro cacti **(Figure 18.31)** enables the plants to expand when they absorb water during wet periods. Some desert mice *never* drink, deriving all their water from the metabolic breakdown of the seeds they eat. Protective adaptations that deter feeding by mammals and insects, such as spines on cacti and poisons in the leaves of shrubs, are common in desert plants. ✓

✓ CHECKPOINT

1. How does the savanna climate vary seasonally?
2. What abiotic factor characterizes deserts?

■ *Answers: 1. Temperature stays about the same year-round, but rainfall varies dramatically. 2. Rainfall is low and unpredictable.*

▼ **Figure 18.30** **Savanna in the Serengeti Plain in Tanzania.**

▼ **Figure 18.31** **Sonoran Desert.**

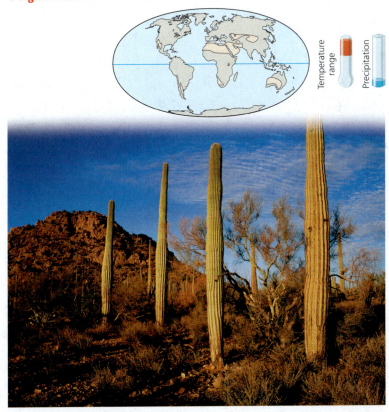

Chaparral

The climate that supports **chaparral** vegetation results mainly from cool ocean currents circulating offshore, producing mild, rainy winters. Summers are hot and dry. This biome is limited to small coastal areas, some in California **(Figure 18.32)**. The largest region of chaparral surrounds the Mediterranean Sea; in fact, Mediterranean is another name for this biome. Dense, spiny, evergreen shrubs dominate chaparral. Annual plants are also common during the wet winter and spring months. Animals characteristic of the chaparral are deer, fruit-eating birds, seed-eating rodents, and lizards and snakes.

Chaparral vegetation is adapted to periodic fires caused by lightning. Many plants contain flammable chemicals and burn fiercely, especially where dead brush has accumulated. After a fire, shrubs use food reserves stored in the surviving roots to support rapid shoot regeneration. Some chaparral plants produce seeds that will germinate only after a hot fire. The ashes of burned vegetation fertilize the soil with mineral nutrients, promoting regrowth of the plant community. Houses do not fare as well. The firestorms that race through the densely populated canyons of Southern California can be devastating to the human inhabitants.

Temperate Grassland

Temperate grasslands have some of the characteristics of tropical savannas, but they are mostly treeless, except along rivers or streams, and are found in regions of relatively cold winter temperatures. Rainfall, averaging between 25 and 75 cm per year (approximately 10 to 30 inches), with frequent severe droughts, is too low to support forest growth. Periodic fires and grazing by large mammals also prevent invasion by woody plants. These grazers include the bison and pronghorn in North America, the wild horses and sheep of the Asian steppes, and kangaroos in Australia. As in the savanna, however, the dominant plant-eaters are invertebrates, especially grasshoppers and soil-dwelling nematodes.

Without trees, many birds nest on the ground. Many small mammals, such as rabbits, voles, ground squirrels, prairie dogs, and pocket gophers, dig burrows to escape predators. Temperate grasslands like the one shown in **Figure 18.33** once covered much of central North America.

Because grassland soil is both deep and rich in nutrients, these habitats provide fertile land for agriculture. Most grassland in the United States has been converted to cropland or pasture, and very little natural prairie exists today. ✅

✅ CHECKPOINT

1. What is one way that homeowners in chaparral areas can protect their neighborhoods from fire?

2. How do people now use most of the North American land that was once temperate grassland?

■ *Answers: 1. by keeping the area clear of dead brush, which is flammable. 2. for farming.*

▼ **Figure 18.32** Chaparral in California.

▼ **Figure 18.33** Temperate grassland in Saskatchewan, Canada.

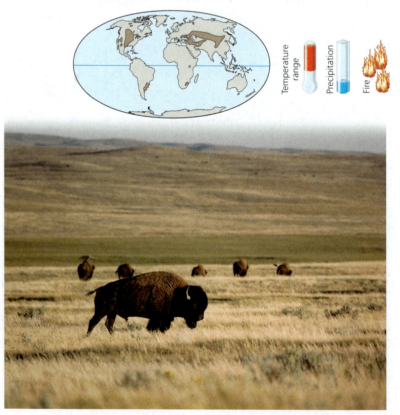

Temperate Broadleaf Forest

Temperate broadleaf forests occur throughout midlatitudes where there is sufficient moisture to support the growth of large trees. Annual precipitation is relatively high at 75 to 150 cm (30 to 60 inches) and typically distributed evenly around the year. Temperature varies seasonally over a wide range, with hot summers and cold winters. In the Northern Hemisphere, dense stands of deciduous trees are trademarks of temperate forests, such as the one pictured in **Figure 18.34**. Deciduous trees drop their leaves before winter, when temperatures are too low for effective photosynthesis and water lost by evaporation is not easily replaced from frozen soil.

Numerous invertebrates live in the soil and the thick layer of leaf litter that accumulates on the forest floor. Some vertebrates, such as mice, shrews, and ground squirrels, burrow for shelter and food, while others, including many species of birds, live in the trees. Predators include bobcats, foxes, black bears, and mountain lions. Many mammals that inhabit these forests enter a dormant winter state called hibernation, and some bird species migrate to warmer climates.

Virtually all the original temperate broadleaf forests in North America were cut for timber or cleared for agriculture or development. These forests tend to recover after disturbance, however, and today we see deciduous trees growing in undeveloped areas over much of their former range.

Coniferous Forest

Cone-bearing evergreen trees such as pine, spruce, fir, and hemlock dominate **coniferous forests** in the Northern Hemisphere. (Other kinds of conifers grow in parts of South America, Africa, and Australia.) The northern coniferous forest, or **taiga (Figure 18.35)**, is the largest terrestrial biome on Earth, stretching in a broad band across North America and Asia south of the Arctic Circle. Taiga is also found at cool, high elevations in more temperate latitudes—for example, in much of the mountainous region of western North America. The taiga is characterized by long, snowy winters and short, wet summers that are sometimes warm. The slow decomposition of conifer needles in the thin, acidic soil makes few nutrients available for plant growth. The conical shape of many conifers prevents too much snow from accumulating on their branches and breaking them. Animals of the taiga include moose, elk, hares, bears, wolves, grouse, and migratory birds. The Asian taiga is home to the dwindling number of Siberian tigers that remain in the wild.

The **temperate rain forests** of coastal North America (from Alaska to Oregon) are also coniferous forests. Warm, moist air from the Pacific Ocean supports this unique biome, which, like most coniferous forests, is dominated by a few tree species, typically hemlock, Douglas fir, and redwood. These forests are heavily logged, and the old-growth stands of trees are rapidly disappearing. ✅

✅ **CHECKPOINT**

1. How does the loss of leaves function as an adaptation of deciduous trees to cold winters?

2. What type of trees are characteristic of the taiga?

spruce, fir, and hemlock.
frozen soil. 2. conifers such as pine,
water from the trees when that
■ Answers: 1. by reducing loss of

▼ **Figure 18.34**
Temperate broadleaf forest in Maine in autumn.

▼ **Figure 18.35**
Northern coniferous forest in Finland, with the sky lit by the northern lights.

Tundra

Tundra covers expansive areas of the Arctic between the taiga and polar ice. **Permafrost** (permanently frozen subsoil), bitterly cold temperatures, and high winds are responsible for the absence of trees and other tall plants in the arctic tundra shown in **Figure 18.36**. The arctic tundra receives very little annual precipitation. However, water cannot penetrate the underlying permafrost, so melted snow and ice accumulate in pools on the shallow topsoil during the short summer.

Tundra vegetation includes small shrubs, grasses, mosses, and lichens. When summer arrives, flowering plants grow quickly and bloom in a rapid burst. Caribou, musk oxen, wolves, and small rodents called lemmings are among the mammals found in the arctic tundra. Many migratory birds use the tundra as a summer breeding ground. During the brief but productive warm season, the marshy ground supports the aquatic larvae of insects, providing food for migratory waterfowl, and clouds of mosquitoes often fill the tundra air.

On very high mountaintops at all latitudes, including the tropics, high winds and cold temperatures create plant communities called alpine tundra. Although these communities are similar to arctic tundra, there is no permafrost beneath alpine tundra.

Polar Ice

Polar ice covers the land at high latitudes north of the arctic tundra in the Northern Hemisphere and in Antarctica in the Southern Hemisphere **(Figure 18.37)**. The temperature in these regions is extremely cold year-round, and precipitation is very low. Only a small portion of these landmasses is free of ice or snow, even during the summer. Nevertheless, small plants, such as mosses and lichens, eke out a living, and invertebrates such as nematodes, mites, and wingless insects called springtails inhabit the frigid soil. Nearby sea ice provides feeding platforms for large animals such as polar bears (in the Northern Hemisphere), penguins (in the Southern Hemisphere), and seals. Seals, penguins, and other marine birds visit the land to rest and breed. The polar marine biome provides the food that sustains these birds and mammals. In the Antarctic, penguins feed at sea, eating a variety of fish, squids, and small shrimplike crustaceans known as krill. Antarctic krill, an important food source for many species of fish, seals, squids, seabirds, and filter-feeding whales as well as penguins, depend on sea ice for breeding and as a refuge from predators. As the amount of sea ice declines and the ice season becomes shorter—consequences of global climate change—krill habitat is shrinking. ✓

✓ CHECKPOINT

1. Global warming is melting permafrost in some areas of the arctic tundra. What biome would you expect to replace tundra in these regions?

2. How does the vegetation found in polar ice regions compare with tundra vegetation?

■ *Answers: 1. taiga. 2. Neither biome is hospitable to plants because of the cold temperatures. However, tundra supports the growth of small shrubs, while polar ice vegetation is limited to mosses and lichens.*

▼ **Figure 18.36** Arctic tundra in Yukon Territory, Canada.

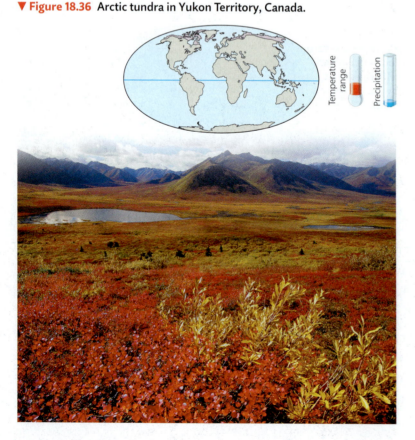

▼ **Figure 18.37** Polar ice in Antarctica.

The Water Cycle

Biomes are not self-contained units. Rather, all parts of the biosphere are linked by the global water cycle, illustrated in **Figure 18.38**, and by nutrient cycles (see Chapter 20). Consequently, events in one biome may reverberate throughout the biosphere.

As you learned earlier in this chapter, water and air move in global patterns driven by solar energy. Precipitation and evaporation continuously move water between the land, oceans, and the atmosphere. Water also evaporates from plants, which pull it from the soil in a process called transpiration.

Over the oceans, evaporation exceeds precipitation. The result is a net movement of water vapor to clouds that are carried by winds from the oceans across the land. On land, precipitation exceeds evaporation and transpiration. The excess precipitation may stay on the surface or it may trickle through the soil to become groundwater. Both surface water and groundwater eventually flow back to the sea, completing the water cycle.

Just as the water draining from your shower carries dead skin cells from your body along with the day's grime, the water washing over and through the ground carries traces of the land and its history. For example, water flowing from land to the sea carries with it silt (fine soil particles) and chemicals such as fertilizers and pesticides.

Erosion from coastal development has caused silt to muddy the waters of some coral reefs, dimming the light available to the photosynthetic algae that power the reef community. Chemicals in surface water may travel hundreds of miles by stream and river to the ocean, where currents then carry them even farther from their point of origin. For instance, traces of pesticides and chemicals from industrial wastes have been found in marine mammals in the Arctic and in deep-sea octopuses and squids. Airborne pollutants such as nitrogen oxides and sulfur oxides, which combine with water to form acid precipitation, are distributed by the water cycle, too.

Human activity also affects the global water cycle itself in a number of important ways. One of the main sources of atmospheric water is transpiration from the dense vegetation making up tropical rain forests. The destruction of these forests changes the amount of water vapor in the air. Pumping large amounts of groundwater to the surface for irrigation increases the rate of evaporation over land and may deplete groundwater supplies. In addition, global warming affects the water cycle in complex ways that will have far-reaching effects on precipitation patterns. We'll consider some of these environmental impacts in the following sections. ✓

POLLUTANTS FROM CAR EXHAUST CAN RETURN TO EARTH AS ACID PRECIPITATION.

✓ CHECKPOINT

What is the main way that living organisms contribute to the water cycle?

■ *Answer: Plants move water from the ground to the air via transpiration.*

Figure Walkthrough

Mastering Biology
goo.gl/FaPcb2

▼ **Figure 18.38** The global water cycle.

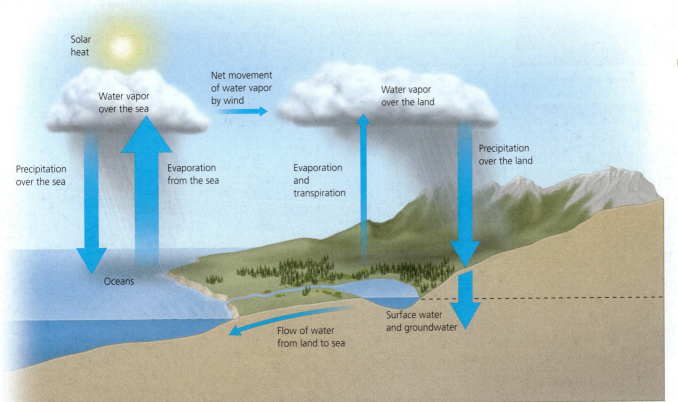

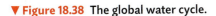

- Solar heat
- Water vapor over the sea
- Net movement of water vapor by wind
- Water vapor over the land
- Precipitation over the sea
- Evaporation from the sea
- Evaporation and transpiration
- Precipitation over the land
- Oceans
- Flow of water from land to sea
- Surface water and groundwater

Human Impact on Biomes

For hundreds of years, people have been using increasingly effective technologies to capture or produce food, to extract resources from the environment, and to build cities. It is now clear that the environmental costs of these enterprises are staggering. In this section, you'll see some examples of how human activities are affecting forest and freshwater resources. Throughout the remainder of this unit, you'll learn about the role of ecological knowledge in achieving **sustainability**, the goal of developing, managing, and conserving Earth's resources in ways that meet the needs of people today without compromising the ability of future generations to meet their needs.

Forests

The map in Figure 18.27 shows the terrestrial biomes that would be expected to flourish under the prevailing climatic conditions. However, about three-quarters of Earth's land surface has been altered by thousands of years of human occupation. Most of the land that we've appropriated is used for agriculture; another hefty chunk is covered by the asphalt and concrete of development. Changes in vegetation are especially dramatic in regions like tropical forests that escaped large-scale human intervention until recently. Satellite photos of a small area in Brazil show how thoroughly a landscape can be altered in a short amount of time (Figure 18.39).

Every year, more and more forested land is cleared for agriculture. You might think that this land is needed to feed new mouths as the human population continues to grow, but that's not entirely the case. Unsustainable agricultural practices have degraded much of the world's cropland so severely that it is unusable. Researchers estimate that replacing worn-out farmland accounts for up to 80% of the deforestation occurring today. Tropical forests, such as the one shown in Figure 18.40, are also being cleared to grow palm oil for products such as cosmetics and a long list of packaged foods, including cookies, crackers, potato chips, chocolate products, and soups. Other forests are being lost to logging, mining, and air pollution, problems that are hitting coniferous forests especially hard. (As we mentioned in the previous section, most temperate broadleaf forests were replaced by human enterprises long ago.) Land that hasn't been directly converted to food production and living space also bears the imprint of our presence. Roads penetrate regions that are otherwise unaltered, bringing pollution to the wilderness, providing avenues for new diseases to emerge, and slicing vast tracts of biome into segments that are too small to support a full array of species.

Land uses that provide resources such as food, fuel, and shelter are clearly beneficial to us. But natural ecosystems also provide services that support the human population—purification of air and water, nutrient cycling, and recreation, to name just a few. (We'll return to the topic of ecosystem services in Chapter 20.)

▼ **Figure 18.39** Satellite photos of the Rondonia area of the Brazilian rain forest.

1986

1986. In 1986, development in this remote region was just beginning.

2001

2001. Same area 15 years later, after loggers and farmers have moved in. The "fishbone" pattern marks the new network of roads carved through the forest.

▼ **Figure 18.40** A tropical forest in Indonesia clear-cut for a palm oil plantation.

Fresh Water

The impact of human activities on freshwater ecosystems may pose an even greater threat to life on Earth—including ourselves—than the damage to terrestrial ecosystems. Freshwater ecosystems are being polluted by large amounts of nitrogen and phosphorus compounds that run off from heavily fertilized farms or from livestock feedlots. A wide variety of other pollutants, such as industrial wastes, also contaminate freshwater habitats, drinking water, and groundwater. Some regions of the world face dire shortages of water as a result of the overuse of groundwater for irrigation, extended droughts (partially caused by global climate change), or poor water management practices.

Las Vegas, the population center of Clark County, Nevada, is one example of a city whose water resources are increasingly stressed by drought and overconsumption. **Figure 18.41a** is a satellite photo of Las Vegas in 1972, when the population of Clark County was 307,421. **Figure 18.41b** shows the same area in 2013, when the population had swelled to more than 2 million. In contrast to the disappearance of greenery in the photos of Brazilian rain forest, the mark of human activities in Figure 18.41b is the notable expansion of greenery—the result of watering lawns and golf courses. Las Vegas is situated in a high valley in the Mojave Desert. Where does it get the water to turn barren desert into green fields?

Las Vegas taps underground aquifers for some water, but its main water supply is Lake Mead. Lake Mead is an enormous reservoir formed by the Hoover Dam on the Colorado River, which in turn receives almost all of its water from snowmelt in the Rocky Mountains. With decreased annual snowfall, attributable largely to climate change, the flow of the Colorado has greatly diminished.

The water level in Lake Mead has dropped drastically **(Figure 18.42)**, and parched cities and farms farther downstream are pleading for more water.

To ensure an adequate water supply for the future, Las Vegas is looking for new sources of water. Among other options, Las Vegas is eyeing the abundant supply of groundwater in the northern end of the valley where it lies. Although sparsely populated, that area is home to many ranchers whose livelihoods depend on the groundwater. It is also home to numerous endangered species. Not surprisingly, environmentalists and residents of the north valley are resisting efforts to pipe its groundwater to Las Vegas.

Nevada is just one of many places where the hard realities of climate change are beginning to affect daily life. Battles over water resources are shaping up throughout the arid West and Southwest of the United States, where changing precipitation patterns due to global warming are projected to continue the drought for many years to come. In other regions of the world, including China, India, and North Africa, the increasing demands of economic, agricultural, and population growth are straining water resources that are already scarce.

While policymakers are dealing with current crises and planning how to manage resources in the future, researchers are seeking methods of sustainable agriculture and water use. Basic ecological research is an essential component of ensuring that sufficient food and water will be available for people now—and for the generations to come. Next, we take a closer look at a major threat to sustainability: climate change. ✅

▼ **Figure 18.42 Low water level in Lake Mead.** The white "bathtub ring" is caused by mineral deposits on rocks that were once submerged.

▼ **Figure 18.41 Satellite photos of Las Vegas, Nevada.**

(a) 1972

(b) 2013

✅ **CHECKPOINT**

Why is decreased snowfall in the Rocky Mountains a concern for people who live in Las Vegas?

Answer: Snowmelt from the Rockies flows into the Colorado River, which supplies water for Las Vegas residents.

Climate Change

Rising concentrations of carbon dioxide (CO_2) and certain other gases in the atmosphere are changing global climate patterns. This was the conclusion of the assessment report released by the Intergovernmental Panel on Climate Change (IPCC) in 2014. Thousands of scientists and policymakers from more than 100 countries participated in producing the report, which is based on data published in thousands of scientific papers. Thus, there is no debate among scientists about whether climate change is occurring.

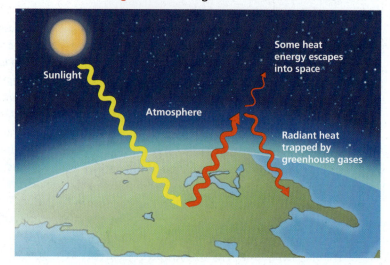

▼ **Figure 18.43** The greenhouse effect.

The Greenhouse Effect and Global Warming

Why is Earth's atmosphere becoming warmer? A familiar and useful analogy is a greenhouse, which is used to grow plants when the weather outside is too cold. Its transparent glass or plastic walls allow solar radiation to pass through and trap some of the heat that accumulates inside the building. Similarly, certain gases in Earth's atmosphere are transparent to solar radiation but absorb or reflect heat. Some of these **greenhouse gases** are natural, including CO_2, water vapor, and methane. Others, such as chlorofluorocarbons (CFCs, found in some aerosol sprays and refrigerants), are synthetic. As **Figure 18.43** shows, greenhouse gases act as a blanket that traps heat in the atmosphere. This heating, often called the **greenhouse effect**, is highly beneficial. Without it, the average air temperature on Earth would be a frigid −18°C (−0.4°F), far too cold for most life as we know it. However, increasing the insulation that the blanket provides is making Earth overly warm.

The effect of increasing greenhouse gases is the increase in the average global temperature, which has risen about 1°C (1.8°F) since 1900 at an accelerating pace. Further increases of 2 to 4.5°C are likely by the end of this century, according to the IPPC. Temperature increases are not distributed evenly around the world. The largest increases are in the northernmost regions of the Northern Hemisphere and parts of Antarctica **(Figure 18.44)**.

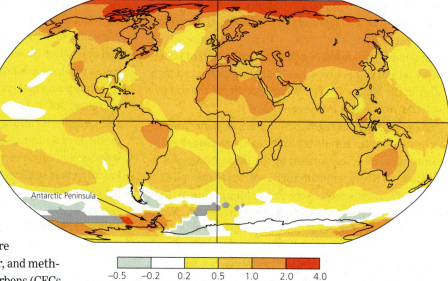

▼ **Figure 18.44** Differences in average temperatures during 2006–2016 compared with long-term averages during 1951–1980, in °C. The largest temperature increases are shown in red. Gray indicates regions for which no data are available.

Antarctic Peninsula

−0.5 −0.2 0.2 0.5 1.0 2.0 4.0

More than 90% of the heat trapped by greenhouse gases is being stored in the ocean. Water expands as it warms, causing sea levels to rise. Melting of the massive ice sheets of Greenland and Antarctica, as well as mountain glaciers, is also contributing to sea level rise. Rising sea levels will cause catastrophic flooding of coastal areas worldwide.

Global warming is causing an increase in extreme weather events. In some regions, a greater proportion of the total precipitation falls in torrential downpours that cause flooding; other regions undergo intense drought. Prolonged heat waves occur more frequently. Higher sea surface temperatures fuel more powerful hurricanes. In addition, warm weather begins earlier each year. ✓

✓ CHECKPOINT

Why are gases such as CO_2 and methane called greenhouse gases?

■ *Answer: They allow solar radiation to pass through the atmosphere but prevent the heat from reflecting back out, much as the glass of a greenhouse retains the sun's heat inside the building.*

The Accumulation of Greenhouse Gases

After many years of data collection and debate, the vast majority of scientists are confident that human activities have caused the rising concentrations of greenhouse gases. Major sources of emissions include agriculture, landfills, and the burning of wood and fossil fuels (oil, coal, and natural gas).

Let's take a closer look at CO_2, the dominant greenhouse gas. For 650,000 years, the atmospheric concentration of CO_2 did not exceed 300 parts per million (ppm); the concentration before the Industrial Revolution was 280 ppm. In 2016, the average atmospheric CO_2 was 404 ppm and continuing to rise **(Figure 18.45)**. The levels of other greenhouse gases have increased dramatically, too. Remember that CO_2 is removed from the atmosphere by the process of photosynthesis and stored in organic molecules such as carbohydrates (see Figure 6.2). These molecules are eventually broken down by cellular respiration, releasing CO_2. Overall, uptake of CO_2 by photosynthesis roughly equals the release of CO_2 by cellular respiration **(Figure 18.46)**. However, extensive deforestation has significantly decreased the incorporation of CO_2 into organic material. At the same time, CO_2 is flooding into the atmosphere from the burning of fossil fuels and wood, a process that releases CO_2 from organic material much more rapidly than cellular respiration.

CO_2 is also exchanged between the atmosphere and the surface waters of the oceans. For decades, the oceans have acted as massive sponges, soaking up considerably more CO_2 than they have released. But now, the excess CO_2 has made the oceans more acidic, a change that could have a profound effect on marine communities. As ocean acidification worsens, many species of plankton and marine animals such as corals and molluscs will be unable to build their shells or exoskeletons. Their demise will remove critical links from marine food webs and ultimately damage marine ecosystems around the world. ✅

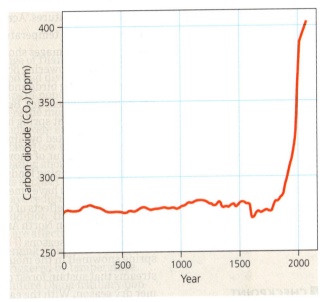

▼ **Figure 18.45 Atmospheric concentration of CO_2.** Notice that the concentration was relatively stable until the Industrial Revolution, which began in the late 1700s.

◄ **Figure 18.46** How CO_2 enters and leaves the atmosphere.

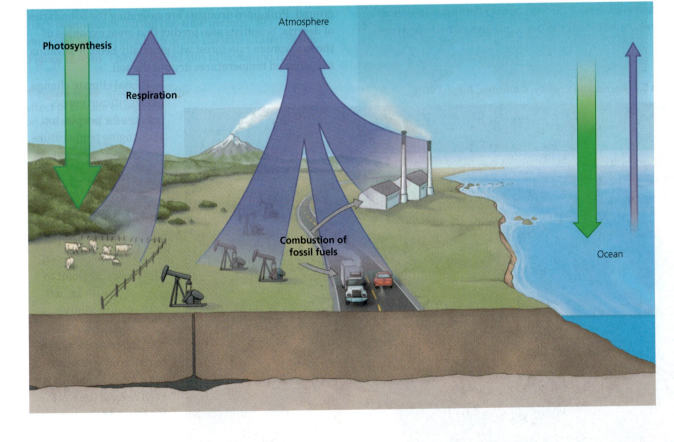

✅ **CHECKPOINT**

What is the major source of CO_2 released by human activities?

■ *Answer: burning fossil fuels*

Living in Earth's Diverse Environments

The biosphere is an environmental patchwork in which abiotic factors affect the distribution and abundance of organisms.

Abiotic Factors of the Biosphere

Abiotic factors include the availability of sunlight, water, nutrients, and temperature. In aquatic habitats, dissolved oxygen, salinity, current, and tides are also important. Additional factors in terrestrial environments include wind and fire.

The Evolutionary Adaptations of Organisms

Adaptation via natural selection results from the interactions of organisms with their environments.

Adjusting to Environmental Variability

Organisms also have adaptations that enable them to cope with environmental variability, including physiological, behavioral, and anatomical responses to changing conditions.

Biomes

A biome is a major terrestrial or aquatic life zone.

Freshwater Biomes

Freshwater biomes include lakes, ponds, rivers, streams, and wetlands. Lakes vary, depending on depth, with regard to light penetration (photic and aphotic zones), temperature, nutrients, oxygen levels, and community structure. Rivers change greatly from their source to the point at which they empty into a lake or ocean. The bottom of an aquatic biome is its benthic realm.

Marine Biomes

Marine life is distributed into distinct realms (benthic and pelagic) and zones (photic, aphotic, and intertidal) according to the depth of the water, degree of light penetration, distance from shore, and open water versus deep-sea bottom. Marine biomes include the pelagic realm and the benthic realm of the oceans, coral reefs, intertidal zones, and estuaries. Coral reefs, which occur in warm tropical waters above the continental shelf, have an abundance of biological diversity. An ecosystem found near hydrothermal vents in the deep ocean is powered by chemical energy from Earth's interior instead of sunlight. Estuaries, located where a freshwater river or stream merges with the ocean, are some of the most biologically productive environments on Earth.

How Climate Affects Terrestrial Biome Distribution

The geographic distribution of terrestrial biomes is based mainly on regional variations in climate. Climate is largely determined by the uneven distribution of solar energy on Earth. Proximity to large bodies of water and the presence of landforms such as mountains also affect climate.

Terrestrial Biomes

Most terrestrial biomes are named for their climate and predominant vegetation. The major terrestrial biomes include tropical forest, savanna, desert, chaparral, temperate grassland, temperate broadleaf forest, coniferous forest, tundra, and polar ice. If the climate in two geographically separate areas is similar, the same type of biome may occur in both.

The Water Cycle

The global water cycle links aquatic and terrestrial biomes. Human activities are disrupting the water cycle.

Human Impact on Biomes

Land use by humans has altered vast tracts of forest and degraded the services provided by natural ecosystems. Unsustainable agricultural practices have depleted cropland fertility. Human activities have polluted freshwater ecosystems, which are vital for life. Agriculture, population growth, drought, and declining snowfall are all factors in the rapid depletion of freshwater resources in some regions.

Climate Change

The Greenhouse Effect and Global Warming

So-called greenhouse gases, including CO_2 and methane, increase the amount of heat retained in Earth's atmosphere. The accumulation of these gases has caused increases in the average global temperature.

The Accumulation of Greenhouse Gases

Human activities, especially the burning of fossil fuels, are responsible for the rise in greenhouse gases over the past century. Release of CO_2 has exceeded the amount that can be absorbed by natural processes.

Effects of Climate Change on Ecosystems

Climate change is disrupting interactions between species. Devastating wildfires are among the effects of climate change in certain ecosystems. Climate change is also shifting biome boundaries.

Looking to Our Future

Each person has a carbon footprint—that person's responsibility for a portion of global greenhouse gas emissions. We can take action to reduce our carbon footprints.

Mastering Biology

For practice quizzes, BioFlix animations, MP3 tutorials, video tutors, and more study tools designed for this textbook, go to Mastering Biology™

SELF-QUIZ

1. Place these levels of ecological study in order from the least to the most comprehensive: community ecology, ecosystem ecology, organismal ecology, population ecology.

2. Name several abiotic factors that might affect the community of organisms living inside a home fish tank.

3. The formation of goose bumps on your skin in cold weather is an example of a(an) _____ response, while seasonal migration is an example of a(an) _____ response.

4. Which of the following sea creatures might be described as a pelagic animal of the aphotic zone?

a. a coral reef fish

b. a giant clam near a deep-sea hydrothermal vent

c. an intertidal snail

d. a deep-sea squid

5. Identify the following biomes on the graph below: tundra, northern coniferous forest, desert, temperate grassland, temperate broadleaf forest, and tropical forest.

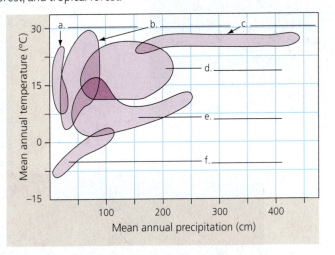

6. We are on a coastal hillside on a hot, dry summer day among evergreen shrubs that are adapted to fire. We are most likely standing in a _____ biome.

7. What three abiotic factors account for the rarity of trees in arctic tundra?

8. What human activity is responsible for the greatest amount of deforestation?

9. What is the greenhouse effect? How is the greenhouse effect related to global warming?

10. The recent increase in atmospheric CO_2 concentration is mainly a result of an increase in

a. plant growth.

b. the absorption of heat radiating from Earth.

c. the burning of fossil fuels and wood.

d. cellular respiration by the increasing human population.

11. What populations of organisms are most likely to survive climate change via evolutionary adaptation?

For answers to the Self-Quiz, see Appendix D.

IDENTIFYING MAJOR THEMES

For each statement below, identify which major theme is evident (the relationship of structure to function, information flow, pathways that transform energy and matter, interactions within biological systems, or evolution) and explain how the statement relates to the theme. If necessary, review the themes (Chapter 1) and review the examples highlighted in blue in this chapter.

12. Reptilian scales and the waxy coating on many leaves reduce water loss.

13. Other organisms may compete with an individual for food and other resources, prey upon it, or change its physical and chemical environment.

14. Solar energy from sunlight captured by chlorophyll during the process of photosynthesis powers most ecosystems.

15. After a period of lower-than-average rainfall, drought-resistant individuals may be more prevalent in a plant population.

For answers to Identifying Major Themes, see Appendix D.

THE PROCESS OF SCIENCE

16. Design a laboratory experiment to measure the effect of water temperature on the population growth of a certain phytoplankton species from a pond.

17. **Interpreting Data** This graph shows average monthly temperature and precipitation for a city in the Northern Hemisphere. Based on the biome descriptions on pages 385–390, in which biome is this city located? Explain your answer.

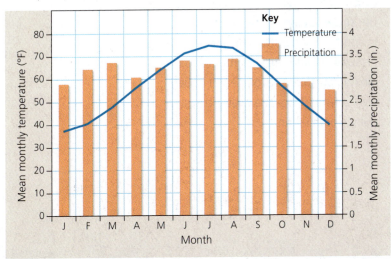

BIOLOGY AND SOCIETY

18. Some people are not convinced that human-induced global climate change is a real phenomenon. Using your knowledge of the scientific process (see Chapter 1) and the information from this chapter, develop arguments you could use to explain the scientific basis for saying that global climate change is truly occurring and that people are responsible for it.

19. Research your country's per capita (per person) carbon emissions. Calculate your own carbon footprint and compare it with the average for your country. Make a list of actions you are willing to take to reduce your personal carbon footprint. What actions could you take to persuade others to reduce their carbon footprints?

19 Population Ecology

CHAPTER CONTENTS
An Overview of Population Ecology 404
Population Growth Models 408
Applications of Population Ecology 412
Human Population Growth 417

DO YOU BUY TOO MUCH STUFF? THE AVERAGE AMERICAN GENERATES ABOUT 40 POUNDS OF TRASH PER WEEK, TWICE AS MUCH AS THE AVERAGE MEXICAN.

Why Population Ecology Matters

Population ecology has many practical applications in such diverse fields as conservation, agriculture, and combating the spread of infectious diseases.
In addition, understanding population ecology provides insight into one of the most important environmental issues we currently face—the ever-increasing human population.

IS THE ABUNDANCE OF NATURE UNLIMITED? THE SAYING "THERE ARE PLENTY OF FISH IN THE SEA" COULD BECOME MEANINGLESS IF OVERHARVESTING CONTINUES.

HOW LARGE IS THE HUMAN POPULATION? COUNTING ONE PERSON PER SECOND, IT WOULD TAKE MORE THAN 238 YEARS TO COUNT ALL 7.5 BILLION PEOPLE ALIVE TODAY.

CHAPTER THREAD

Biological Invasions

BIOLOGY AND SOCIETY   Invasion of the Lionfish 403

THE PROCESS OF SCIENCE   Can Fences Stop Cane Toads? 415

EVOLUTION CONNECTION   Humans as an Invasive Species 421

BIOLOGY AND SOCIETY   Biological Invasions

The red lionfish, a beautiful but deadly invader, is a threat to coral reef communities. This lionfish was photographed in the Caribbean, halfway around the world from its native habitat.

Invasion of the Lionfish

Lionfish, with their graceful, flowing fins, bold stripes, and eye-catching array of spines, are striking members of tropical reef communities. They are also favorites of saltwater aquarium enthusiasts—especially the red lionfish, a native to the coral reefs of the South Pacific and Indian Oceans. There are a few drawbacks to owning a red lionfish, however. The spines are venomous and can inflict an intensely painful sting. Lionfish are merciless predators, so any tankmates must be chosen with care. And they are large. A 2-inch juvenile can rapidly become an 18-inch adult that requires, at minimum, a 120-gallon tank. Apparently, some aquarium owners who regretted their purchase released their lionfish into the wild.

Freed from the competitors and predators of their native reefs as well as their tanks, red lionfish have multiplied exponentially. Within a few years of the first sightings off the southeastern coast of Florida, lionfish populations had spread up the East Coast. They have since invaded islands and coastlines throughout the Atlantic and Caribbean regions and the Gulf of Mexico. The speed of the onslaught has stunned scientists, who are just beginning to document its devastating effects on native ecosystems. Lionfish consume prodigious numbers of other fish, including species that are key to maintaining the legendary diversity of reef communities and juveniles of economically important fishes such as grouper and snapper. Some biologists think our best hope of stopping the lionfish invasion is for *us* to consume *them*. The National Oceanic and Atmospheric Administration (NOAA) has launched an "Eat Lionfish" campaign to encourage human predation on the tasty fish.

Lionfish are not the only invasive species that have had devastating effects. For as long as people have traveled the world, they have carried—intentionally or accidentally—thousands of species to new habitats. Many of these non-native species have established populations that spread far and wide, leaving environmental havoc in their wake. We humans, too, have multiplied and spread far from our point of origin, radically changing our environment in the process. As you explore population ecology in this chapter, you'll also learn about trends in human population growth and other applications of this area of ecological research.

Life Tables and Survivorship Curves

Life tables track survivorship, the chance of an individual in a given population surviving to various ages. The life insurance industry uses life tables to predict how long, on average, a person of a given age will live. Starting with a population of 100,000 people, **Table 19.1** shows the number of people who are expected to be alive at the beginning of each age interval, based on death rates in 2012. For example, 94,351 out of 100,000 people are expected to live to age 50. Their chance of surviving to age 60, shown in the last column of the same row, is 0.941; 94% of 50-year-olds will reach the age of 60. The chance of 80-year-olds surviving to age 90, however, is only 0.418. Population ecologists have adopted this technique and constructed life tables to help them understand the structure and dynamics of various plant and animal species. By identifying the most vulnerable stage of the life cycle, life table data may also help conservationists develop effective measures to protect species whose populations are declining.

Ecologists represent life table data graphically in a **survivorship curve**, a plot of the number of individuals still alive at each age in the maximum life span **(Figure 19.4)**. By using a percentage scale instead of actual ages on the *x*-axis, we can compare species with widely varying life spans, such as humans and squirrels, on the same graph. The curve for the human population (red) shows that most people survive to the older age intervals. Ecologists refer to the shape of this curve as Type I survivorship. Species that exhibit a Type I curve—humans

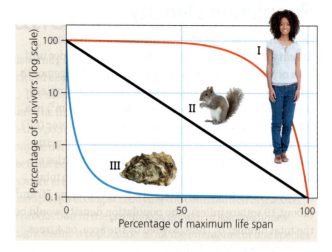

▲ **Figure 19.4** Three idealized types of survivorship curves.

and many other large mammals—usually produce few offspring but give them good care, increasing the likelihood that they will survive to maturity.

In contrast, a Type III curve (blue) indicates low survivorship for the very young, followed by a period when survivorship is high for those few individuals who live to a certain age. Species with this type of survivorship curve usually produce very large numbers of offspring but provide little or no care for them. Some species of fish, for example, produce millions of eggs at a time, but most of these offspring die as larvae from predation or other causes. Many invertebrates, including oysters, also have Type III survivorship curves.

A Type II curve (black) is intermediate, with survivorship constant over the life span. That is, individuals are no more vulnerable at one stage of the life cycle than another. This type of survivorship has been observed in some invertebrates, lizards, and rodents. ✓

Life History Traits as Adaptations

A population's pattern of survivorship is an important feature of its **life history**, the set of traits that affect an organism's schedule of reproduction and survival. Some key life history traits are the age at first reproduction, the frequency of reproduction, the number of offspring, and the amount of parental care given. As you might expect from the different types of survivorship curves, life history traits vary among organisms. Let's take a closer look at how natural selection shapes these traits.

As you may recall, reproductive success is key to evolutionary success (see Chapter 13). Accordingly,

Table 19.1	Life Table for the U.S. Population in 2012		
	Number Living at Start of Age Interval	Number Dying During Interval	Chance of Surviving Interval
Age Interval	(*N*)	(*D*)	$1 - (D/N)$
0–10	100,000	759	0.992
10–20	99,241	301	0.997
20–30	98,940	900	0.991
30–40	98,040	1,234	0.987
40–50	96,805	2,454	0.975
50–60	94,351	5,546	0.941
60–70	88,805	10,465	0.882
70–80	78,340	20,485	0.739
80–90	57,855	33,653	0.418
90+	24,202	24,202	0.000

you may wonder why all organisms don't simply produce a large number of offspring. One reason is that reproduction is expensive in terms of time, energy, and nutrients—resources that are available in limited amounts. An organism that gives birth to a large number of offspring will not be able to provide a great deal of parental care. Consequently, the combination of life history traits represents trade-offs that balance the demands of reproduction and survival. **In other words, life history traits, like anatomical features, are shaped by evolution.**

Because selective pressures vary, life histories are very diverse. Nevertheless, ecologists have observed some patterns that are useful for understanding how natural selection influences life history characteristics.

One life history pattern is typified by small-bodied, short-lived species (for example, insects, small rodents, and dandelions) that develop and reach sexual maturity rapidly, have a large number of offspring, and offer little or no parental care. In plants, "parental care" is measured by the amount of nutritional material stocked in each seed. Many small, nonwoody plants produce thousands of tiny seeds. Such organisms have an **opportunistic life history**, one that enables the plant or animal to take immediate advantage of favorable conditions. In general, populations with this life history pattern exhibit a Type III survivorship curve.

In contrast, organisms with an **equilibrial life history** develop and reach sexual maturity slowly and produce a few well-cared-for offspring. Organisms that have an equilibrial life history are typically larger-bodied, longer-lived species (for example, bears and elephants). Populations with this life history pattern exhibit a Type I survivorship curve. Plants with comparable life history traits include certain trees. For example, coconut palms produce relatively few seeds, but those seeds are well stocked with nutrient-rich material. **Table 19.2** compares key traits of opportunistic and equilibrial life history patterns.

What accounts for the differences in life history patterns?

Some ecologists hypothesize that the potential survival rate of the offspring and the likelihood that the adult will live to reproduce again are the critical factors. In a harsh, unpredictable environment, an adult may have just one good shot at reproduction, so it may be an advantage to invest in quantity rather than quality. On the other hand, in an environment where favorable conditions are more dependable, an adult is more likely to survive to reproduce again. Seeds are more likely to fall on fertile ground, and newly emerged animals are more likely to survive to adulthood. In that case, it may be more advantageous for the adult to invest its energy in producing a few well-cared-for offspring at a time.

Of course, there is much more diversity in life history patterns than the two extremes described here. Nevertheless, the contrasting patterns are useful for understanding the interactions between life history traits and our next topic, population growth. ✓

✓ **CHECKPOINT**

How does the term *opportunistic* capture the key characteristics of that life history pattern?

■ Answer: An opportunistic life history is characterized by an ability to produce a large number of offspring very rapidly when the environment affords a temporary opportunity for a burst of reproduction.

Dandelions have an opportunistic life history.

Elephants have an equilibrial life history.

Table 19.2	Some Life History Characteristics of Opportunistic and Equilibrial Populations	
Characteristic	Opportunistic Populations (such as many wildflowers)	Equilibrial Populations (such as many large mammals)
Climate	Relatively unpredictable	Relatively predictable
Maturation time	Short	Long
Life span	Short	Long
Death rate	Often high	Usually low
Number of offspring per reproductive episode	Many	Few
Number of reproductions per lifetime	Usually one	Often several
Timing of first reproduction	Early in life	Later in life
Size of offspring or eggs	Small	Large
Parental care	Little or none	Often extensive

Regulation of Population Growth

Now let's take a closer look at how population growth is regulated in nature. What stops a population from continuing to increase after reaching carrying capacity?

Density-Dependent Factors

Several **density-dependent factors**—limiting factors whose intensity is related to population density—can limit growth in natural populations. The most obvious is **intraspecific competition**, the competition between individuals of the same species for the same limited resources. As a limited food supply is divided among more and more individuals, birth rates may decline as

individuals have less energy available for reproduction. Density-dependent factors may also depress a population's growth by increasing the death rate. For example, in a population of song sparrows, both factors reduced the number of offspring that survived and left the nest **(Figure 19.8a)**. As the number of competitors for food increased, female song sparrows laid fewer eggs. In addition, the death rate of eggs and nestlings increased with increasing population density.

Plants that grow close together may experience an increased death rate as intraspecific competition for resources increases. And those that do survive will produce fewer flowers, fruits, and seeds than uncrowded individuals. After seeds sprout, gardeners often pull out some of the seedlings to allow sufficient resources for the remaining plants. Intraspecific competition is also the reason that plants purchased from a nursery come with instructions to space the plants a certain distance apart.

A limited resource may be something other than food or nutrients. Like a game of musical chairs, the number of safe hiding places may limit a prey population by exposing some individuals to a greater risk of predation. For example, young kelp perch hide from predators in "forests" of the large seaweed known as kelp (see Figure 15.25). In the experiment shown in **Figure 19.8b**, the proportion of perch eaten by a predator increased with increasing perch density. In many animals that defend a territory, the availability of space may limit reproduction. For instance, the number of nesting sites on rocky islands may limit the population size of oceanic birds such as gannets, which maintain breeding territories **(Figure 19.9)**.

In addition to competition for resources, other factors may cause density-dependent deaths in a population. For example, the death rate may climb as a result of increased disease transmission under crowded conditions or the accumulation of toxic waste products.

▼ **Figure 19.8** Density-dependent regulation of population growth.

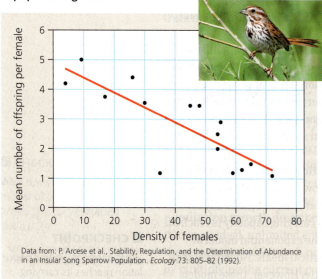

Data from: P. Arcese et al., Stability, Regulation, and the Determination of Abundance in an Insular Song Sparrow Population. *Ecology* 73: 805–82 (1992).

(a) Declining reproductive success of song sparrows (inset) with increasing population density.

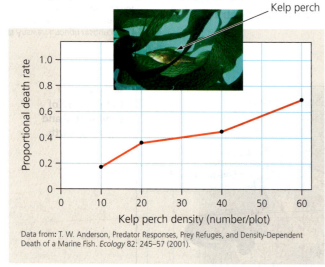

Data from: T. W. Anderson, Predator Responses, Prey Refuges, and Density-Dependent Death of a Marine Fish. *Ecology* 82: 245–57 (2001).

(b) Increasing death rate of kelp perch (inset) with increasing population density.

▼ **Figure 19.9** Space as a limiting resource in a population of gannets.

Density-Independent Factors

In many natural populations, abiotic factors such as weather may limit or reduce population size well before other limiting factors become important. A population-limiting factor whose intensity is unrelated to population density is called a **density-independent factor**. If we look at the growth curve of such a population, we see something like exponential growth followed by a rapid decline rather than a leveling off. **Figure 19.10** shows this effect for a population of aphids, insects that feed on the sugary sap of plants. These and many other insects undergo virtually exponential growth in the spring and then rapidly die off when the weather turns hot and dry in the summer. A few individuals may remain, allowing population growth to resume if favorable conditions return. In some populations of insects—many mosquitoes and grasshoppers, for instance—the adults die off entirely, leaving behind eggs that will initiate population growth the following year. In addition to seasonal changes in the weather, environmental disturbances such as fire, floods, and storms can affect a population's size regardless of its density.

Over the long term, most populations are probably regulated by a complex interaction of density-dependent and density-independent factors. Although some populations remain fairly stable in size and are presumably close to a carrying capacity that is determined by biotic factors such as competition or predation, most populations for which we have long-term data do fluctuate.

Population Cycles

Some populations of insects, birds, and mammals undergo dramatic fluctuations in density with remarkable regularity. "Booms" characterized by rapid exponential growth are followed by "busts," during which the population falls back to a minimal level. Lemmings, small rodents that live in the tundra, are a striking example. In lemming populations, boom-and-bust growth cycles occur every three to four years. Some researchers hypothesize that natural changes in the lemmings' food supply may be the underlying cause. Another hypothesis is that stress from crowding during the "boom" triggers hormonal changes that may cause the "bust" by reducing birth rates.

Population cycles of the snowshoe hare and the lynx illustrate interconnections within biological systems (Figure 19.11). The lynx is one of the main predators of the snowshoe hare in the far northern forests of Canada and Alaska. About every ten years, both hare and lynx populations show a rapid increase followed by a sharp decline. What causes these boom-and-bust cycles? Since ups and downs in the two populations seem to almost match each other on the graph, does this mean that changes in one directly affect the other? For the hare cycles, there are three main hypotheses. First, cycles may be caused by winter food shortages that result from overgrazing. Second, cycles may be due to predator-prey interactions. Many predators other than lynx, such as coyotes, foxes, and great-horned owls, eat hares, and together these predators might overexploit their prey. Third, cycles may be affected by a combination of food resource limitation and excessive predation. Recent field studies support the hypothesis that the ten-year cycles of the snowshoe hare are largely driven by excessive predation but are also influenced by fluctuations in the hare's food supplies. Long-term studies are the key to unraveling the complex causes of such population cycles. ✅

✅ CHECKPOINT

List some density-dependent factors that limit population growth.

■ *Answer: food and nutrient limitations, insufficient space for territories or nests, increase in disease and predation, accumulation of toxins*

▼ Figure 19.10 Weather change as a density-independent factor limiting growth of an aphid population.

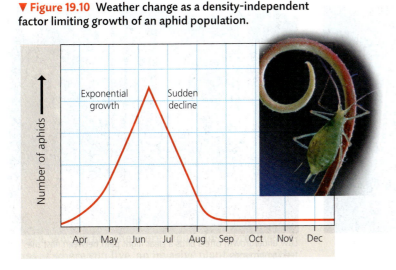

▼ Figure 19.11 Population cycles of the snowshoe hare and the lynx.

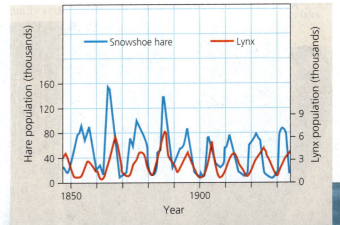

the Saint Lawrence Seaway. They quickly spread to rivers and lakes beyond the Great Lakes region and are now widely distributed. The economic damage caused by zebra mussels results from their astronomically large populations. Adult zebra mussels form thick layers that clog pipes and the water intakes of cities, power plants, and factories. As many as 70,000 individuals can be found in 1 square meter! In addition, these tiny molluscs frequently attach to larger native bivalves, interfering with their ability to feed and reproduce, and they compete with native species for food and spaces. As a result, populations of native bivalves have declined in areas colonized by zebra mussels.

For a non-native organism like zebra mussels to become invasive, the biotic and abiotic factors of the new environment must be compatible with the organism's needs and tolerances. For example, Burmese pythons set loose in South Florida—either accidentally released by damaging storms or deliberately freed by disenchanted pet owners—found a hot, humid climate similar to their native area. Prey such as birds, mammals, reptiles, and amphibians are readily available, especially in the Everglades. As a result, South Florida is now home to a burgeoning population of the giant reptiles **(Figure 19.15)**. Burmese pythons released in a less favorable environment might survive for a short time but would not be able to establish a population. ✅

▼ **Figure 19.15** **A Burmese python.** Researchers from the U.S. Geological Survey captured this enormous snake in the Everglades in 2017.

✅ **CHECKPOINT**

What distinguishes invasive species from organisms that are introduced to non-native habitats but do not become invasive?

■ *Answer: Invasive species spread far from where they are introduced, and they cause environmental or economic damage.*

Biological Control of Pests

The absence of biotic factors that limit population growth, such as pathogens, predators, or herbivores, may contribute to the success of invasive species. Accordingly, efforts to eliminate or control these troublesome organisms often focus on **biological control**, the intentional release of a natural enemy to attack a pest population. Agricultural researchers have long been interested in identifying potential biological agents to control insects, weeds, and other organisms that reduce crop yield.

Biological control has been effective in numerous instances, especially with invasive insects and plants. In one classic success story, beetles were brought in to combat St. John's wort, a perennial (long-lived) European weed that invaded the western United States. By the 1940s, St. John's wort (also known as Klamath weed) had overgrown millions of acres of rangeland and pasture, leaving few edible plants for grazing livestock. Researchers imported leaf beetles from the plant's native region that feed exclusively on St. John's wort. The shiny, pea-sized insects reduced the weed to less than 5% of its former abundance, restoring the land's value to ranchers.

One potential pitfall of biological control is the danger that an imported control agent may be as invasive as its target. One cautionary tale comes from introducing the mongoose **(Figure 19.16)** to control rats. Rats that originated in India and northern Asia were accidentally transported around the world and became invasive in many places. For sugarcane growers, rat infestation meant massive crop damage. Cane planters imported the small Indian mongoose, a fierce little carnivore, to deal with the problem. In time, mongooses were introduced to dozens of natural habitats, including all of the largest Caribbean and Hawaiian islands—and became invasive themselves. Mongooses are not picky eaters, and they have voracious appetites. On island after island, populations of reptiles, amphibians, and ground-nesting birds have declined or vanished as mongoose populations have grown and spread. They also prey on domestic poultry and ruin crops, costing millions of dollars a year. Clearly, rigorous research is needed to assess the safety and effectiveness of potential biological control agents.

◀ **Figure 19.16** A small Indian mongoose.

Can Fences Stop Cane Toads?

BACKGROUND

In 1935, Australian agricultural experts imported the cane toad **(Figure 19.17a)** to control invasive beetles that were destroying sugar cane, a major crop. Unfortunately, the beetles were usually underground or in the air, safe from being eaten by cane toads. Instead of recruiting an ally in the battle against beetles, the agriculturalists had introduced a super-invader. From the original introduction of 102 cane toads, the amphibians spread rapidly. They now occupy more than 1 million km^2 (over 386,000 mi^2) of Australia, an area larger than Texas and Arizona, with a total population estimated at 1.5 billion. The toads' voracious appetite threatens many native species. However, the greatest danger is to predators, such as snakes, lizards, and marsupials, which risk death when they try to eat the toxic toads.

Humans have unwittingly aided the cane toad invasion. To survive the dry season on the semi-arid plains of Australia's Northern Territory, the toads gather near permanent water sources. Cattle farmers in this region have expanded suitable cane toad habitat by providing additional watering holes—artificial stock ponds built to sustain their livestock through the dry season. The stock ponds are rings of packed dirt holding water that can be directed to troughs. Researchers hypothesized that blocking access to the ponds could limit the cane toad invasion.

METHOD

During the dry season, researchers selected nine stock ponds that were at least 40 yards apart. Three of the ponds were left as unfenced controls **(Figure 19.17b)**. Three ponds were surrounded by closed, toad-proof fences made of stakes and mesh from the ground to several feet above water. Three ponds had open fences (raised up so that toads could hop under the fence to access water, but otherwise identical to the closed fences). Researchers removed the toads from all of the ponds. Then, the researchers waited to see if the ponds in the immediate vicinity of each would be reinvaded by toads. On day 70 (early wet season), the researchers counted live and dead toads in and around each pond. They repeated this on day 190 (early dry season) and day 365 (late dry season).

RESULTS

There were almost no dead toads at either the unfenced or open-fence ponds at any time in the year, but toad mortality was high near the closed-fenced ponds late in the dry season **(Figure 19.17c)**. Almost no live toads were found at the closed-fenced ponds, while there were 30—100 toads per survey at the open-fence and unfenced ponds. The closed-fenced ponds acted as ecological traps: Toads were attracted to the scent of water but died when they could not get inside fences to access the ponds. Enclosing stock ponds, the researchers concluded, could protect the dry regions of Australia from cane toad invasion.

▼ **Figure 19.17** An experiment to test a strategy for limiting the spread of cane toads.

Data from: M. Letnic et al. Restricting Access to Invasion Hubs Enables Sustained Control of an Invasive Vertebrate. Journal of Applied Ecology 52(2): 341– 47 (2015).

(a) The cane toad has spread across large parts of Australia.

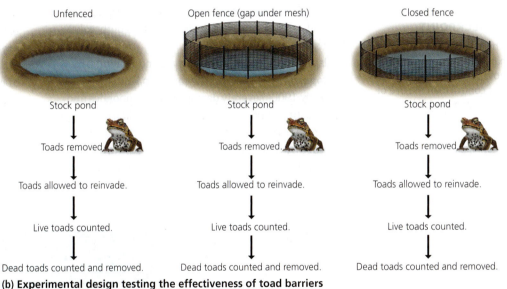

(b) Experimental design testing the effectiveness of toad barriers

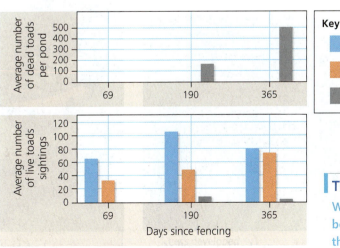

(c) Influence of fencing on toad reinvasions of ponds

Key
Unfenced
Open fence
Closed fence

Thinking Like a Scientist

Why did researchers bother to put up fences that toads could go under?

For the answer, see Appendix D.

Life History Traits as Adaptations

Life history traits are shaped by evolutionary adaptation. Most populations probably fall between the extreme opportunistic life histories (reach sexual maturity rapidly; produce many offspring; little or no parental care) of many insects and the equilibrial life histories (develop slowly and produce few, well-cared-for offspring) of many larger-bodied species.

Population Growth Models

The Exponential Population Growth Model: The Ideal of an Unlimited Environment

Exponential population growth is the accelerating increase that occurs when growth is unlimited. The exponential model predicts that the larger a population becomes, the faster it grows.

The Logistic Population Growth Model: The Reality of a Limited Environment

Logistic population growth occurs when growth is slowed by limiting factors. The logistic model predicts that a population's growth rate will be low when the population size is either small or large and highest when the population is at an intermediate level relative to the carrying capacity.

Regulation of Population Growth

Over the long term, most population growth is limited by a mixture of density-independent factors, which affect the same percentage of individuals regardless of population size, and density-dependent factors, which intensify as a population increases in density. Some populations have regular boom-and-bust cycles.

Applications of Population Ecology

Conservation of Endangered Species

Endangered and threatened species are characterized by very small population sizes. One approach to conservation is identifying and attempting to supply the critical combination of habitat factors needed by the population.

Sustainable Resource Management

Resource managers apply principles of population ecology to help determine sustainable harvesting practices.

Invasive Species

Invasive species are non-native organisms that spread far beyond their original point of introduction and cause environmental and economic damage. Typically, invasive species have an opportunistic life history pattern.

Biological Control of Pests

Biological control, the intentional release of a natural enemy to attack a pest population, is sometimes effective against invasive species. However, prospective control agents can themselves become invasive.

Integrated Pest Management

Crop scientists have developed integrated pest management (IPM) strategies—combinations of biological, chemical, and cultural methods—to deal with agricultural pests.

Human Population Growth

The History of Human Population Growth

The human population grew rapidly during the 1900s and is currently more than 7.5 billion. A shift from high birth and death rates to low birth and death rates has lowered the rate of growth in more developed countries. In developing nations, death rates have dropped, but birth rates are still high.

Age Structures

The age structure of a population affects its future growth. The wide base of the age structure of Mexico in 1990—the 0–14 age-group—predicts continued population growth in the next generation. Population momentum is the continued growth that occurs after a population's high fertility rate has been reduced to replacement rate; it is a result of girls in the 0–14 age-group reaching their childbearing years. Age structures may also indicate social and economic trends, as in the age structure on the right below.

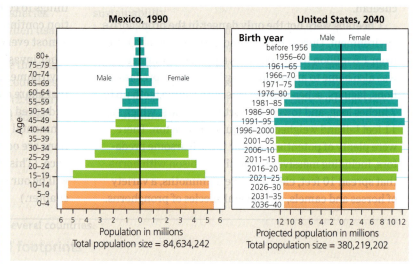

Our Ecological Footprint

An ecological footprint represents the amount of land and water needed to produce the resources used by an individual or nation. There is a huge disparity between resource consumption by individuals in more developed and less developed nations.

Mastering Biology

For practice quizzes, BioFlix animations, MP3 tutorials, video tutors, and more study tools designed for this textbook, go to Mastering Biology™

SELF-QUIZ

1. What two values would you need to know to figure out the human population density of your community?

2. If members of a species produce a large number of offspring but provide minimal parental care, then a Type _____ survivorship curve is expected. In contrast, if members of a species produce few offspring and provide them with long-standing care, then a Type _____ survivorship curve is expected.

3. Use this graph of the idealized exponential and logistic growth curves to complete the following.

   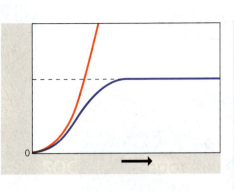

   a. Label the axes and curves on the graph.
   b. What does the dotted line represent?
   c. For each curve, indicate and explain where population growth is the most rapid.
   d. Which of these curves better represents global human population growth?

4. Which of the following describes the effects of a density-dependent limiting factor?
   a. A forest fire kills all the pine trees in a patch of forest.
   b. Early rainfall triggers the explosion of a locust population.
   c. Drought decimates a wheat crop.
   d. Rabbits multiply, and their food supply begins to dwindle.

5. Which life history pattern is typical of invasive species?

6. Skyrocketing growth of the human population since the beginning of the Industrial Revolution appears to be mainly a result of
   a. migration to thinly settled regions of the globe.
   b. better nutrition boosting the birth rate.
   c. a drop in the death rate due to better nutrition and health care.
   d. the concentration of humans in cities.

7. According to data on ecological footprints,
   a. the carrying capacity of the world is 10 billion.
   b. Earth's resources are sufficient to sustain future generations at current levels of consumption.
   c. the ecological footprint of the United States is more than twice the world average.
   d. nations with the largest ecological footprints have the fastest population growth rates.

For answers to the Self-Quiz, see Appendix D.

IDENTIFYING MAJOR THEMES

For each statement below, identify which major theme is evident (the relationship of structure to function, information flow, pathways that transform energy and matter, interactions within biological systems, or evolution) and explain how the statement relates to the theme. If necessary, review the themes (Chapter 1) and review the examples highlighted in blue in this chapter.

8. The mussel secretes a combination of molecules that hardens into exceptionally sticky threads.

9. The lynx is one of the main predators of the snowshoe hare in the far northern forests of Canada and Alaska. About every ten years, both lynx and hare populations show a rapid increase followed by a sharp decline.

10. Life history traits, like anatomical features, are shaped by trade-offs that balance the demands of reproduction and survival, both of which determine fitness.

For answers to Identifying Major Themes, see Appendix D.

THE PROCESS OF SCIENCE

11. One hypothesis suggests that population cycles observed in snowshoe hares are caused by winter food shortages. Describe an experiment to test this hypothesis.

12. **Interpreting Data** The graph below shows data for population trends in Mexico from 1890 to 2016, with projected trends for 2016–2050. How has Mexico's rate of population growth changed over this time period? How is it expected to change by midcentury? Describe Mexico's projected age structure in 2050.

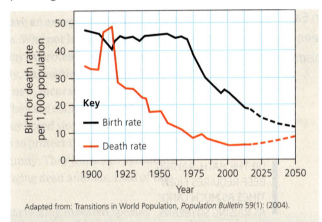

Adapted from: Transitions in World Population, *Population Bulletin* 59(1): (2004).

BIOLOGY AND SOCIETY

13. Experts are enlisting the help of citizen scientists in the battle against invasive species. For example, in 2014, the Florida Fish and Wildlife Commission released a phone app for reporting lionfish sightings, data that will help target fish for removal. A phone app to identify and report the locations of invasive species at more than 100 sites across the United States and Canada is available at www.whatsinvasive.org. Other citizen science projects are listed at www.birds.cornell.edu/citscitoolkit/projects/find/projects-invasive-species. Using these sites or other resources, determine which invasive species are of greatest concern in your area. Learn how to identify them, and find out what control measures are being taken.

14. Affluent, more developed nations consume a disproportionate amount of Earth's resources and produce a disproportionate amount of wastes, including carbon dioxide and other greenhouse gases that are causing global climate change. The consequences of climate change include rising sea level, drought, and extreme weather events. Do you think affluent countries should be responsible for helping poorer, less developed countries bear the financial burden of these consequences?

The Nitrogen Cycle

As an ingredient of proteins and nucleic acids, nitrogen is essential to the structure and functioning of all organisms. Nitrogen has two abiotic reservoirs, the atmosphere and the soil. The atmospheric reservoir is huge; almost 80% of the atmosphere is nitrogen gas (N_2). However, plants cannot use nitrogen gas. Like someone dying of thirst in the middle of the ocean because they can't remove salt from the water, plants are surrounded by nitrogen, a critical resource, in an unusable form. The process of **nitrogen fixation** converts gaseous N_2 to ammonia (NH_3). NH_3 then picks up another H^+ to become ammonium (NH_4^+), which plants can assimilate. Most of the nitrogen available in natural ecosystems comes from biological fixation performed by certain bacteria. Without these organisms, the reservoir of usable soil nitrogen would be extremely limited.

Figure 20.34 illustrates the actions of two types of nitrogen-fixing bacteria. **1** Some bacteria live symbiotically in the roots of certain species of plants, supplying their hosts with a direct source of usable nitrogen. The largest group of plants with this mutualistic relationship is the legumes, including peanuts and soybeans. Many farmers improve soil fertility by alternating crops of legumes, which add nitrogen to the soil, with plants such as corn that require nitrogen fertilizer. **2** Free-living bacteria in soil or water fix nitrogen, resulting in NH_4^+. **3** After nitrogen is fixed, some of the ammonium is taken up and used by plants. **4** Nitrifying bacteria in the soil also convert some of the ammonium to nitrate (NO_3^-),

5 which is more readily acquired by plants. Plants use this nitrogen to make molecules such as amino acids, which are then incorporated into proteins.

6 When an herbivore (represented here by a rabbit) eats a plant, it digests the proteins into amino acids and then uses the amino acids to build the proteins it needs. Higher-order consumers get nitrogen from the organic molecules of their prey. Because animals form nitrogen-containing waste products during protein metabolism, consumers excrete some nitrogen into soil or water. The urine that rabbits and other mammals excrete contains urea, a nitrogen compound that is widely used as fertilizer.

Organisms that are not consumed eventually die and become detritus, which is decomposed by bacteria and fungi. **7** The decomposition of organic compounds releases ammonium into the soil, replenishing that abiotic reservoir. Under low-oxygen conditions, however, **8** soil bacteria known as denitrifying bacteria strip the oxygen atoms from nitrates, releasing N_2 back into the atmosphere and depleting the soil of usable nitrogen.

Human activities are disrupting the nitrogen cycle by adding extra nitrogen to the biosphere. Combustion of fossil fuels and modern agricultural practices are two major sources of nitrogen. Many farmers apply enormous amounts of synthetic nitrogen fertilizer. However, less than half that fertilizer is actually used by the crop plants. Some nitrogen escapes to the atmosphere, where it forms nitrous oxide (N_2O), a gas that contributes to global warming. And as you'll learn next, nitrogen fertilizers also pollute aquatic systems. ✓

☑ **CHECKPOINT**

What are the abiotic reservoirs of nitrogen? In what form does nitrogen occur in each reservoir?

■ *Answer: atmosphere: N_2; soil: NH_4^+ and NO_3^-*

▼ **Figure 20.34** The nitrogen cycle.

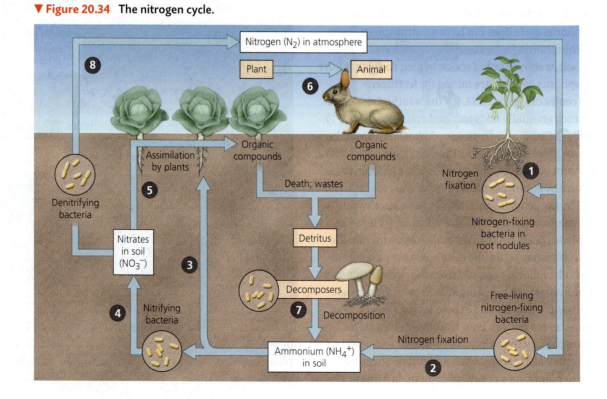

Nutrient Pollution

Low nutrient levels, especially of phosphorus and nitrogen, often limit the growth of producers, such as algae and cyanobacteria in aquatic ecosystems. Nutrient pollution occurs when human activities add excess amounts of these chemicals to aquatic ecosystems, increasing populations of producers and disrupting the ecosystem in which they live.

In many areas, phosphate and nitrogen pollution comes from the large amount of inorganic fertilizers routinely applied to crops, lawns, and golf courses and runoff of animal waste from livestock feedlots (where hundreds of animals are penned together). Phosphates are also a common ingredient in dishwasher detergents, making the outflow from sewage treatment facilities—which also contains phosphorus from human waste—a major source of phosphate pollution. Pollution may also come from sewage treatment facilities when extreme conditions (such as unusual storms) or malfunctioning equipment prevents them from meeting water quality standards. Nutrient pollution of lakes and rivers results in heavy growth of algae and cyanobacteria (Figure 20.35).

In an example of how far-reaching this problem can be, nutrient runoff from Midwestern farm fields has been linked to an annual summer "dead zone" in the Gulf of Mexico that is nearly devoid of animals (Figure 20.36). Vast algal blooms extend outward from where the Mississippi River deposits its nutrient-laden waters. It might seem that producers would add oxygen to the water through photosynthesis, allowing consumers to thrive. However, producers consume oxygen when they do not get sufficient light, such as at night or when they are at the bottom of thick mats of algae. Algae have a short life span. As the algae die, bacteria decompose the huge quantities of biomass. Cellular respiration by bacteria diminishes the supply of dissolved oxygen over an area that ranges from 13,000 km^2 to 22,000 km^2 (from roughly the size of Connecticut to the size of New Jersey). Oxygen depletion disrupts benthic (aquatic, bottom dwelling) communities, displacing fish and invertebrates that move along or near the substrate and killing organisms that are attached to it. More than 400 recurring or permanent coastal dead zones totaling approximately 245,000 km^2 (about the area of Michigan) have been documented worldwide. Some algal blooms are caused by toxic algae, greatly increasing the ecological and economic damage. ✅

▼ **Figure 20.35** Producer growth in Lake Erie resulting from nutrient pollution.

▼ **Figure 20.36** The Gulf of Mexico dead zone.

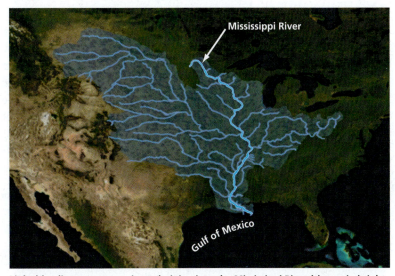

Light blue lines represent rivers draining into the Mississippi River (shown in bright blue). Nitrogen runoff carried by these rivers ends up in the Gulf of Mexico. In the images below, red and orange indicate high concentrations of phytoplankton. Bacteria feeding on dead phytoplankton deplete the water of oxygen, creating a "dead zone."

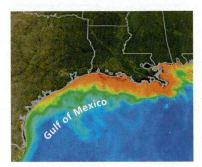

Summer

Winter

✅ CHECKPOINT

How does the excessive addition of mineral nutrients to a pond eventually result in the loss of most fish in the pond?

Answer: Excessive mineral nutrients initially cause population explosions of producers and the organisms that feed on them. The respiration of so much life, especially of the microbes decomposing all the organic refuse, consumes most of the pond's oxygen, which the fish require.

Conservation and Restoration Biology

As we have seen in this unit, many of the environmental problems facing us today have been caused by human enterprises. But the science of ecology is not just useful for telling us how things have gone wrong. Ecological research is also the foundation for finding solutions to these problems and for reversing the negative consequences of ecosystem alteration. Thus, we end the ecology unit by highlighting these beneficial applications of ecological research.

Conservation biology is a goal-oriented science that seeks to understand and counter the loss of biodiversity. Conservation biologists recognize that biodiversity can be sustained only if we maintain the ecosystems and processes that shaped the evolution of the species. Thus, the goal is not simply to preserve individual species but to sustain ecosystems, where natural selection can continue to function, and to maintain the genetic variability on which natural selection acts. The expanding field of **restoration ecology** uses ecological principles to develop methods of returning degraded areas to their natural state.

Biodiversity "Hot Spots"

☑ **CHECKPOINT**

What is a biodiversity hot spot?

■ *Answer: a relatively small area with a disproportionately large number of species, including endangered species.*

Conservation biologists are applying their understanding of population, community, and ecosystem dynamics in establishing parks, wilderness areas, and other legally protected nature reserves. Choosing locations for these protected zones often focuses on **biodiversity hot spots**. These relatively small areas have a large number of endangered and threatened species and an exceptional concentration of **endemic species**, species that are found nowhere else. Together, the "hottest" of Earth's biodiversity hot spots, shown in **Figure 20.37**, total less than 1.5% of Earth's land surface but are home to a third of all species of plants and vertebrates. For example, all of the many species of lemurs on Earth—more than 50 species—are endemic to Madagascar, a large island off the eastern coast of Africa. In fact, almost all of the mammals, reptiles, amphibians, and plants that inhabit Madagascar are endemic. There are also hot spots in aquatic ecosystems, such as certain river systems and coral reefs. Because biodiversity hot spots can also be hot spots of extinction, they rank high on the list of areas demanding strong global conservation efforts.

Concentrations of species provide an opportunity to protect many species in very limited areas. We will explore this more in the Evolution Connection essay. However, preserving biodiversity requires more than protecting the hot spots, which are usually defined using only the most noticeable organisms, especially vertebrates and plants. Invertebrates and microorganisms are often overlooked. Furthermore, species endangerment is a global problem, and focusing on hot spots should not detract from efforts to conserve habitats and species diversity in other areas. Finally, even the protection of a nature reserve does not shield organisms from the effects of climate change or other threats, such as invasive species or infectious disease. To stem the tide of biodiversity loss, we will have to address environmental problems globally as well as locally. ☑

▶ **Figure 20.37** Earth's terrestrial (purple) and marine (red) biodiversity hot spots.

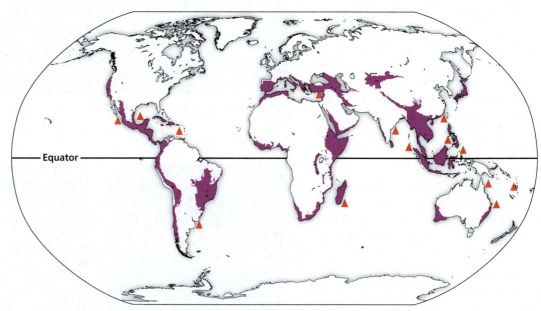

Adapted from: Critical Ecosystem Partnership Fund, *Annual Report 2014*, Conservation International.

Conservation at the Ecosystem Level

In the past, most conservation efforts focused on saving individual species, and this work continues. (You have already learned about one example, the red-cockaded woodpecker; see Figure 19.12.) More and more, however, conservation biology aims at sustaining the biodiversity of entire communities and ecosystems. On an even broader scale, conservation biology considers the biodiversity of whole landscapes. Ecologically, a **landscape** is a regional assemblage of interacting ecosystems, such as an area with forest, adjacent fields, wetlands, streams, and streamside habitats. **Landscape ecology** is the application of ecological principles to the study of land-use patterns. Its goal is to make ecosystem conservation a functional part of the planning for land use.

▼ **Figure 20.38** **Edges between ecosystems within a landscape.**

Natural edges. Forests border grassland ecosystems in Lake Clark National Park, Alaska.

Edges created by human activity. Forest edges surround farmland in the Cotswolds region of south central England.

Edges between ecosystems are prominent features of landscapes, whether natural or altered by human activity **(Figure 20.38)**. Edges have their own sets of physical conditions—such as soil type and surface features—that differ from the ecosystems on either side of them. Edges also may have their own type and amount of disturbance. For instance, the edge of a forest often has more blown-down trees than a forest interior because the edge is less protected from strong winds. Because of their specific physical features, edges also have their own communities of organisms. Some organisms thrive in edges because those organisms require resources found only there. For instance, white-tailed deer browse on woody shrubs found in edges between woods and fields, so deer populations often expand as edges do when forests are logged or fragmented by development.

Edges can have both positive and negative effects on biodiversity. A recent study in a tropical rain forest in western Africa indicated that natural edge communities are important sites of speciation. On the other hand, landscapes where human activities have produced edges often have fewer species.

Another important landscape feature, especially where habitats have been severely fragmented, is the **movement corridor**, a narrow strip or series of small clumps of suitable habitat connecting otherwise isolated patches. In places where there is extremely heavy human impact, artificial corridors are sometimes constructed **(Figure 20.39)**. Corridors can promote dispersal of individuals from a population and thereby help sustain populations. Corridors are especially important to species that migrate between different habitats seasonally. But a corridor can also be harmful—as, for example, in the spread of disease, especially among small subpopulations in closely situated habitat patches. ✅

✅ CHECKPOINT

How is a landscape different from an ecosystem?

Answer: A landscape is more inclusive in that it consists of several interacting ecosystems in the same region.

▼ **Figure 20.39** **An artificial corridor.** This bridge over a road provides an artificial corridor for animals in Canada.

THE PROCESS OF SCIENCE | **Importance of Biodiversity**

Does Biodiversity Protect Human Health?

BACKGROUND

As humans extract natural resources, build roads, and convert forests to agricultural land, they are fragmenting many landscapes and decreasing biodiversity in those communities. Species that require large areas to survive are disappearing as their habitat is destroyed. Species that thrive in small fragments are increasing in numbers. The loss of biodiversity due to fragmentation can be seen in the northern forests of the United States, where populations of the white-footed mouse (*Peromyscus leucopus*) are exploding **(Figure 20.40a)**. White-footed mice carry Lyme disease.

BIODIVERSITY CAN PROTECT YOU FROM CATCHING LYME DISEASE FROM THIS TICK.

In humans, Lyme disease can be fatal if it spreads to the heart or nervous system. The bacterium that causes Lyme disease (*Borrelia burgdorferi*) is carried by black-legged ticks (*Ixodes scapularis*) in the northern forests of the United States and in similar biomes worldwide. The tick, a parasite, requires more than one host species to complete its life cycle **(Figure 20.40b)**. Adult ticks feed on the blood of deer and humans, then females lay eggs. Larvae hatch and feed on the blood of an intermediate host. Then the larvae grow into nymphs. Once a nymph, a tick attaches to another intermediate host. As the tick feeds, the Lyme disease bacteria is spread, either from the tick to the host or vice versa. Host species differ in their ability to infect ticks with bacteria. White-footed mice are particularly likely to infect ticks. The higher biodiversity found in large forest fragments would potentially help break the chain of disease transmission because more biologically diverse communities include many intermediate hosts, for example, birds and lizards, that do not spread the bacteria, even if they are infected.

In recent years, the number of cases of Lyme disease in the United States has spiked. To find out if fragmentation is increasing the risk of Lyme disease, researchers conducted a study to test the hypothesis that smaller forest fragments would have an increased density of infected ticks.

METHOD

Researchers studied 14 maple forest fragments of different sizes in southeastern New York. They chose a county with an extremely high number of Lyme disease cases. They identified fragments from aerial photographs and collected ticks by dragging cloths along the ground in a standard pattern. A subset of all the ticks collected was analyzed for presence of Lyme disease bacteria.

RESULTS

The smallest forest fragments had much higher densities of infected black-legged tick nymphs **(Figure 20.40c)**. In the small fragments, white-footed mice are extremely common, and other intermediate hosts may be rare or absent. White-footed mice are also the most likely intermediate host in the landscapes studied to infect ticks with Lyme disease. These findings suggest that high biodiversity provides people some protection from Lyme disease. Future studies conducted on other parasite-transmitted diseases may show a similar pattern.

▼ **Figure 20.40** Habitat fragmentation, decreased biodiversity, and Lyme disease transmission by ticks. Data from: B. F. Allan et al., Effect of Forest Fragmentation on Lyme Disease Risk. *Conservation Biology* 17(1): 267–72 (2003), Figure 1c, p. 271.

(a) The white-footed mouse spreads Lyme disease in forest communities.

Thinking Like a Scientist

Researchers ensured that the fragments they studied were a minimum distance apart. Why was this important?

For answer, see Appendix D.

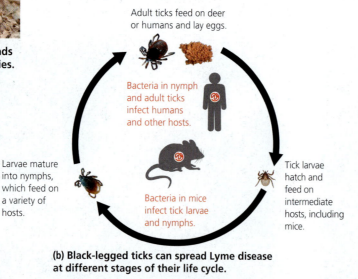

Adult ticks feed on deer or humans and lay eggs.

Bacteria in nymph and adult ticks infect humans and other hosts.

Tick larvae hatch and feed on intermediate hosts, including mice.

Bacteria in mice infect tick larvae and nymphs.

Larvae mature into nymphs, which feed on a variety of hosts.

(b) Black-legged ticks can spread Lyme disease at different stages of their life cycle.

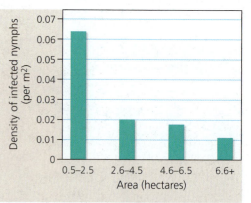

(c) Larger forest fragments have a lower density of infected tick nymphs. (A hectare is an area of 100 × 100 m.)

Restoring Ecosystems

One of the major strategies in restoration ecology is **bioremediation**, the use of living organisms to detoxify polluted ecosystems. For example, bacteria have been used to clean up old mining sites and oil spills. Researchers are also investigating the potential of using plants to remove toxic substances such as heavy metals and organic pollutants (for example, PCBs) from contaminated soil **(Figure 20.41)**.

Some restoration projects have the broader goal of returning ecosystems to their natural state. Such projects may involve replanting vegetation, fencing out non-native animals, or removing dams that restrict water flow. Hundreds of restoration projects are currently under way in the United States. One of the most ambitious endeavors is the Kissimmee River Restoration Project in south central Florida.

The Kissimmee River was once a shallow, meandering river that wound its way from Lake Kissimmee southward into Lake Okeechobee. During about half of the year, the river flooded into a wide floodplain, creating wetlands that provided habitat for large numbers of birds, fishes, and invertebrates. And as the floods deposited the river's load of nutrient-rich silt on the floodplain, they boosted soil fertility and maintained the water quality of the river.

Between 1962 and 1971, the U.S. Army Corps of Engineers converted the 166-km wandering river to a straight canal 9 m deep, 100 m wide, and 90 km long. This project, designed to allow development on the floodplain, drained approximately 31,000 acres of wetlands, with significant negative impacts on fish and wetland bird populations. Without the marshes to help filter and reduce agricultural runoff, the river transported phosphates and other excess nutrients from Lake Okeechobee farther south to the Everglades.

The restoration project involves removing water-control structures such as dams, reservoirs, and channel modifications and filling in about 35 km of the canal **(Figure 20.42)**. The first phase of the restoration project was completed in 2004. The time line for completion has been extended repeatedly, but the latest end date is 2019. The photo in Figure 20.42 shows a section of the Kissimmee canal that has been plugged, diverting flow into the remnant river channels. Birds and other wildlife have returned in unexpected numbers to the 11,000 acres of wetlands that have been restored. The marshes are filled with native vegetation, and game fishes again swim in the river channels. ☑

☑ **CHECKPOINT**

The water in the Kissimmee River eventually flows into the Everglades. How will the Kissimmee River Restoration Project affect water quality in the Everglades ecosystem?

■ *Answer: Wetlands filter agricultural runoff, which prevents nutrient pollution from flowing downstream. By restoring this ecosystem service, the project will improve water quality in the Everglades.*

▼ **Figure 20.41 Bioremediation using plants.** A researcher from the U.S. Department of Agriculture investigates the use of canola plants to reduce toxic levels of selenium in contaminated soil.

▼ **Figure 20.42** **The Kissimmee River Restoration Project.**

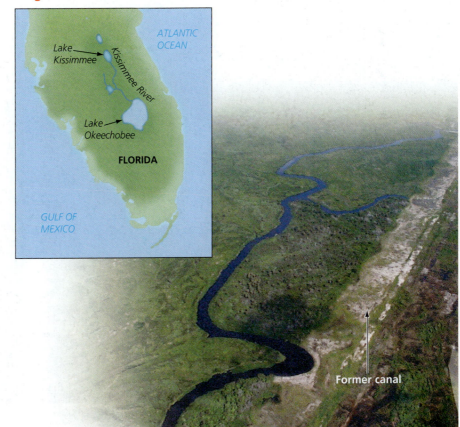

The Goal of Sustainable Development

As the world population grows and becomes more affluent, the demand for the "provisioning" services of ecosystems, such as food, wood, and water, is increasing. Although these demands are currently being met for much of the world, they are satisfied at the expense of other critical ecosystem services, such as climate regulation and protection against natural disasters. Clearly, we have set ourselves and the rest of the biosphere on a precarious path into the future. How can we achieve **sustainable development**—development that meets the needs of people today without limiting the ability of future generations to meet their needs?

Many nations, scientific associations, corporations, and private foundations have embraced the concept of sustainable development. The Ecological Society of America, the world's largest organization of ecologists, endorses a research agenda called the Sustainable Biosphere Initiative. The goal of this initiative is to acquire the ecological information necessary for the responsible development, management, and conservation of Earth's resources. The research agenda includes the search for ways to sustain the productivity of natural and artificial ecosystems and studies of the relationship between biological diversity, global climate change, and ecological processes.

Sustainable development depends on more than continued research and application of ecological knowledge. It also requires that we connect the life sciences with the social sciences, economics, and humanities. Conservation of biodiversity is only one side of sustainable development; the other side is improving the human condition. Public education and the political commitment and cooperation of nations are essential to the success of this endeavor.

An awareness of our unique ability to alter the biosphere and jeopardize the existence of other species, as well as our own, may help us choose a path toward a sustainable future. The risk of a world without adequate natural resources for all its people is not a vision of the distant future. It is a prospect for your children's lifetime, or perhaps even your own. But although the current state of the biosphere is grim, the situation is far from hopeless. Now is the time to take action by aggressively pursuing greater knowledge about the diversity of life on our planet and by joining with others in working toward long-term sustainability **(Figure 20.43)**. ✅

✅ CHECKPOINT

What is meant by sustainable development?

■ *Answer: development that meets current needs while ensuring an adequate supply of natural and economic resources for future generations*

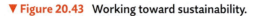

▼ **Figure 20.43** **Working toward sustainability.**

Students at the University of Virginia sorted trash from dumpsters to promote recycling. Recycling just one aluminum can saves enough energy to power a laptop for five hours.

A student at California State University, Fresno, pulls weeds in a plot of mustard and kale plants at the university's organic farm. Weeding by hand is more sustainable than herbicide use.

Saving the Hot Spots

As we have learned in this chapter, islands can be hot spots of both diversity and extinction. For example, the island of Guam was once home to several endemic bird species that are now extinct in the wild **(Figure 20.44)**. Guam's story is not unique. Geographical separation often leads to evolution of new species (see Figure 14.7). Different conditions and random events shape species in ways that are specific to their islands. Eventually, endemic species and subspecies evolve.

On an island with no predators, investing in mechanical, physiological, or behavioral strategies to avoid predation is a waste. Random mutations that decrease genes for anti-predation traits actually leave individuals on the island better adapted to their environment. These individuals produce more offspring than others in their population. In fact, on such islands, there is no selection pressure to even maintain fear responses to predators. However, when predators arrive, native species have no defenses and may go extinct before any evolve. With no native competitors or predators to control their numbers, invasive predators can devastate island communities. This happened on Guam, which has no native snakes.

When one or more brown treesnakes hitchhiked to Guam in cargo soon after World War II, no one noticed. The snake is nocturnal and tends to stay hidden. At first, people found a few shed snake skins and reported the occasional snake sighting. Then native species started disappearing. Eventually, about half of the native birds and several native species of bats and lizards became extinct in the wild. Forests that once rang with birdsong fell silent. Guam's biodiversity continues to be affected. The island has seen a decline in trees dependent on pollination by the missing bird species. Brown treesnakes also cause economic damage, eating poultry and causing frequent power outages by shorting out high-voltage wires.

The isolation of islands provides humans a unique opportunity to protect them. On Guam, government agencies use trained dogs to prevent snakes from leaving to infest other areas. Snake traps are placed along

fences. Brown treesnakes are unusually vulnerable to the common painkiller acetaminophen (Tylenol), so mouse carcasses laced with the drug are attached to tiny parachutes dropped from helicopters. A 2012 study showed that Guam's efforts resulted in treesnake reductions of over 80%. Someday, it may be possible to reintroduce into the wild some endemic bird species that are currently maintained in captive populations. Worldwide, over 230 island species have benefited from the eradication of pests.

Understanding the structure and function of communities and ecosystems provides us with tools for reversing some of the damage done by human activities. With this knowledge, we may ensure that we leave a diverse and healthy world for future generations.

Guam bridled white-eye

Guam kingfisher

Guam rail

(a) Some of Guam's endemic birds that are now extinct in the wild.

◄ **Figure 20.44** Brown treesnakes as a threat to Guam's biodiversity.

(b) Guam uses detection dogs, traps, and poisoned bait to control brown treesnakes.

Chapter Review

SUMMARY OF KEY CONCEPTS

Biodiversity

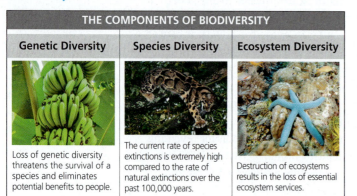

THE COMPONENTS OF BIODIVERSITY

Genetic Diversity	Species Diversity	Ecosystem Diversity
Loss of genetic diversity threatens the survival of a species and eliminates potential benefits to people.	The current rate of species extinctions is extremely high compared to the rate of natural extinctions over the past 100,000 years.	Destruction of ecosystems results in the loss of essential ecosystem services.

Causes of Declining Biodiversity

Habitat destruction is the leading cause of extinctions. Invasive species, overexploitation, and pollution are also significant factors.

Community Ecology

Interspecific Interactions

Populations in a community interact in a variety of ways that can be generally categorized as being beneficial (**+**), harmful (**–**), or neutral (**0**) to individuals of different species. Because **+/–** interactions (exploitation of one species by another species) may have such a negative impact on the individual that is harmed, defensive evolutionary adaptations are common.

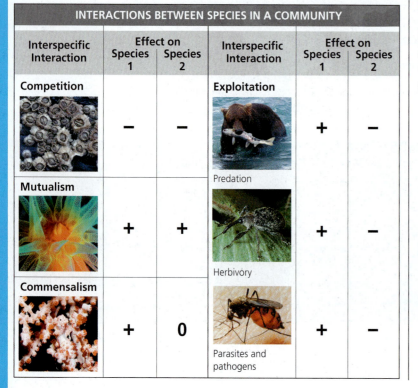

INTERACTIONS BETWEEN SPECIES IN A COMMUNITY

Interspecific Interaction	Effect on Species 1	Effect on Species 2	Interspecific Interaction	Effect on Species 1	Effect on Species 2
Competition	–	–	Exploitation	+	–
			Predation		
Mutualism	+	+		+	–
			Herbivory		
Commensalism	+	0		+	–
			Parasites and pathogens		

Trophic Structure

The trophic structure of a community defines the feeding relationships among organisms. In the process of biological magnification, toxins become more concentrated as they are passed up a food chain to the top predators.

Increasing PCB concentration

Species Diversity in Communities

Diversity within a community includes species richness and relative abundance of different species. A keystone species is a species that has a great impact on the composition of the community despite a relatively low abundance or biomass.

Disturbances and Succession in Communities

Disturbances are episodes that damage communities, at least temporarily, by destroying organisms or altering the availability of resources such as mineral nutrients and water. People are the most significant cause of disturbances today.

Ecological Succession

The sequence of changes in a community after a disturbance is called ecological succession. Primary succession occurs where a community arises in a virtually lifeless area with no soil. Secondary succession occurs where a disturbance has destroyed an existing community but left the soil intact.

Ecosystem Ecology

Energy Flow in Ecosystems

An ecosystem is a biological community and the abiotic factors with which the community interacts. Energy must be added continuously to an ecosystem, because as it flows from producers to consumers and

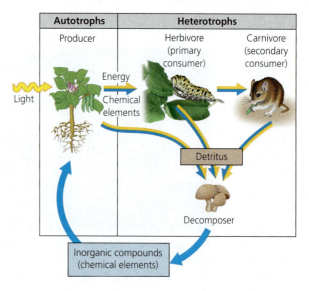

decomposers, it is constantly being lost. Chemical elements can be recycled between an ecosystem's living community and the abiotic environment. Trophic relationships determine an ecosystem's routes of energy flow and chemical cycling.

Primary production is the rate at which plants and other producers build biomass. Ecosystems vary considerably in their productivity. Primary production sets the spending limit for the energy budget of the entire ecosystem because consumers must acquire their organic fuels from producers. In a food chain, only about 10% of the biomass at one trophic level is available to the next, resulting in an energy pyramid.

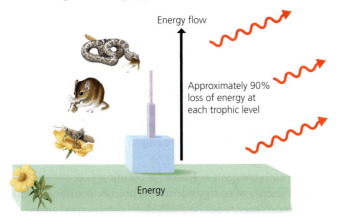

Energy flow

Approximately 90% loss of energy at each trophic level

Energy

When people eat producers instead of consumers, less photosynthetic production is required, which reduces the impact on the environment.

Chemical Cycling in Ecosystems

Biogeochemical cycles involve biotic and abiotic components. Each circuit has an abiotic reservoir through which the chemical cycles. Some chemical elements require "processing" by certain microorganisms before those chemical elements are available to plants as inorganic nutrients. A chemical's specific route through an ecosystem varies, depending on the element and the trophic structure of the ecosystem. Phosphorus is not very mobile and is cycled locally. Carbon and nitrogen spend part of their time in gaseous form and are cycled globally. Runoff of nitrogen and phosphorus, especially from agricultural land, causes algal blooms in aquatic ecosystems, lowering water quality and sometimes depleting the water of oxygen.

Conservation and Restoration Biology

Biodiversity "Hot Spots"

Conservation biology is a goal-oriented science that seeks to counter the loss of biodiversity. The front lines for conservation biology are biodiversity "hot spots," relatively small geographic areas that are especially rich in endangered species.

Conservation at the Ecosystem Level

Increasingly, conservation biology aims at sustaining the biodiversity of entire communities, ecosystems, and landscapes. Edges between ecosystems are prominent features of landscapes, with positive and negative effects on biodiversity. Corridors can promote dispersal and help sustain populations.

Restoring Ecosystems

In some cases, ecologists use microbes or plants to remove toxic substances, such as heavy metals, from ecosystems. Ecologists are working to revitalize some ecosystems by planting native vegetation, removing barriers to wildlife, and other means. The Kissimmee River Restoration Project is an attempt to undo the ecological damage done when the river was engineered into straight channels.

The Goal of Sustainable Development

Balancing the needs of people with the health of the biosphere, sustainable development has the goal of long-term prosperity of human societies and the ecosystems that support them.

Mastering Biology

For practice quizzes, BioFlix animations, MP3 tutorials, video tutors, and more study tools designed for this textbook, go to Mastering Biology™

SELF-QUIZ

1. Currently, the number one cause of biodiversity loss is _____.

2. According to the concept of competitive exclusion,
   a. two species cannot coexist in the same habitat.
   b. extinction or emigration is the only possible result of competitive interactions.
   c. intraspecific competition results in the success of the best-adapted individuals.
   d. two species cannot share exactly the same niche in a community.

3. The concept of trophic structure emphasizes the
   a. prevalent form of vegetation.
   b. keystone species concept.
   c. feeding relationships within a community.
   d. species richness of the community.

4. Match each organism with its trophic level (you may choose a level more than once).
   a. alga   1. decomposer
   b. grasshopper   2. producer
   c. zooplankton   3. tertiary consumer
   d. eagle   4. secondary consumer
   e. fungus   5. primary consumer

5. Why are the top predators in food chains most severely affected by pesticides such as DDT?

6. Over a period of many years, grass grows on a sand dune, then shrubs grow, and then eventually trees grow. This is an example of ecological _____.

7. According to the energy pyramid, why is eating grain-fed beef a relatively inefficient means of obtaining the energy trapped by photosynthesis?

APPENDIX B The Periodic Table

Atomic number (number of protons) → 6 C 12.01 ← Atomic mass (number of protons plus number of neutrons averaged over all isotopes)

Element symbol

Metals Metalloids Nonmetals

Representative elements

Groups: Elements in a vertical column have the same number of electrons in their valence (outer) shell and thus have similar chemical properties.

Periods: Each horizontal row contains elements with the same total number of electron shells. Across each period, elements are ordered by increasing atomic number.

Alkali metals — Alkaline earth metals — Halogens — Noble gases — Transition elements

Period number — Group 1A — Group 2A — Group 3A — Group 4A — Group 5A — Group 6A — Group 7A — Group 8A

*Lanthanides:
Ce 140.1, Pr 140.9, Nd 144.2, Pm (145), Sm 150.4, Eu 152.0, Gd 157.3, Tb 158.9, Dy 162.5, Ho 164.9, Er 167.3, Tm 168.9, Yb 173.0, Lu 175.0

†Actinides:
Th 232.0, Pa 231.0, U 238.0, Np (237), Pu (244), Am (243), Cm (247), Bk (247), Cf (251), Es 252, Fm 257, Md 258, No 259, Lr 260

Name (Symbol)	Atomic Number	Name (Symbol)	Atomic Number	Name (Symbol)	Atomic Number	Name (Symbol)	Atomic Number	Name (Symbol)	Atomic Number
Actinium (Ac)	89	Copper (Cu)	29	Iron (Fe)	26	Osmium (Os)	76	Silicon (Si)	14
Aluminum (Al)	13	Curium (Cm)	96	Krypton (Kr)	36	Oxygen (O)	8	Silver (Ag)	47
Americium (Am)	95	Darmstadtium (Ds)	110	Lanthanum (La)	57	Palladium (Pd)	46	Sodium (Na)	11
Antimony (Sb)	51	Dubnium (Db)	105	Lawrencium (Lr)	103	Phosphorus (P)	15	Strontium (Sr)	38
Argon (Ar)	18	Dysprosium (Dy)	66	Lead (Pb)	82	Platinum (Pt)	78	Sulfur (S)	16
Arsenic (As)	33	Einsteinium (Es)	99	Lithium (Li)	3	Plutonium (Pu)	94	Tantalum (Ta)	73
Astatine (At)	85	Erbium (Er)	68	Livermorium (Lv)	116	Polonium (Po)	84	Technetium (Tc)	43
Barium (Ba)	56	Europium (Eu)	63	Lutetium (Lu)	71	Potassium (K)	19	Tellurium (Te)	52
Berkelium (Bk)	97	Fermium (Fm)	100	Magnesium (Mg)	12	Praseod`ymium (Pr)	59	Tennessine (Ts)	117
Beryllium (Be)	4	Flerovium (Fl)	114	Manganese (Mn)	25	Promethium (Pm)	61	Terbium (Tb)	65
Bismuth (Bi)	83	Fluorine (F)	9	Meitnerium (Mt)	109	Protactinium (Pa)	91	Thallium (Tl)	81
Bohrium (Bh)	107	Francium (Fr)	87	Mendelevium (Md)	101	Radium (Ra)	88	Thorium (Th)	90
Boron (B)	5	Gadolinium (Gd)	64	Mercury (Hg)	80	Radon (Rn)	86	Thulium (Tm)	69
Bromine (Br)	35	Gallium (Ga)	31	Molybdenum (Mo)	42	Rhenium (Re)	75	Tin (Sn)	50
Cadmium (Cd)	48	Germanium (Ge)	32	Moscovium (Mc)	115	Rhodium (Rh)	45	Titanium (Ti)	22
Calcium (Ca)	20	Gold (Au)	79	Neodymium (Nd)	60	Roentgenium (Rg)	111	Tungsten (W)	74
Californium (Cf)	98	Hafnium (Hf)	72	Neon (Ne)	10	Rubidium (Rb)	37	Uranium (U)	92
Carbon (C)	6	Hassium (Hs)	108	Neptunium (Np)	93	Ruthenium (Ru)	44	Vanadium (V)	23
Cerium (Ce)	58	Helium (He)	2	Nickel (Ni)	28	Rutherfordium (Rf)	104	Xenon (Xe)	54
Cesium (Cs)	55	Holmium (Ho)	67	Nihonium (Nh)	113	Samarium (Sm)	62	Ytterbium (Yb)	70
Chlorine (Cl)	17	Hydrogen (H)	1	Niobium (Nb)	41	Scandium (Sc)	21	Yttrium (Y)	39
Chromium (Cr)	24	Indium (In)	49	Nitrogen (N)	7	Seaborgium (Sg)	106	Zinc (Zn)	30
Cobalt (Co)	27	Iodine (I)	53	Nobelium (No)	102	Selenium (Se)	34	Zirconium (Zr)	40
Copernicium (Cn)	112	Iridium (Ir)	77	Oganesson (Og)	118				

Eric Isselee/Shutterstock; Ashley Toone/Alamy; **8.3** Ed Reschke/Photolibrary/Getty Images; **8.4** Biophoto Associates/Science Source; **8.7** Conly L. Rieder; **8.8a** Don W. Fawcett/Science Source; **8.8b** Kent Wood/Science Source; **8.10** Portra/Digital Vision/Getty Images; **8.11** CNRI/Science Source; **8.12** Iofoto/Shutterstock; **p. 133** Ed Reschke/Photolibrary/Getty Images; **8.17** David M. Phillips/Science Source; **8.19a** Scenics & Science/Alamy; **8.22 top** Lauren Shear/Science Source; **8.22 bottom** CNRI/Science Source; **8.23** Nhpa/Superstock; **p. 143** Ed Reschke/Photolibrary/Getty Images.

CHAPTER 9:
Chapter opening photo top right Petrenko Andriy/Shutterstock; **chapter opening photo left** Africa Studio/Shutterstock; **chapter opening photo bottom** KidStock/Getty Images; **p. 145** Eric Simon; **9.1** Science Source; **9.4** Patrick Lynch/Alamy; **9.5** James King-Holmes/Science Source; **9.8** Martin Shields/Science Source; **9.9 left to right** Tracy Morgan/Dorling Kindersley, Ltd.; Tracy Morgan/Dorling Kindersley, Ltd.; Eric Isselee/Shutterstock; Victoria Rak/Shutterstock; **9.10 left** Tracy Morgan/Dorling Kindersley, Ltd.; **9.10 right** Eric Isselee/Shutterstock; **9.12 left** James Woodson/Getty Images; **9.12 right** Blend Images/KidStock/Getty Images; **Table 9.1 top to bottom** David Terrazas Morales/Corbis; Editorial Image, LLC/Alamy; Eye of Science/Science Source; Science Source; **9.13 left** Ostill/Shutterstock; **9.13 right** Image Source/Getty Images; **9.16a left to right** Kuznetsov Alexey/Shutterstock; Erik Lam/123RF; Irina Oxilixo Danilova/Shutterstock; Purplequeue/Shutterstock; Markos86/Shutterstock; Eric Isselee/Shutterstock; **9.17** Saturn Stills/Science Source; **9.20** Mauro Fermariello/Science Source; **9.21** Eye of Science/Science Source; **9.23** Jean Dickey; **p. 161** Rootstock/Shutterstock; **9.25 clockwise from left** In Green/Shutterstock; Rido/Shutterstock; Andrew Syred/Science Source; Dave King/Dorling Kindersley, Ltd.; Jo Foord/Dorling Kindersley, Ltd.; **9.28** Archive Pics/Alamy; **9.29 top left to bottom right** Jerry Young/Dorling Kindersley, Ltd.; Gelpi/Fotolia; Dave King/Dorling Kindersley, Ltd.; Dave King/Dorling Kindersley, Ltd.; Tracy Morgan/Dorling Kindersley, Ltd.; Dave King/Dorling Kindersley, Ltd.; Jerry Young/Dorling Kindersley, Ltd.; Dave King/Dorling Kindersley, Ltd.; Dave King/Dorling Kindersley, Ltd.; Tracy Morgan/Dorling Kindersley, Ltd.; Irina Oxilixo Danilova/Shutterstock; **p. 165** Eric J. Simon; **p. 168** Arco/G. Lacz/GmbH/Alamy; **p. 169** Eric J. Simon.

CHAPTER 10:
Chapter opening photo top right JLPH/Cultura/Getty Images; **chapter opening photo center** Volker Steger/Science Source; **chapter opening photo bottom left** Eye of Science/Science Source; **chapter opening photo bottom right** Thomas Deerinck/NCMIR/Science Source; **p. 171** Kateryna Kon/Science Photo Library/Getty Images; **10.3 left** Barrington Brown/Science Source; **10.3 right** Library of Congress Print and photographs Division; **10.11** Eye of Science/Science Source; **10.22** Georgette Apol/Steve Bloom Images/Alamy; **10.23** Koji Niino/Mixa/Alamy; **10.24** Oliver Meckes/Science Source; **10.26** Norm Thomas/Science Source; **10.28** Hazel Appleton/Health Protection Agency Centre for Infections/Science Source; **10.31** NIBSC/Science Photo Library/Science Source; **10.32** Will & Deni McIntyre/Science Source; **10.33 clockwise from left** Scott Camazine/Alamy; Cynthia Goldsmith/CDC; Cynthia Goldsmith/CDC; Cultura Creative (RF)/Alamy; Cynthia Goldsmith/CDC; National Institute of Allergy and Infectious Diseases (NIAD); Cynthia Goldsmith/CDC; Phanie/Alamy.

CHAPTER 11:
Chapter opening photo top Peathegee Inc/Blend Image/Getty Images; **chapter opening photo center** Alila Medical Media/Shutterstock; **chapter opening photo center right** S4svisuals/Shutterstock; **chapter opening photo bottom left** Akkharat Jarusilawong/Shutterstock; **p. 197** SPL/Science Source; **11.1 left to right** Steve Gschmeissner/Science Source; Steve Gschmeissner/Science Photo Library/Alamy; Ed Reschke/Oxford Scientific/Getty Images; **11.4** Iuliia Lodia/Fotolia; **11.9** Dr. Thomas Kaufman Dept. of Biology Indiana University; **11.10** Alila Medical Media/Shutterstock; **11.11** Videowokart/Shutterstock; **p. 205** Joseph T. Collins/Science Source; **11.13a** Courtesy of the Roslin Institute, Edinburgh; **11.13b** Randall S. Prather, PhD/Courtesy of Jim Curley; **11.13c** Robert Lanza; **11.15 left** Mauro Fermariello/Science Source; **11.15 right** Klaus Guldbrandsen/Science Source; **p. 210** Eric Isselee/Shutterstock; **11.19** Geo Martinez/Shutterstock; **11.21** Alastair Grant/Pool/AP Images; **11.23** Biophoto Associates/Science Source.

CHAPTER 12:
Chapter opening photo top right Photolinc/Shutterstock; **chapter opening photo left** Volker Steger/Science Source; **chapter opening photo bottom right** Dinodia Photos/Alamy; **p. 217** Isak55/Shutterstock; **12.1** STR/AP Images; **12.2 left** Prof. S. Cohen/Science Source; **12.2 right** Huntington Potter/University of South Florida College of Medicine; **12.3** Dmitry Lobanov/Fotolia; **12.6** Eric Carr/Alamy; **12.7** Volker Steger/Science Source; **12.8** Inga Spence/Alamy; **12.9 top** Christopher Gable and Sally Gable/Dorling Kindersley, Ltd.; **12.9 bottom** U.S. Department of Agriculture (USDA); **12.10 left** International Rice Research Institute (IRRI); **12.10 right** Fotosearch RM/AGE Fotostock; **p. 223** Andy Manis/Bloomberg/Getty Images; **12.13** Eric J. Simon; **21.15** Andrew Brookes/National Physical Laboratory/Science Source; **12.17** Steve Helber/AP Images; **12.18** Fine Art Images/Heritage Image Partnership Ltd/Alamy; **12.19** Volker Steger/Science Source; **12.21 top** Pikselstock/Shutterstock; **12.21 bottom** David Parker/Science Photo Library/Science Source; **12.24 left** Scott Camazine/Science Source; **12.24 right** FBI UPI Photo Service/Newscom; **12.25** James King-Holmes/Science Source; **12.26** Science Source; **12.27** Alex Milan Tracy/NurPhoto/Sipa U/Newscom; **12.28** Isak55/Shutterstock; **12.29** Image Point Fr/Shutterstock; **12.30** Pictures From History/Newscom; **12.31** Jekesai Njikizana/AFP/GettyImages.

CHAPTER 13:
Chapter opening photo top BW Folsom/Shutterstock; **chapter opening photo bottom left** Lapina/Shutterstock; **chapter opening photo bottom right** John Bryant/Gallo Images/Getty Images; **p. 243** George Mulala/Reuters; **13.1** Aditya Singh/Moment/Getty Images; **13.2a** Chrispo/Shutterstock; **13.2b** Sabena Jane Blackbird/Alamy; **13.3 left** Science Source; **13.3 right** Classic Image/Alamy; **13.4a** Celso Diniz/Shutterstock; **13.4b** Tim Laman/National Geographic/Getty Images; **13.5 clockwise from left** Francois Gohier/Science Source; Francois Gohier/Science Source; Science Source; Sergei Cherkashin/Reuters; Pixtal/SuperStock; **13.7 left** Dr.Keith Wheeler/Science Source; **13.7 right** Lennart Nilsson/TT Nyhetsbyrån; **13.9** Blickwinkel/Hecker/Alamy; **13.10** Matthew Oldfield Underwater Photography/Alamy; **13.11** Philippe Wojazer/Reuters; **13.12** Our Wild Life Photography/Alamy; **13.13** Adam Jones/The Image Bank/Getty Images; **13.14** Andy Levin/Science Source; **13.19** Steve Bloom Images/Alamy; **13.20** Planetpix/Alamy; **13.21** Heather Angel/Natural Visions/Alamy; **13.22** Mariko Yuki/Shutterstock; **13.23a** Peter Grant; **13.23b** Gabrielle Therin-Weise/Photographer's Choice RF/Getty Images; **13.24a** Reinhard/ARCO/Nature Picture Library; **13.24b** Paolo-manzi/Shutterstock.

CHAPTER 14:
Chapter opening photo top right Sappington Todd/Getty Images; **chapter opening photo center left** Oberhaeuser/Agencja Fotograficzna Caro/Alamy; **chapter opening photo center right** GL Archive/Alamy; **chapter opening photo bottom** Keneva Photography/Shutterstock; **p. 269** Michael Collier; **14.1 left** Cathleen A Clapper/Shutterstock; **14.1 right** Peter Scoones/Nature Picture Library; **14.2 clockwise from left** Bill Draker/Rolf Nussbaumer Photography/Alamy; David Kjaer/Nature Picture Library; Jupiterimages/Stockbyte/Getty images; Photos/Getty Images; Phil Date/Shutterstock; Blvdone/Shutterstock; Comstock/Stockbyte/Getty Images; Robert Kneschke/Shutterstock; **14.4 left to right** APHIS Animal and Plant health inspection Service/USDA; Jared Hobbs/SuperStock; McDonald/Photoshot; Joe McDonald/Corbis Documentary/Getty Images; J & C Sohns/Tier und Naturfotografie/

AGE Fotostock; Oyvind Martinsen Wildlife Collection/Alamy; Jon G. Fuller/VWPics/Alamy; Danita Delimont/Gallo Images/AGE Fotostock; **14.5 left to right** Chuck Brown/Science Source; Dogist/Shutterstock; Alistair Duncan/Dorling Kindersley, Ltd.; Dorling Kindersley, Ltd.; Dr. Kazutoshi Okuno; **14.6 left to right** John Shaw/Photoshot; Morey Milbradt/Stockbyte/Getty Images; Ron Niebrugge/Alamy; **14.8** Michelle Gilders/Alamy; **14.9** Artemyeva/Shutterstock; **14.11a** Florida Museum photo by Jeff Gage; **14.11b clockwise from left** Bill Brooks/Alamy; Arco Images GmbH/Alamy; David Chapman/Alamy; **14.12 top to bottom** Interfoto/Alamy; Mary Plage/Oxford Scientific/Getty Images; Ralph Lee Hopkins/National Geographic RF/Getty Images; **14.13** Barbara Rich/Moment/Getty Images; **14.17 left to right** Jean-Paul Ferrero/Auscape International Pty Ltd/Alamy; Eugene Sergeev/Shutterstock; Kamonrat/Shutterstock; **14.19** George Atsametakis/Alamy; **14.20** Chris Hellier/Science Source; **14.21 left to right** Image Quest Marine; Christophe Courteau/Science Source; Reinhard Discherl/Alamy; Image Quest Marine; James Watt/Image Quest Marine; **p. 286** Florida Stock/Shutterstock; **14.23** Mark Thiessen/National Geographic Creative/Alamy; **14.27a** John Sullivan/Alamy; **14.27b** MikeLane45/iStock/Getty Images.

CHAPTER 15:

Chapter opening photo top right MintImages/Shutterstock; **chapter opening photo center left** Harry Vorsteher/Cultura Creative (RF)/Alamy; **chapter opening photo center right** Odilon Dimier/PhotoAlto sas/Alamy; **chapter opening photo bottom** Floris van Breugel/Nature Picture Library/Alamy; **p. 293** Stephanie Schuller/Science Source; **15.2** Mark Garlick/Science Photo Library/Corbis; **15.5** B. Murton/Southampton Oceanography Centre/Science Source; **15.6 left** Imagedb/Shutterstock; **15.6 right** Dr. Tony Brain and David Parker/Science Photo Library/Science Source; **15.7 left to right** Scimat/Science Source; Niaid/CDC/Science Source; CNRI/SPL/Science Source; **15.8a** David M. Phillips/Science Source; **15.8b** John Walsh/Science Source; **15.8c** Esther R. Angert; **15.9** Science Photo Library - Steve Gschmeissner/Brand X Pictures/Getty Images; **15.10** Eye of Science/Science Source; **15.11a** Sinclair Stammers/Science Source; **15.11b** Dr. Gary Gaugler/Science Source; **15.12** Peter Batson/Image Quest Marine; **15.13 top** Huetter, C./Arco Images/Alamy; **15.13 bottom** Nigel Cattlin/Alamy; **15.15** Sipa USA/Newscom; **15.16** Eric J. Simon; **15.17** Jim West/Alamy; **15.18** SPL/Science Source; **15.19 top** Centers for Disease Control and Prevention (CDC); **15.19 center** Scott Camazine/Science Source; **15.19 bottom left** Sarah2/Shutterstock; **15.19 bottom right** David M. Phillips/Science Source; **15.21a** Carol Buchanan/F1online/AGE Fotostock; **15.21b** Eye of Science/Science Source; **15.21c** Blickwinkel/NaturimBild/Alamy; **15.22 left to right** Eye of Science/Science

Source; David M. Phillips/Science Source; Biophoto Associates/Science Source; Claude Carre/Science Source; Eye of Science/Science Source; Michael Abbey/Science Source; **15.23a** Nigel Downer/Science Source; **15.23b** Eye of Science/Science Source; **15.24a** Eye of Science/Science Source; **15.24b** Steve Gschmeissner/Science Source; **15.24c** Manfred Kage/Science Source; **15.25 left to right** Marevision/AGE Fotostock; Marevision/AGE Fotostock; David Hall/Science Source; **15.26** Lucidio Studio, Inc/Moment/Getty Images.

CHAPTER 16:

Chapter opening photo top right Webphotographeer/E+/Getty Images; **chapter opening photo center** ITS AL Dente/Shutterstock; **chapter opening photo bottom** National Geographic Creative/Alamy; **p. 315** Valentyn Volkov/Shutterstock; **16.2** Science Source; **16.3** Steve Gorton/Dorling Kindersley, Ltd.; **16.4** Garry DeLong/Oxford Scientific/Getty Images; **16.5 left** Bob Gibbons/Alamy; **16.5 right** Gerd Guenther/Science Source; **16.7 left to right** Kristin Piljay; James Randklev/Photographer's Choice RF/Getty Images; V. J. Matthew/Shutterstock; Dale Wagler/Shutterstock; **16.8** Duncan Shaw/Science Source; **16.9** John Serrao/Science Source; **16.11 clockwise from left** Jon Bilous/Shutterstock; Biophoto Associates/Science Source; Mothy20/Getty Images; Ed Reschke/Photolibrary/Getty Images; **16.12** Field Museum Library/Premium Archive/Getty Images; **16.13** Danilo Donadoni/Marka/AGE Fotostock; **16.15 left to right** Stephen P. Parker/Science Source; Morales/AGE Fotostock; Gunter Marx/Alamy; **16.16** Gene Cox/Science Source; **16.18 left to right** Jean Dickey; Tyler Boyes/Shutterstock; Christopher Marin/Shutterstock; Jean Dickey; **16.20 clockwise from left** Jean Dickey; Scott Camazine/Science Source; Sonny Tumbelaka/AFP Creative/Getty Images; **16.21** Prill/Shutterstock; **Table 16.1 top to bottom** Steve Gorton/Dorling Kindersley, Ltd.; Richard Griffin/Shutterstock; Photogal/Shutterstock; Dionisvera/Fotolia; Radu Razvan/Shutterstock; Dorling Kindersley/Getty Images; Photogal/Shutterstock; National Tropical Botanical Garden; Alle/Shutterstock; **16.22 clockwise from left** Jean Dickey; Stan Rohrer/Alamy; Science Source; Astrid & Hanns-Frieder Michler/Science Source; VEM/Science Source; Pabkov/Shutterstock; **16.23 top** Jupiterimages/Photos/Getty Images; **16.23 bottom** Blickwinkel/Alamy; **16.24a** Wolfness72/Shutterstock; **16.24c** Raymond Louis/Alamy; **16.25** Bedrich Grunzweig/Science Source; **16.26 top left** Mikeledray/Shutterstock; **16.26 bottom left** Imagebroker/Food-Drinks/SuperStock; **16.26 right** Will Heap/Dorling Kindersley, Ltd.; **16.27** Christine Case; **16.28** Wildlife GmbH/Alamy.

CHAPTER 17:

Chapter opening photo top right Gazimal/Getty Images; **chapter opening photo center** Dropu/

Shutterstock; **chapter opening photo bottom** Sebastien Plailly/Science Source; **p. 337** Sabena Jane Blackbird/Alamy; **17.1** Gunter Ziesler/Photolibrary/Getty Images; **17.4** Sinclair Stammers/Science Source; **17.5 left** Christian Jegou/Publiphoto/Science Source; **17.5 right** Lorraine Hudgins/Shutterstock; **17.9** James Watt/Image Quest Marine; **17.10 top left** Joe Belanger/Shutterstock; **17.10 top right** Sue Daly/Nature Picture Library/Alamy; **17.10 center** Marek Mis/Science Source; **17.10 bottom** Pavlo Vakhrushev/Fotolia; **17.13 left to right** Georgette Douwman/Nature Picture Library; Jez Tryner/Image Quest Marine; Christophe Courteau/Nature Picture Library; Marevision/AGE Fotostock; Reinhard Dirscherl/Alamy; **17.14 top right** CMB/AGE Fotostock; **17.14 bottom left** Geoff Brightling/Gary Stabb/Dorling Kindersley Ltd.; **17.14 bottom right** Eye of Science/Science Source; **17.16 left** Schulz, H./Juniors Bildarchiv GmbH/Alamy; **17.16 right** Blue-Sea.cz/Shutterstock; **17.17a** Steve Gschmeissner/Science Source; **17.17b** Eye of Science/Science Source; **17.17c** Eye of Science/Science Source; **p. 346** Astrid & Hanns-Frieder Michler/Science Source; **17.18 top to bottom** Snowleopard1/Getty Images; Maximilian Weinzierl/Alamy; Jean L. Dickey; **17.19** Dave King/Dorling Kindersley, Ltd.; **17.20 clockwise from left** Nenad Druzic/Moment Open/Getty Images; Dave King/Dorling Kindersley Ltd.; Andrew Syred/Science Source; Larry West/Science Source; Snowleopard1/Getty Images; **17.21 top** Dave King/Dorling Kindersley Ltd.; **17.21 center left** Maximilian Weinzierl/Alamy; **17.21 center** Tom McHugh/Science Source; **17.21 center right** Nature's Images/Science Source; **17.21 bottom left** Dietmar Nill/Nature Picture Library; **17.21 bottom right** Nancy Sefton/Science Source; **17.22 left** Jean L. Dickey; **17.22 right** Tom McHugh/Science Source; **17.23** Radius Images/Alamy; **17.24 clockwise from left** NH/Shutterstock; Stuart Wilson/Science Source; Huayang/Moment/Getty Images; Jean L. Dickey; Jean L. Dickey; Lkpro/Shutterstock; David Oberholzer/Prisma by Dukas Presseagentur GmbH/Alamy; Stuart Wilson/Science Source; **17.25 top** Thomas Kitchin & Victoria Hurst/Design Pics Inc./Alamy; **17.25 bottom** Keith Dannemiller/Alamy; **17.26 clockwise from left** Roger Steene/Image Quest Marine; Andrew J. Martinez/Science Source; Jose B. Ruiz/Nature Picture Library; Image Quest Marine; Roger Steene/Image Quest Marine; Tbkmedia.de/Alamy Alamy; **17.27** Colin Keates/Courtesy of the Natural History Museum/Dorling Kindersley Ltd.; **17.29 left** Heather Angel/Natural Visions/Alamy; **17.29 right** Image Quest Marine; **17.31a** Tom McHugh/Science Source; **17.31b** Marevision/AGE Fotostock; **17.31b inset** A Hartl/Blickwinkel/AGE Fotostock; **17.31c** George Grall/National Geographic/Getty Images; **17.31d** Christian Vinces/Shutterstock; **17.32a left** Tom McHugh/Science Source; **17.32a right** Jack Goldfarb/AGE Fotostock; **17.32b** LeChatMachine/Fotolia; **17.32c left** Gary Meszaros/Science

APPENDIX D Selected Answers

CHAPTER 1

1. b (some living organisms are single-celled)
2. atom, molecule, cell, tissue, organ, organism, population, ecosystem, biosphere; the cell
3. Photosynthesis cycles nutrients by converting the carbon in carbon dioxide to sugar, which is then consumed by other organisms. Additionally, the oxygen in water is released as oxygen gas. Photosynthesis contributes to energy flow by converting sunlight to chemical energy, which is then also consumed by other organisms, and by producing heat.
4. On average, those individuals with heritable traits best suited to the local environment produce the greatest number of offspring that survive and reproduce. This increases the frequency of those traits in the population over time. The result is the accumulation of evolutionary adaptations.
5. d
6. c
7. evolution
8. A fact should be confirmed to be the same by all observers, whereas opinions can differ from one observer to the next.
9. a3, b2, c1, d4
10. Information flow: Genes encode information to make proteins, which then produce physical traits.
11. Pathways that transform energy and matter: Sea turtles obtain both energy and molecular building blocks from the grass they consume.
12. Interactions within biological systems: Chemicals released in one part of the world can interact with ocean water in another part, affecting the life found there.

Thinking Like a Scientist

The buckets acted as a control group. By comparing the movement of the sea turtles with the movement of the floating buckets, the researchers could assign any difference to be the result of swimming by the turtles.

CHAPTER 2

1. electrons; neutrons
2. protons

3. Nitrogen-14 has an atomic number of 7 and a mass number of 14. The radioactive isotope, nitrogen-16, has an atomic number of 7 and a mass number of 16.
4. Organisms incorporate radioactive isotopes of an element into their molecules just as they do nonradioactive isotopes, and researchers can detect the presence of the radioactive isotopes.
5. Each carbon atom has only three covalent bonds instead of the required four.
6. The positively charged hydrogen regions would repel each other.
7. d
8. a
9. The positive and negative poles cause adjacent water molecules to become attracted to each other, forming hydrogen bonds. The properties of water such as cohesion, temperature regulation, and water's ability to act as a solvent all arise from this atomic "stickiness."
10. Because a nonpolar molecule cannot form hydrogen bonds, it would not have the properties that allow water to act as the basis of life, such as the ability to dissolve substances and water's cohesive properties.
11. The cola is an aqueous solution, with water as the solvent, sugar as the main solute, and the CO_2 making the solution acidic.
12. Pathways that transform energy and matter: The rearrangement of molecules within cells is an example of matter being transformed.
13. The relationship between structure and function: The polar structure of a water molecule helps enable it to carry out the function of supporting life.
14. Interactions within biological systems: Since there are interactions between the atmosphere and ocean, release of CO_2 into the atmosphere can affect ocean life.

Thinking Like a Scientist

The rat study could include a placebo-controlled group (rats who receive nonradioactive seed implantation). That would be unethical in a human trial.

CHAPTER 3

1. Isomers have different structures, or shapes, and the shape of a molecule usually helps determine the way it functions in the body.
2. dehydration reactions; water
3. hydrolysis
4. b
5. $C_6H_{12}O_6 + C_6H_{12}O_6 \rightarrow C_{12}H_{22}O_{11} + H_2O$
6. fatty acid; glycerol; triglyceride
7. b
8. c
9. If the change does not affect the shape of the protein in any way, then that change would not affect the function of the protein.
10. Hydrophobic amino acids are most likely to be found within the interior of a protein, far from the watery environment.
11. a
12. starch (or glycogen or cellulose); nucleic acid
13. Both DNA and RNA are polynucleotides; both have the same phosphate group along the backbone; and both use A, C, and G bases. But DNA uses T while RNA uses U as a base; the sugar differs between them; and DNA is usually double-stranded, while RNA is usually single-stranded.
14. Structurally, a gene is a long stretch of DNA. Functionally, a gene contains the information needed to produce a protein.
15. Pathways that transform energy and matter: The energy within a molecule of glucose can be transformed to promote cellular work, and the matter within glucose can be used to build large molecules such as starch.
16. The relationship of structure to function: The specific structure of a polysaccharide (how the monosaccharide monomers are joined together) affects the function of that polysaccharide.
17. Information flow: Nucleic acids are information storage molecules, and the information of genes is stored as a precise nucleotide sequence.

Thinking Like a Scientist

A change in the nucleotide sequence of a chromosome outside of a gene may affect a protein that interacts with

that gene, such as a protein that turns a gene on or off.

CHAPTER 4

1. b
2. A membrane is fluid because its components are not locked into place. A membrane is mosaic because it contains a variety of suspended proteins.
3. endomembrane system
4. smooth ER; rough ER
5. because many bacterial cells have walls, but no human cells do
6. Both organelles use membranes to organize enzymes and both provide energy to the cell. But chloroplasts capture energy from sunlight during photosynthesis, whereas mitochondria release energy from glucose during cellular respiration. Chloroplasts are only in photosynthetic plants and protists, whereas mitochondria are in almost all eukaryotic cells.
7. a3, b1, c5, d2, e4
8. nucleus, nuclear pores, ribosomes, rough ER, Golgi apparatus
9. Both are appendages that aid in movement and that extend from the surface of a cell. Cells with flagella typically have one long flagellum that propels the cell in a whiplike motion; cilia are usually shorter, are more numerous, and beat in a coordinated fashion.
10. Information flow: The precise DNA nucleotide sequence of a gene contains the information necessary to build a protein.
11. Interactions within biological systems: Only by acting together can organelles such as the nucleus, ER, and ribosomes express a genetic message.
12. Pathways that transform energy and matter: The energy in sunlight is transformed into the chemical energy of sugar molecules.

Thinking Like a Scientist

Most antibiotics were discovered in nature. Therefore, we need to preserve natural habitats because we depend upon such areas for new drug discoveries.

CHAPTER 5

1. You transform the chemical energy from food to the kinetic

8. Chop the genome into fragments using restriction enzymes, clone and sequence each fragment, and reassemble the short sequences into a continuous sequence for every chromosome.
9. c, b, a, d
10. Evolution: Analysis of DNA sequences can provide insight into evolutionary history.
11. Interactions within biological systems: New functions arise through the interactions of many smaller parts, such as the many proteins that are involved in metabolism.
12. Information flow: Genes are the units of genetic information that are conveyed from one generation to the next.

Thinking Like a Scientist
Random mutations happen all the time. If a sequence is the same in very different species, it is evidence that the exact sequence is essential to survival.

CHAPTER 13
1. species, genus, family, order, class, phylum, kingdom, domain
2. c
3. Lyell and other geologists presented evidence for the gradual change of geologic features over millions of years. Darwin applied this idea to suggest that species evolve through the slow accumulation of small changes over long periods of time.
4. *Bb*: 0.42; *BB*: 0.49; *bb*: 0.09
5. The fitness of an individual (or of a particular genotype) is measured by the relative number of alleles that it contributes to the gene pool of the next generation compared with the contribution of others. Thus, the number of fertile offspring produced determines an individual's fitness.
6. b
7. c
8. Both effects result in populations small enough for significant sampling error in the gene pool for the first few generations. A bottleneck event reduces the size of an existing population in a given location. The founder effect occurs when a new, small population colonizes a new territory.
9. stabilizing selection

10. Information flow: The evolutionary history of a species is documented in the genetic information inherited from its ancestral species.
11. The relationship between structure and function: The bird's shape minimizes friction as it dives into water from a height of 75 feet.
12. Evolution: Natural selection is the process by which life evolves.

Thinking Like a Scientist
To perform a controlled experiment, the Grants would have to artificially manipulate food resources for a subset of the finch population. Such an experiment would be extremely difficult to carry out in the field, and capturing birds for a laboratory experiment would be disruptive to the island's ecology. An observational study allowed the Grants to test their hypothesis under natural conditions with the entire medium ground finch population of Daphne.

CHAPTER 14
1. Microevolution is a change in the gene pool of a population, often associated with adaptation. Speciation is an evolutionary process in which one species splits into two or more species. Macroevolution is evolutionary change above the species level, for example, the origin of evolutionary novelty and new taxonomic groups and the impact of mass extinctions on the diversity of life and its subsequent recovery. Macroevolution is marked by major changes in the history of life, and these changes are often noticeable enough to be evident in the fossil record.
2. prezygotic: a, b, c, e; postzygotic: d
3. because a small gene pool is more likely to be changed substantially by genetic drift and natural selection
4. exaptations
5. d
6. d
7. 2.6
8. Homologies reflected a shared evolutionary history, whereas analogies do not. Analogies result from convergent evolution.
9. Archaea and Bacteria
10. Information flow: Genetic changes that affect the rate of developmental events can have a profound effect on body form.

11. The relationship of structure to function: The structure of the skin between the "finger" bones of the wing provides the surface area needed for flight.
12. Pathways that transform energy and matter: In the process of photosynthesis, plants convert the sun's energy into chemical energy that is stored in the bonds of organic molecules.

Thinking Like a Scientist
New species of *Tragopogon* originated very recently (within the past several decades). This gives researchers an opportunity to study the adaptive and genetic changes that occur early in a species' existence.

CHAPTER 15
1. g, a, d, f, c, b, e
2. DNA polymerase is a protein, which must be transcribed from a gene. But a DNA gene requires DNA polymerase to be replicated. This creates a paradox about which came first—DNA or protein. But RNA can act as both an information-storage molecule and an enzyme, suggesting that dual-role RNA may have preceded both DNA and proteins.
3. Exotoxins are poisons secreted by pathogenic bacteria; endotoxins are components of the outer membrane of pathogenic bacteria.
4. food (products of photosynthesis); usable form of nitrogen
5. They can form endospores.
6. Prokaryotes in soil or water decompose the organic matter in leaves and other plant and animal remains, returning the elements to the environment in inorganic form. Prokaryotes in a sewage treatment facility decompose the organic matter in sewage, converting it to an inorganic form.
7. They are eukaryotes that are not plants, animals, or fungi.
8. c
9. b
10. Information flow: Cells store information in DNA, which is used to make RNA, which is used to make proteins.
11. Pathways that transform energy and matter: When prokaryotes break down complex molecules to simpler ones, they transform energy and matter.

12. Relationship of structure to function: The web form increases the surface area, which facilitates absorption of necessary resources.

Thinking Like a Scientist
Extract microbiota from an obese "donor," then introduce equal amounts of these microbes into the intestinal tracts of lean and obese recipients. Measure the body composition of recipients before microbial transplant and measure again at a pre-determined amount of time after the transplant. Recipients would adhere to a prescribed diet during the experiment. Drawbacks: In practical terms, human subjects are more variable than lab-raised mice, which can be bred and raised in standardized conditions. In addition, each person has his own unique microbiota that will interact with the transplanted microbes. Solving these problems creates ethical issues. For example, it would be unethical for the researchers to treat subjects with large doses of antibiotics to kill off their existing microbiota. Also, the transplanted microbes could result in health problems for the recipients.

CHAPTER 16
1. cuticle
2. flowers
3. a. sporophyte b. cones; angiosperms c. fruit
4. b
5. b
6. a fern
7. Because the plants do not lose their leaves during autumn and winter, the leaves are already fully developed for photosynthesis when the short growing season begins in spring.
8. vascular plant
9. a
10. a ripened ovary of a flower that protects and aids in the dispersal of seeds contained in the fruit
11. plant roots; fungi
12. A fungus digests its food externally by secreting digestive juices into the food and then absorbing the small nutrients that result from digestion. In contrast, humans and most other animals ingest relatively large pieces of food and digest the food within their bodies.

13. Information Flow: DNA from different gametophytes combines and is passed to the sporophyte generation. The spores contain the DNA for the next generation of gametophytes.
14. The Relationship of Structure to Function: The fine branches of roots increase the surface area, which facilitates absorption of necessary resources.
15. Evolution: Natural selection led to the development of vascular systems in ferns, giving them an advantage in a variety of habitats. This could also be an example of the relationship of structure to function because vascular tissue functions to move materials throughout the plant.

Thinking Like a Scientist
So they could tell whether disturbing the soil influenced pine growth, even if mycorrhizae were not added.

CHAPTER 17
1. c
2. arthropod
3. amphibians
4. b
5. chordata; notochord; cartilage disks between your vertebrae
6. a
7. *Australopithecus* species, *Homo habilis, Homo erectus, Homo sapiens*
8. a4, b5, c1, d2, e3
9. Information flow: The information needed to direct the development of diverse body forms is contained in DNA, which is transmitted between generations.
10. Relationship of structure to function: The suckers and hooks function as effective tools that allow tapeworms to attach to hosts.
11. Interactions within biological systems: Insects have multiple types of interactions with people. Some interactions are beneficial to both humans and insects. Others are harmful to humans, insects, or both.

Thinking Like a Scientist
Mutations result in genetic differences between populations that don't

interbreed. Analyzing more genes provides multiple, independent lines of evidence.

CHAPTER 18
1. organismal ecology, population ecology, community ecology, ecosystem ecology
2. light, water temperature, chemicals added
3. physiological; behavioral
4. d
5. a. desert; b. temperate grassland; c. tropical forest; d. temperate broadleaf forest; e. northern coniferous forest; f. tundra
6. chaparral
7. permafrost, very cold winters, and high winds
8. agriculture
9. Carbon dioxide and other gases in the atmosphere absorb heat energy radiating from Earth and reflect it back toward Earth. This is called the greenhouse effect. As the carbon dioxide concentration in the atmosphere increases, more heat is retained, causing global warming.
10. c
11. populations of organisms that have high genetic variability and short life spans
12. The relationship of structure to function: The structures of molecules like the keratin in reptilian scales and the wax that coats plant leaves create a water-tight barrier.
13. Interactions within biological systems: Organisms in an ecosystem are connected in many ways, including competition and predation.
14. Pathways that transform energy and matter: Photosynthesis is a pathway that uses light energy to produce energy-rich molecules from carbon dioxide and water.
15. Evolution: As a result of random mutation, plants in a population vary in their resistance to drought. During dry periods, individuals that are more drought resistant leave more offspring than other individuals in the population. Therefore, the drought-resistant individuals

pass along their genes. This process of natural selection leads to evolution of increased drought resistance.

Thinking Like a Scientist
The researchers tested their hypothesis with an observational study: They collected data from historical and present-day population records and compared these observations to determine changes in the butterfly's range.

CHAPTER 19
1. the number of people and the land area in which they live
2. III; I
3. a. The x-axis is time; the y-axis is the number of individuals; the red curve represents exponential growth; the blue curve represents logistic growth.
   b. carrying capacity
   c. In exponential growth, the size of the population increases more and more rapidly. In logistic growth, the population grows fastest when it is about one-half the carrying capacity.
   d. exponential growth curve, though the worldwide growth rate is slowing
4. d
5. opportunistic
6. c
7. c
8. The Relationship of Structure to Function: The structure of the molecules in this combination causes them to function as effective anchors.
9. Interactions within Biological Systems: Changes in population size can be caused by interactions between species (such as predation and herbivory) as well as between organisms and their environment.
10. Evolution: Individuals have limited resources to invest in reproduction and survival. Alleles that lead to a better balance between survival and reproduction will lead to greater fitness, and those alleles

will become more common in the population.

Thinking Like a Scientist
A valid control must match the experimental treatment in every way except for the variable in question. Researchers were testing whether blocking access would limit the toads. The open-fence treatment allowed researchers to determine whether some other aspect of the fence (such as giving predators a place to perch) was responsible for any reduction in toad numbers.

CHAPTER 20
1. habitat destruction
2. d
3. c
4. a2, b5, c5, d3 or d4, e1
5. because the pesticides become concentrated in their prey
6. succession
7. Only about 10% of the energy trapped by photosynthesis is turned into biomass by the plant, and only about 10% of that energy is turned into the meat of a grazing animal. Therefore, grain-fed beef provides only about 1% of the energy captured by photosynthesis.
8. Many nutrients come from the soil, but carbon comes from CO_2 in the air.
9. landscape
10. a
11. The relationship of structure to function: The nature of the structures (sharp or hard) allows them to function for defense.
12. Interactions within biological systems: Competition between different species is a type of interaction.
13. Pathways that transform energy and matter: As energy flows through an ecosystem, some is lost as heat. Matter is conserved.

Thinking Like a Scientist
If fragments are too close together, they may act as one large fragment for species that can move between them.

Glossary

A

abiotic factor (ā′bī-ot′-ik)
A nonliving component of an ecosystem, such as air, water, light, minerals, or temperature.

abiotic reservoir
The part of an ecosystem where a chemical, such as carbon or nitrogen, accumulates or is stockpiled outside of living organisms.

ABO blood groups
Genetically determined classes of human blood that are based on the presence or absence of carbohydrates A and B on the surface of red blood cells. The ABO blood group phenotypes, also called blood types, are A, B, AB, and O.

absorption
The uptake of small nutrient molecules by an organism's own body. In animals, absorption is the third main stage of food processing, following digestion; in fungi, it is acquisition of nutrients from the surrounding medium.

acclimation (ak′-li-mā′-shun)
Physiological adjustment that occurs gradually, though still reversibly, in response to an environmental change.

acid
A substance that increases the hydrogen ion (H+) concentration in a solution.

activation energy
The amount of energy that reactants must absorb before a chemical reaction will start. An enzyme lowers the activation energy of a chemical reaction, allowing it to proceed faster.

activator
A protein that switches on a gene or group of genes by binding to DNA.

active site
The part of an enzyme molecule where a substrate molecule attaches—typically, a pocket or groove on the enzyme's surface.

active transport
The movement of a substance across a biological membrane against its concentration gradient, aided by specific transport proteins and requiring the input of energy (often as ATP).

adenine (A) (ad′-uh-nēn)
A double-ring nitrogenous base found in DNA and RNA.

ADP (adenosine diphosphate)
(a-den′-ō-sēn dī-fos′-fāt)
A molecule composed of adenosine and two phosphate groups. The molecule ATP is made by combining a molecule of ADP with a third phosphate in an energy-consuming reaction.

adult stem cell
A cell present in adult tissues that generates replacements for nondividing differentiated cells.

aerobic (ār-ō′-bik)
Containing or requiring molecular oxygen (O_2).

age structure
The relative number of individuals of each age in a population.

AIDS
Acquired immunodeficiency syndrome; the late stages of HIV infection, characterized by a reduced number of T cells; usually results in death caused by infections that would be defeated by a properly functioning immune system.

alga (al′-guh)
(plural, **algae**) An informal term that describes a great variety of photosynthetic protists, including unicellular, colonial, and multicellular forms. Prokaryotes that are photosynthetic autotrophs are also regarded as algae.

allele (uh-lē′-ul)
An alternative version of a gene.

allopatric speciation
The formation of a new species in populations that are geographically isolated from one another. *See also* sympatric speciation.

alternation of generations
A life cycle in which there is both a multicellular diploid form, the sporophyte, and a multicellular haploid form, the gametophyte; a characteristic of plants and multicellular green algae.

alternative RNA splicing
A type of regulation at the RNA-processing level in which different mRNA molecules are produced from the same primary transcript, depending on which RNA segments are treated as exons and which as introns.

amino acid (uh-mēn′-ō)
An organic molecule containing a carboxyl group, an amino group, a hydrogen atom, and a variable side chain (also called a radical group, or R group); serves as the monomer of proteins.

amniote (am′-nē-ōt)
Member of a clade of tetrapods that has an amniotic egg containing specialized membranes that protect the embryo. Amniotes include mammals and reptiles (including birds).

amniotic egg (am′-nē-ot′-ik)
A shelled egg in which an embryo develops within a fluid-filled amniotic sac and is nourished by yolk. Produced by reptiles (including birds) and egg-laying mammals, it enables them to complete their life cycles on dry land.

amoeba (uh-mē′-buh)
A general term for a protozoan (animal-like protist) characterized by great structural flexibility and the presence of pseudopodia.

amphibian
Member of a class of vertebrate animals that includes frogs and salamanders.

anaerobic (an'-ār-ō'-bik)
Lacking or not requiring molecular oxygen (O_2).

analogy
The similarity between two species that is due to convergent evolution rather than to descent from a common ancestor with the same trait.

anaphase
The third stage of mitosis, beginning when sister chromatids separate from each other and ending when a complete set of daughter chromosomes has arrived at each of the two poles of the cell.

anecdotal evidence
An assertion based on a single or just a few examples. Anecdotal evidence is not considered valid proof of a generalized conclusion.

angiosperm (an'-jē-ō-sperm)
A flowering plant, which forms seeds inside a protective chamber called an ovary.

animal
A eukaryotic, multicellular, heterotrophic organism that obtains nutrients by ingestion.

annelid (an'-uh-lid)
A segmented worm. Annelids include earthworms, polychaetes, and leeches.

anther
A sac in which pollen grains develop, located at the tip of a flower's stamen.

anthropoid (an'-thruh-poyd)
Member of a primate group made up of the apes (gibbons, orangutans, gorillas, chimpanzees, and bonobos), monkeys, and humans.

anticodon (an'-tī-kō'-don)
On a tRNA molecule, a specific sequence of three nucleotides that is complementary to a codon triplet on mRNA.

aphotic zone (ā-fō'-tik)
The region of an aquatic ecosystem beneath the photic zone, where light levels are too low for photosynthesis to take place.

apicomplexan (ap'-ē-kom-pleks'-un)
A type of parasitic protozoan (animal-like protist). Some apicomplexans cause serious human disease.

aqueous solution
A solution in which water is the solvent.

arachnid
Member of a major arthropod group that includes spiders, scorpions, ticks, and mites.

Archaea (ar-kē'-uh)
One of two prokaryotic domains of life, the other being Bacteria.

archaean (ar-kē'-uhn)
(plural, **archaea**) An organism that is a member of the domain Archaea.

arthropod (ar'-thruh-pod)
Member of the most diverse phylum in the animal kingdom; includes the horseshoe crab, arachnids (for example, spiders, ticks, scorpions, and mites), crustaceans (for example, crayfish, lobsters, crabs, and barnacles), millipedes, centipedes, and insects. Arthropods are characterized by a chitinous exoskeleton, molting, jointed appendages, and a body formed of distinct groups of segments.

artificial selection
The selective breeding of domesticated plants and animals to promote the occurrence of desirable traits in the offspring.

asexual reproduction
The creation of genetically identical offspring by a single parent, without the participation of gametes (sperm and egg).

atom
The smallest unit of matter that retains the properties of an element.

atomic mass
The total mass of an atom.

atomic number
The number of protons in each atom of a particular element. Elements are ordered by atomic number in the periodic table of the elements.

ATP (adenosine triphosphate)
(a-den'-ō-sēn trī-fos'-fāt)
A molecule composed of adenosine and three phosphate groups; the main energy source for cells. A molecule of ATP can be broken down to a molecule of ADP (adenosine diphosphate) and a free phosphate; this reaction releases energy that can be used for cellular work.

ATP synthase
A protein cluster, found in a cellular membrane (including the inner membrane of mitochondria, the thylakoid membrane of chloroplasts, and the plasma membrane of prokaryotes), that uses the energy of a hydrogen ion concentration gradient to make ATP from ADP. An ATP synthase provides a port through which hydrogen ions (H^+) diffuse.

autosome
A chromosome not directly involved in determining the sex of an organism; in mammals, for example, any chromosome other than X or Y.

autotroph (ot'-ō-trōf)
An organism that makes its own food from inorganic ingredients, thereby sustaining itself without eating other organisms or their molecules. Plants, algae, and photosynthetic bacteria are autotrophs.

B

bacillus (buh-sil'-us)
(plural, **bacilli**) A rod-shaped prokaryotic cell.

Bacteria
One of two prokaryotic domains of life, the other being Archaea.

bacteriophage (bak-tēr'-ē-ō-fāj)
A virus that infects bacteria; also called a phage.

bacterium
(plural, **bacteria**) An organism that is a member of the domain Bacteria.

base
A substance that decreases the hydrogen ion (H^+) concentration in a solution.

benign tumor
An abnormal mass of cells that remains at its original site in the body.

benthic realm
A seafloor or the bottom of a freshwater lake, pond, river, or stream. The benthic realm is occupied by communities of organisms known as benthos.

bilateral symmetry
An arrangement of body parts such that an organism can be divided equally by a single cut passing longitudinally through it. A bilaterally symmetric organism has mirror-image right and left sides.

binary fission
A means of asexual reproduction in which a parent organism, often a single cell, divides into two individuals of about equal size.

binomial
The two-part format for naming a species; for example, *Homo sapiens*.

biocapacity
Earth's capacity to produce the resources such as food, water, and fuel consumed by humans and to absorb human-generated waste.

biodiversity
The variety of living things; includes genetic diversity, species diversity, and ecosystem diversity.

biodiversity hot spot
A small geographic area that contains a large number of threatened or endangered species and an exceptional concentration of endemic species (those found nowhere else).

biofilm
A surface-coating cooperative colony of prokaryotes.

biogeochemical cycle
Any of the various chemical circuits occurring in an ecosystem, involving both biotic and abiotic components of the ecosystem.

biogeography
The study of the past and present distribution of organisms.

bioinformatics
A scientific field of study that uses mathematics to develop methods for organizing and analyzing large sets of biological data.

biological control
The intentional release of a natural enemy to attack a pest population.

biological magnification
The accumulation of persistent chemicals in the living tissues of consumers in food chains.

biological species concept
The definition of a species as a population or group of populations the members of which have the potential in nature to interbreed and produce fertile offspring.

biology
The scientific study of life.

biomass
The amount, or mass, of living organic material in an ecosystem.

biome (bī'-ōm)
A major terrestrial or aquatic life zone, characterized by vegetation type in terrestrial biomes and the physical environment in aquatic biomes.

bioremediation
The use of living organisms to detoxify and restore polluted and degraded ecosystems.

biosphere
The global ecosystem; the entire portion of Earth inhabited by life; all of life and where it lives.

biotechnology
The manipulation of living organisms to perform useful tasks.

biotic factor (*bī-ot'-ik*)
A living component of a biological community; any organism that is part of an individual's environment.

bird
Member of a group of reptiles with feathers and adaptations for flight.

bivalve
Member of a group of molluscs that includes clams, mussels, scallops, and oysters.

blastula (*blas'-tyū-luh*)
An embryonic stage that marks the end of cleavage during animal development; a hollow ball of cells in many species.

body cavity
A fluid-filled space separating the digestive tract from the outer body wall.

body segmentation
Subdivision of an animal's body into a series of repeated parts called segments.

bony fish
A fish that has a stiff skeleton reinforced by calcium salts.

bottleneck effect
Genetic drift resulting from a drastic reduction in population size. Typically, the surviving population is no longer genetically representative of the original population.

bryophyte (*brī'-uh-fīt*)
A type of plant that lacks xylem and phloem; a nonvascular plant. Bryophytes include mosses and their close relatives.

buffer
A chemical substance that decreases the hydrogen ion (H⁺) concentration in a solution.

C

calorie
The amount of energy that raises the temperature of 1 g of water by 1°C. Commonly reported as Calories, which are kilocalories (1,000 calories).

Calvin cycle
The second of two stages of photosynthesis; a cyclic series of chemical reactions that occur in the stroma of a chloroplast, using the carbon in CO_2 and the ATP and NADPH produced by the light reactions to make the energy-rich sugar molecule G3P, which is later used to produce glucose.

cancer
A malignant growth or tumor caused by abnormal and uncontrolled cell division.

carbohydrate (*kar'-bō-hī'-drāt*)
A biological molecule consisting of a simple sugar (a monosaccharide), two monosaccharides joined into a double sugar (a disaccharide), or a chain of monosaccharides (a polysaccharide).

carbon fixation
The initial incorporation of carbon from CO_2 into organic compounds by autotrophic organisms such as photosynthetic plants, algae, or bacteria.

carbon footprint
The amount of greenhouse gas emitted as a result of the actions of a person, nation, or other entity.

carcinogen (*kar-sin'-uh-jin*)
A cancer-causing agent, either high-energy radiation (such as X-rays or UV light) or a chemical.

carnivore
An animal that mainly eats other animals. *See also* herbivore; omnivore.

carpel (*kar'-pul*)
The egg-producing part of a flower, consisting of a stalk with an ovary at the base and a stigma, which traps pollen, at the tip.

carrier
An individual who is heterozygous for a recessively inherited disorder and who therefore does not show symptoms of that disorder.

carrying capacity
The maximum population size that a particular environment can sustain.

cartilaginous fish (*kar-ti-laj'-uh-nus*)
A fish that has a flexible skeleton made of cartilage.

cell cycle
An ordered sequence of events (including interphase and the mitotic phase) that extends from the time a eukaryotic cell is first formed from a dividing parent cell until its own division into two cells.

cell cycle control system
A cyclically operating set of proteins that triggers and coordinates events in the eukaryotic cell cycle.

cell division
The reproduction of a cell.

cell plate
A membranous disk that forms across the midline of a dividing plant cell. During cytokinesis, the cell plate grows outward, accumulating more cell wall material and eventually fusing into a new cell wall.

cell theory
The theory that all living things are composed of cells and that all cells come from earlier cells.

cellular respiration
The aerobic harvesting of energy from food molecules; the energy-releasing chemical breakdown of food molecules, such as glucose, and the storage of potential energy in a form that cells can use to perform work; involves glycolysis, the citric acid cycle, the electron transport chain, and chemiosmosis.

cellulose (*sel'-yū-lōs*)
A large polysaccharide composed of many glucose monomers linked into cable-like fibrils that provide structural support in plant cell walls. Because cellulose cannot be digested by animals, it acts as fiber, or roughage, in the diet.

centipede
A carnivorous terrestrial arthropod that has one pair of long legs for each of its numerous body segments, with the front pair modified as poisonous claws.

central vacuole (*vak'-yū-ōl*)
A membrane-enclosed sac occupying most of the interior of a mature plant cell, having diverse roles in reproduction, growth, and development.

centromere (*sen'-trō-mer*)
The region of a chromosome where two sister chromatids are joined and where spindle microtubules attach during mitosis and meiosis. The centromere divides at the onset of anaphase during mitosis and anaphase II of meiosis.

cephalopod
Member of a group of molluscs that includes squids and octopuses.

chaparral (*shap-uh-ral'*)
A terrestrial biome limited to coastal regions where cold ocean currents circulate offshore, creating mild, rainy winters and long, hot, dry summers; also known as the Mediterranean biome. Chaparral vegetation is adapted to fire.

character
A heritable feature that varies among individuals within a population, such as flower color in pea plants or eye color in humans.

charophyte (*kār'-uh-fīt'*)
A member of the green algal group that shares features with land plants. Charophytes are considered the closest relatives of land plants;

modern charophytes and modern plants likely evolved from a common ancestor.

chemical bond
An attraction between two atoms resulting from a sharing of outer-shell electrons or the presence of opposite charges on the atoms.

chemical cycling
The use and reuse of chemical elements such as carbon within an ecosystem.

chemical energy
Energy stored in the chemical bonds of molecules; a form of potential energy.

chemical reaction
A process leading to chemical changes in matter, involving the making and/or breaking of chemical bonds. A chemical reaction involves rearranging atoms, but no atoms are created or destroyed.

chemotherapy (*kē'-mō-ther'-uh-pē*)
Treatment for cancer in which drugs are administered to disrupt cell division of the cancer cells.

chlorophyll (*klor'-ō-fil*)
A light-absorbing pigment in chloroplasts that plays a central role in converting solar energy to chemical energy.

chloroplast (*klō'-rō-plast*)
An organelle found in plants and photosynthetic protists. Enclosed by two membranes, a chloroplast absorbs sunlight and uses it to power the synthesis of organic food molecules (sugars).

chordate (*kōr'-dāt*)
An animal that at some point during its development has a dorsal, hollow nerve cord, a notochord, pharyngeal slits, and a post-anal tail. Chordates include lancelets, tunicates, and vertebrates.

chromatin (*krō'-muh-tin*)
The combination of DNA and proteins that constitutes chromosomes; often used to refer to the diffuse, very extended form taken by the chromosomes when a eukaryotic cell is not dividing.

chromosome (*krō'-muh-sōm*)
A gene-carrying structure found in the nucleus of a eukaryotic cell and most visible when compacted during mitosis and meiosis; also, the main gene-carrying structure of a prokaryotic cell. Each chromosome consists of one very long threadlike DNA molecule and associated proteins. *See also* chromatin.

embryonic stem cell (ES cell)
Any of the cells in the early animal embryo that differentiate during development to give rise to all the kinds of specialized cells in the body.

emerging virus
A virus that has appeared suddenly or has recently come to the attention of medical scientists.

endangered species
As defined in the U.S. Endangered Species Act, a species that is in danger of extinction throughout all or a significant portion of its range.

endemic species
A species whose distribution is limited to a specific geographic area.

endocytosis (*en'-dō-sī-tō'-sis*)
The movement of materials from the external environment into the cytoplasm of a cell via vesicles or vacuoles.

endomembrane system
A network of organelles that partitions the cytoplasm of eukaryotic cells into functional compartments. Some of the organelles are structurally connected to each other, whereas others are structurally separate but functionally connected by the traffic of vesicles among them.

endoplasmic reticulum (ER)
(*reh-tik'-yuh-lum*)
An extensive membranous network in a eukaryotic cell, continuous with the outer nuclear membrane and composed of ribosome-studded (rough) and ribosome-free (smooth) regions. *See also* rough ER; smooth ER.

endoskeleton
A hard interior skeleton located within the soft tissues of an animal; found in all vertebrates and a few invertebrates (such as echinoderms).

endosperm
In flowering plants, a nutrient-rich mass formed by the union of a sperm cell with the diploid central cell of the embryo sac during double fertilization; provides nourishment to the developing embryo in the seed.

endospore
A thick-coated, protective cell produced within a prokaryotic cell exposed to harsh conditions.

endotherm
An animal that derives most of its body heat from its own metabolism.

endotoxin
A poisonous component of the outer membrane of certain bacteria.

energy
The capacity to cause change, or to move matter in a direction it would not move if left alone.

energy flow
The passage of energy through the components of an ecosystem.

energy pyramid
A diagram depicting the cumulative loss of energy with each transfer in a food chain.

enhancer
A eukaryotic DNA sequence that helps stimulate the transcription of a gene at some distance from it. An enhancer functions by means of a transcription factor called an activator, which binds to it and then to the rest of the transcription apparatus.

entropy (*en'-truh-pē*)
A measure of disorder, or randomness. One form of disorder is heat, which is random molecular motion.

enzyme (*en'-zīm*)
A molecule (usually a protein, but sometimes RNA) that serves as a biological catalyst, changing the rate of a chemical reaction without itself being changed in the process.

enzyme inhibitor
A chemical that interferes with an enzyme's activity by changing the enzyme's shape, either by plugging up the active site or by binding to another site on the enzyme.

epigenetic inheritance
Inheritance of traits transmitted by mechanisms not directly involving the nucleotide sequence of a genome; frequently involves chemical modification of DNA bases and/or histone proteins.

equilibrial life history
(*ē-kwi-lib'-rē-ul*)
The pattern of reaching sexual maturity slowly and producing few offspring but caring for the young; often seen in long-lived, large-bodied species.

errantian
A member of a major annelid lineage that includes mostly marine worms with an active lifestyle.

estuary (*es'-chuh-wār-ē*)
The area where a freshwater stream or river merges with seawater.

Eukarya (*yū-kār'-yuh*)
The domain of eukaryotes, organisms made up of eukaryotic cells; includes all of the protists, plants, fungi, and animals.

eukaryote (*yū-kār'-ē-ōt*)
An organism characterized by eukaryotic cells. *See also* eukaryotic cell.

eukaryotic cell (*yū-kār'-ē-ot'-ik*)
A type of cell that has a membrane-enclosed nucleus and other membrane-enclosed organelles. All organisms except bacteria and archaea (including protists, plants, fungi, and animals) are composed of eukaryotic cells.

eutherian (*yū-thēr'-ē-un*)
Mammal whose young complete their embryonic development in the uterus, nourished via the mother's blood vessels in the placenta; also called a placental mammal.

evaporative cooling
A property of water whereby a body becomes cooler as water evaporates from it.

evolution
Descent with modification; the idea that living species are descendants of ancestral species that were different from present-day ones; also, defined more narrowly as the change in the genetic composition of a population from generation to generation. Evolution is the central unifying theme of biology.

evolutionary adaptation
An inherited characteristic that enhances an organism's ability to survive and reproduce in a particular environment.

evolutionary tree
A branching diagram that reflects a hypothesis about evolutionary relationships among groups of organisms.

exaptation
A structure that evolves in one context and gradually becomes adapted for other functions.

exocytosis (*ek'-sō-sī-tō'-sis*)
The movement of materials out of the cytoplasm of a cell via membranous vesicles or vacuoles.

exon (*ek'-son*)
In eukaryotes, a coding portion of a gene. *See also* intron.

exoskeleton
A hard, external skeleton that protects an animal and provides points of attachment for muscles.

exotoxin
A poisonous protein secreted by certain bacteria.

experiment
A scientific test, often carried out under controlled conditions, that involves changing just one factor (the variable) at a time.

experimental group
A set of subjects that has (or receives) the specific factor being tested in a controlled experiment. Ideally, the experimental group should be identical to the control group for all other factors.

exponential population growth
A model that describes the expansion of a population in an ideal, unlimited environment.

extracellular matrix
The meshwork that surrounds animal cells, consisting of a web of protein and polysaccharide fibers embedded in a liquid, jelly, or solid.

F

F_1 generation
The offspring of two parental (P generation) individuals. F_1 stands for first filial.

F_2 generation
The offspring of the F_1 generation. F_2 stands for second filial.

facilitated diffusion
The passage of a substance across a biological membrane down its concentration gradient aided by specific transport proteins.

fact
A piece of information that is correct based on all current information. A fact is not subject to opinion and can be independently verified.

family
In classification, the taxonomic category above genus.

fat
A large lipid molecule made from an alcohol called glycerol and three fatty acids; a triglyceride. Most fats function as energy-storage molecules.

fermentation
The anaerobic harvest of energy from food by some cells. Different pathways of fermentation can produce different end products, including ethanol and lactic acid.

fern
Any of a group of seedless vascular plants.

fertilization
The union of a haploid sperm cell with a haploid egg cell, producing a zygote.

flagella (*fla-jel′-uh*)
Extensions from a eukaryotic cell that propel the cell with an undulating, whiplike motion.

flagellate (*flaj′-uh-lit*)
A protozoan (animal-like protist) that moves by means of one or more flagella.

flatworm
A bilateral animal with a thin, flat body form, a gastrovascular cavity with a single opening, and no body cavity. Flatworms include planarians, flukes, and tapeworms.

flower
In an angiosperm, a short stem with four sets of modified leaves, bearing structures that function in sexual reproduction.

fluid mosaic
A description of membrane structure, depicting a cellular membrane as a mosaic of diverse protein molecules suspended in a fluid bilayer of phospholipid molecules.

food chain
The sequence of food transfers between the trophic levels of a community, beginning with the producers.

food web
A network of interconnecting food chains.

foram
A marine protozoan (animal-like protist) that secretes a shell and extends pseudopodia through pores in its shell.

forensics
The scientific analysis of evidence for crime scene investigations and other legal proceedings.

fossil
A preserved imprint or remains of an organism that lived in the past.

fossil fuel
An energy deposit formed from the fossilized remains of long-dead plants and animals.

fossil record
The ordered sequence of fossils as they appear in rock layers, marking the passing of geologic time.

founder effect
Genetic drift resulting from the establishment of a new, small population whose gene pool represents only a sample of the genetic variation present in the original population.

fruit
A ripened, thickened ovary of a flower, which protects dormant seeds and aids in their dispersal.

functional group
A group of atoms that form the chemically reactive part of an organic molecule. A particular functional group usually behaves similarly in different chemical reactions.

fungus
(plural, **fungi**) A heterotrophic eukaryote that digests its food externally and absorbs the resulting small nutrient molecules. Most fungi consist of a netlike mass of filaments called hyphae. Molds, mushrooms, and yeasts are examples of fungi.

G

gamete (*gam′-ēt*)
A sex cell; a haploid egg or sperm. The union of two gametes of opposite sex (fertilization) produces a zygote.

gametophyte (*guh-mē′-tō-fīt*)
The multicellular haploid form in the life cycle of organisms undergoing alternation of generations; results from a union of spores and mitotically produces haploid gametes that unite and grow into the sporophyte generation.

gastropod
Member of the largest group of molluscs, including snails and slugs.

gastrovascular cavity
A digestive compartment with a single opening that serves as both the entrance for food and the exit for undigested wastes; may also function in circulation, body support, and gas exchange. Jellies and hydras are examples of animals with a gastrovascular cavity.

gastrula (*gas′-trū-luh*)
The embryonic stage resulting from gastrulation in animal development. Most animals have a gastrula made up of three layers of cells: ectoderm, endoderm, and mesoderm.

gel electrophoresis
(*jel e-lek′-trō-fōr-ē′-sis*)
A technique for sorting macromolecules. A mixture of molecules is placed on a gel between a positively charged electrode and a negatively charged one; negatively charged molecules migrate toward the positive electrode. The molecules separate in the gel according to their rates of migration.

gene
A unit of inheritance in DNA (or RNA, in some viruses) consisting of a specific nucleotide sequence that programs the amino acid sequence of a polypeptide. Most of the genes of a eukaryote are located in its chromosomal DNA; a few are carried by the DNA of mitochondria and chloroplasts.

gene cloning
The production of multiple copies of a gene.

gene expression
The process whereby genetic information flows from genes to proteins; the flow of genetic information from the genotype to the phenotype: DNA → RNA → protein.

gene flow
The gain or loss of alleles from a population by the movement of individuals or gametes into or out of the population.

gene pool
All copies of every type of allele at every locus in all members of a population at any one time.

gene regulation
The turning on and off of specific genes within a living organism.

genetically modified (GM) organism
An organism that has acquired one or more genes by artificial means. If the gene is from another organism, typically of another species, the recombinant organism is also known as a transgenic organism. *See also* transgenic organism.

genetic code
The set of rules giving the correspondence between nucleotide triplets (codons) in mRNA and amino acids in protein.

genetic drift
A change in the gene pool of a population due to chance. Effects of genetic drift are most pronounced in small populations.

genetic engineering
The direct manipulation of genes for practical purposes.

genetics
The scientific study of heredity (inheritance).

genomics
The study of whole sets of genes and their interactions.

genotype (*jē′-nō-tīp*)
The genetic makeup of an organism.

genus (*jē′-nus*)
(plural, **genera**) In classification, the taxonomic category above species; the first part of a species' binomial; for example, *Homo*.

geologic time scale
A time scale established by geologists that divides Earth's history into a sequence of geologic periods, grouped into four divisions: Precambrian, Paleozoic, Mesozoic, and Cenozoic.

germinate
To initiate growth, as in a plant seed or a plant or fungal spore.

glycogen (*glī′-kō-jen*)
A complex, extensively branched polysaccharide made up of many glucose monomers; serves as a temporary energy-storage molecule in liver and muscle cells.

glycolysis (*glī-kol′-uh-sis*)
The multistep chemical breakdown of a molecule of glucose into two molecules of pyruvic acid; the first stage of cellular respiration in all organisms; occurs in the cytoplasmic fluid.

Golgi apparatus (*gol′-jē*)
An organelle in eukaryotic cells consisting of stacks of membranous sacs that modify, store, and ship products of the endoplasmic reticulum.

granum (*gran′-um*)
(plural, **grana**) A stack of hollow disks formed of thylakoid membrane in a chloroplast. Grana are the sites where light energy is trapped by chlorophyll and converted to chemical energy during the light reactions of photosynthesis.

green alga (*al′-guh*)
One of a group of photosynthetic protists that includes unicellular, colonial, and multicellular species. Green algae are the photosynthetic protists most closely related to plants.

greenhouse effect
The warming of the atmosphere caused by CO_2, CH_4, and other gases that absorb heat radiation and slow its escape from Earth's surface.

greenhouse gas
Any of the gases in the atmosphere that absorb heat radiation, including CO_2, methane, water vapor, and synthetic chlorofluorocarbons.

growth factor
A protein secreted by certain body cells that stimulates other cells to divide.

guanine (G) (*gwa′-nēn*)
A double-ring nitrogenous base found in DNA and RNA.

gymnosperm (*jim′-nō-sperm*)
A naked-seed plant. Its seed is said to be naked because it is not enclosed in an ovary.

H

habitat
A place where an organism lives; a specific environment in which an organism lives.

half-life
The amount of time it takes for 50% of a sample of a radioactive isotope to decay.

haploid
Containing a single set of chromosomes; referring to an *n* cell.

Hardy-Weinberg equilibrium
The condition describing a nonevolving population (one that is in genetic equilibrium).

heat
The amount of kinetic energy contained in the movement of the atoms and molecules in a body of matter. Heat is energy in its most random form.

herbivore
An animal that eats mainly plants, algae, or phytoplankton. *See also* carnivore; omnivore.

herbivory
The consumption of plant parts or algae by an animal.

heredity
The transmission of traits from one generation to the next.

heterotroph (*het'-er-ō-trōf*)
An organism that cannot make its own organic food molecules from inorganic ingredients and must obtain them by consuming other organisms or their organic products; a consumer (such as an animal) or a decomposer (such as a fungus) in a food chain.

heterozygous (*het'-er-ō-zī'-gus*)
Having two different alleles for a given gene.

histone (*his'-tōn*)
A small protein molecule associated with DNA and important in DNA packing in the eukaryotic chromosome.

HIV
Human immunodeficiency virus; the retrovirus that attacks the human immune system and causes AIDS.

homeotic gene (*hō'-mē-ot'-ik*)
A master control gene that determines the identity of a body structure of a developing organism, presumably by controlling the developmental fate of groups of cells. (In plants, such genes are called organ identity genes.)

hominin (*hah'-mi-nin*)
Any anthropoid on the human branch of the evolutionary tree, more closely related to humans than to chimpanzees.

homologous chromosomes
(*hō-mol'-uh-gus*)
The two chromosomes that make up a matched pair in a diploid cell. Homologous chromosomes are of the same length, centromere position, and staining pattern and possess genes for the same characteristics at corresponding loci. One homologous chromosome is inherited from the organism's father, the other from the mother.

homology (*hō-mol'-uh-jē*)
Similarity in characteristics resulting from a shared ancestry.

homozygous (*hō'-mō-zī'-gus*)
Having two identical alleles for a given gene.

host
An organism that is exploited by a parasite or pathogen.

human gene therapy
A recombinant DNA procedure intended to treat disease by altering an afflicted person's genes.

Human Genome Project
An international collaborative effort that sequenced the DNA of the entire human genome.

hybrid
The offspring of parents of two different species or of two different varieties of one species; the offspring of two parents that differ in one or more inherited traits; an individual that is heterozygous for one or more pairs of genes.

hydrogen bond
A type of weak chemical bond formed when a partially positive hydrogen atom from one polar molecule is attracted to the partially negative atom in another molecule (or in another part of the same molecule).

hydrogenation
The artificial process of converting unsaturated fats to saturated fats by adding hydrogen.

hydrolysis (*hī-drol'-uh-sis*)
A chemical process in which macromolecules are broken down by the chemical addition of water molecules to the bonds linking their monomers; an essential part of digestion. A hydrolysis reaction is essentially the opposite of a dehydration reaction.

hydrophilic (*hī'-drō-fil'-ik*)
"Water-loving"; pertaining to polar, or charged, molecules (or parts of molecules), which are soluble in water.

hydrophobic (*hī'-drō-fō'-bik*)
"Water-fearing"; pertaining to nonpolar molecules (or parts of molecules), which do not dissolve in water.

hypertonic
In comparing two solutions, referring to the one with the greater concentration of solutes.

hypha (*hī'-fuh*)
(plural, **hyphae**) One of many filaments making up the body of a fungus.

hypothesis (*hī-poth'-uh-sis*)
(plural, **hypotheses**) A tentative explanation that a scientist proposes for a specific phenomenon that has been observed.

hypotonic
In comparing two solutions, referring to the one with the lower concentration of solutes.

I

incomplete dominance
A type of inheritance in which the phenotype of a heterozygote (*Aa*) is intermediate between the phenotypes of the two types of homozygotes (*AA* and *aa*).

independent variable
A factor whose value is manipulated or changed during an experiment to reveal possible effects on another factor (the dependent variable).

induced fit
The interaction between a substrate molecule and the active site of an enzyme, which changes shape slightly to embrace the substrate and catalyze the reaction.

insect
An arthropod that usually has three body segments (head, thorax, and abdomen), three pairs of legs, and one or two pairs of wings.

interphase
The phase in the eukaryotic cell cycle when the cell is not actually dividing. During interphase, cellular metabolic activity is high, chromosomes and organelles are duplicated, and cell size may increase. Interphase accounts for 90% of the cell cycle. *See also* mitosis.

interspecific competition
Competition between individuals of two or more species that require similar limited resources.

interspecific interaction
Any interaction between members of different species.

intertidal zone (*in'-ter-tīd'-ul*)
A shallow zone where the waters of an estuary or ocean meet land.

intraspecific competition
Competition between individuals of the same species for the same limited resources.

intron (*in'-tron*)
In eukaryotes, a nonexpressed (noncoding) portion of a gene that is excised from the RNA transcript. *See also* exon.

invasive species
A non-native species that has spread far beyond the original point of introduction and causes environmental or economic damage by colonizing and dominating suitable habitats.

invertebrate
An animal that does not have a backbone.

ion
An atom or molecule that has gained or lost one or more electrons, thus acquiring an electrical charge.

ionic bond
An attraction between two ions with opposite electrical charges. The electrical attraction of the opposite charges holds the ions together.

isomer (*ī'-sō-mer*)
One of two or more molecules with the same molecular formula but different structures and thus different properties.

isotonic (*ī-sō-ton'-ik*)
Having the same solute concentration as another solution.

isotope (*ī'-sō-tōp*)
A variant form of an atom. Different isotopes of an element have the same number of protons and electrons but different numbers of neutrons.

K

karyotype (*kār'-ē-ō-tīp*)
A display of micrographs of the metaphase chromosomes of a cell, arranged by size and centromere position.

keystone species
A species whose impact on its community is much larger than its biomass or abundance indicates.

kinetic energy (*kuh-net'-ik*)
Energy of motion. Moving matter performs work by transferring its motion to other matter, such as leg muscles pushing bicycle pedals.

kingdom
In classification, the broad taxonomic category above phylum.

L

lancelet
One of a group of bladelike invertebrate chordates.

landscape
A regional assemblage of interacting ecosystems.

landscape ecology
The application of ecological principles to the study of land-use patterns; the scientific study of the biodiversity of interacting ecosystems.

larva
An immature individual that looks different from the adult animal.

lateral line system
A row of sensory organs along each side of a fish's body. Sensitive to changes in water pressure, it enables a fish to detect minor vibrations in the water.

law of independent assortment
A general rule of inheritance, first proposed by Gregor Mendel, that states that when gametes form during meiosis, each pair of alleles for a particular character segregates (separates) independently of each other pair.

law of segregation
A general rule of inheritance, first proposed by Gregor Mendel, that states that the two alleles in a pair segregate (separate) into different gametes during meiosis.

life cycle
The entire sequence of generation-to-generation stages in the life of an organism, from fertilization to the production of its own offspring

life history
The traits that affect an organism's schedule of reproduction and survival.

life table
A listing of survivals and deaths in a population in a particular time period and predictions of how long, on average, an individual of a given age will live.

light reactions
The first of two stages in photosynthesis; the steps in which solar energy is absorbed and converted to chemical energy in the form of ATP and NADPH. The light reactions power the sugar-producing Calvin cycle but produce no sugar themselves.

lignin (*lig'-nin*)
A chemical that hardens the cell walls of plants. Lignin makes up most of what we call wood.

limiting factor
An environmental factor that restricts the number of individuals that can occupy a particular habitat, thus holding population growth in check.

linked genes
Genes located close enough together on a chromosome that they are usually inherited together.

lipid
An organic compound consisting mainly of carbon and hydrogen atoms linked by nonpolar covalent bonds and therefore mostly hydrophobic and insoluble in water. Lipids include fats, waxes, phospholipids, and steroids.

lobe-finned fish
A bony fish with strong, muscular fins supported by bones. *See also* ray-finned fish.

locus
(plural, **loci**) The particular site where a gene is found on a chromosome. Homologous chromosomes have corresponding gene loci.

logistic population growth
A model that describes population growth that decreases as population size approaches carrying capacity.

lysogenic cycle (*lī-sō-jen'-ik*)
A bacteriophage reproductive cycle in which the viral genome is incorporated into the bacterial host chromosome as a prophage. New phages are not produced, and the host cell is not killed or lysed unless the viral genome leaves the host chromosome.

lysosome (*lī'-sō-sōm*)
A digestive organelle in eukaryotic cells; contains enzymes that digest the cell's food and wastes.

lytic cycle (*lit'-ik*)
A viral reproductive cycle resulting in the release of new viruses by lysis (breaking open) of the host cell.

M

macroevolution
Evolutionary change above the species level. Examples of macroevolutionary change include the origin of a new group of organisms through a series of speciation events, the impact of mass extinctions on the diversity of life, and the origin of key adaptations.

macromolecule
A giant molecule formed by joining smaller molecules. Examples of macromolecules include proteins, polysaccharides, and nucleic acids.

malignant tumor
An abnormal tissue mass that spreads into neighboring tissue and to other parts of the body; a cancerous tumor.

mammal
Member of a class of endothermic amniotes that possesses mammary glands and hair.

mantle
In molluscs, the outgrowth of the body surface that drapes over the animal. The mantle produces the shell and forms the mantle cavity.

marsupial (*mar-sū'-pē-ul*)
A pouched mammal, such as a kangaroo, opossum, or koala. Marsupials give birth to embryonic offspring that complete development while housed in a pouch and attached to nipples on the mother's abdomen.

mass
A measure of the amount of matter in an object.

mass number
The sum of the number of protons and neutrons in an atom's nucleus.

matter
Anything that occupies space and has mass.

medusa (*med-ū'-suh*)
(plural, **medusae**) One of two types of cnidarian body forms; a floating, umbrella-like body form; also called a jelly.

meiosis (*mī-ō'-sis*)
In a sexually reproducing organism, the process of cell division that produces haploid gametes from diploid cells within the reproductive organs.

messenger RNA (mRNA)
The type of ribonucleic acid that encodes genetic information from DNA and conveys it to ribosomes, where the information is translated into amino acid sequences.

metabolism (*muh-tab'-uh-liz-um*)
The total of all the chemical reactions in an organism.

metamorphosis (*met'-uh-mōr'-fuh-sis*)
The transformation of a larva into an adult.

metaphase (*met'-eh-fāz*)
The second stage of mitosis. During metaphase, the centromeres of all the cell's duplicated chromosomes are lined up along the center line of the cell.

metastasis (*muh-tas'-tuh-sis*)
The spread of cancer cells beyond their original site.

microbiome
The collection of genomes of individual microbial species present in a particular environment, such as the human intestinal tract.

microbiota
The community of microorganisms that live in and on the body of an animal.

microevolution
A change in a population's gene pool over a succession of generations.

microRNA (miRNA)
A small, single-stranded RNA molecule that associates with one or more proteins in a complex that can degrade or prevent translation of an mRNA with a complementary sequence.

microtubule
The thickest of the three main kinds of fibers making up the cytoskeleton of a eukaryotic cell; a straight, hollow tube made of globular proteins called tubulins. Microtubules form the basis of the structure and movement of cilia and flagella.

millipede
A terrestrial arthropod that has two pairs of short legs for each of its numerous body segments and that eats decaying plant matter.

mitochondrion (*mī'-tō-kon'-drē-on*)
(plural, **mitochondria**) An organelle in eukaryotic cells where cellular respiration occurs. Enclosed by two concentric membranes, it is where most of the cell's ATP is made.

mitosis (*mī-tō'-sis*)
The division of a single nucleus into two genetically identical daughter nuclei. Mitosis and cytokinesis make up the mitotic (M) phase of the cell cycle.

mitotic (M) phase
The phase of the cell cycle when mitosis divides the nucleus and distributes its chromosomes to the daughter nuclei and cytokinesis divides the cytoplasm, producing two daughter cells.

mitotic spindle
A spindle-shaped structure formed of microtubules and associated proteins that is involved in the movement of chromosomes during mitosis and meiosis. (A spindle is shaped roughly like a football.)

molecular biology
The study of biological structures, functions, and heredity at the molecular level.

molecule
A group of two or more atoms held together by covalent bonds.

mollusc (*mol′-lusk*)
A soft-bodied animal characterized by a muscular foot, mantle, mantle cavity, and radula. Molluscs include gastropods (snails and slugs), bivalves (clams, oysters, and scallops), and cephalopods (squids and octopuses).

monohybrid cross
A mating of individuals that are heterozygous for the character being followed.

monomer (*mon′-uh-mer*)
A chemical subunit that serves as a building block of a polymer.

monosaccharide (*mon′-uh-sak′-uh-rīd*)
The smallest kind of sugar molecule; a single-unit sugar; also known as a simple sugar.

monotreme (*mon′-uh-trēm*)
An egg-laying mammal, such as the duck-billed platypus.

moss
Any of a group of seedless nonvascular plants.

movement corridor
A series of small clumps or a narrow strip of quality habitat (usable by organisms) that connects otherwise isolated patches of quality habitat.

mutagen (*myū′-tuh-jen*)
A chemical or physical agent that interacts with DNA and causes a mutation.

mutation
Any change to the genetic information of a cell or virus.

mutualism
An interspecific interaction in which both partners benefit.

mycelium (*mī-sē′-lē-um*)
(plural, **mycelia**) The densely branched network of hyphae in a fungus.

mycorrhiza (*mī′-kō-rī′-zuh*)
(plural, **mycorrhizae**) A mutually beneficial symbiotic association of a plant root and fungus.

N

NADH
An electron carrier (a molecule that carries electrons) involved in cellular respiration and photosynthesis. NADH carries electrons from glucose and other fuel molecules and deposits them at the top of an electron transport chain. NADH is generated during glycolysis and the citric acid cycle.

NADPH
An electron carrier (a molecule that carries electrons) involved in photosynthesis. Light drives electrons from chlorophyll to NADP$^+$, forming NADPH, which provides the high-energy electrons for the reduction of carbon dioxide to sugar in the Calvin cycle.

natural selection
A process in which individuals with certain inherited traits are more likely to survive and reproduce than are individuals that do not have those traits.

nematode (*nēm′-uh-tōd*)
See roundworm.

neutron
An electrically neutral particle (a particle having no electrical charge) found in the nucleus of an atom.

nitrogen fixation
The conversion of atmospheric nitrogen (N_2) to ammonia (NH_3). NH_3 then picks up another H$^+$ to become NH_4^+ (ammonium), which plants can absorb and use.

nondisjunction
An accident of meiosis or mitosis in which a pair of homologous chromosomes or a pair of sister chromatids fails to separate at anaphase.

notochord (*nō′-tuh-kord*)
A flexible, cartilage-like, longitudinal rod located between the digestive tract and nerve cord in chordate animals, present only in embryos in many species.

nuclear envelope
A double membrane, perforated with pores, that encloses the nucleus and separates it from the rest of the eukaryotic cell.

nuclear transplantation
A technique in which the nucleus of one cell is placed into another cell that already has a nucleus or in which the nucleus has been previously destroyed. The cell is then stimulated to grow, producing an embryo that is a genetic copy of the nucleus donor.

nucleic acid (*nū-klā′-ik*)
A polymer consisting of many nucleotide monomers; serves as a blueprint for proteins and, through the actions of proteins, for all cellular structures and activities. The two types of nucleic acids are DNA and RNA.

nucleoid
A non—membrane-enclosed region in a prokaryotic cell where the DNA is concentrated.

nucleolus (*nū-klē′-ō-lus*)
A structure within the nucleus of a eukaryotic cell where ribosomal RNA is made and assembled with proteins to make ribosomal subunits; consists of parts of the chromatin DNA, RNA transcribed from the DNA, and proteins imported from the cytoplasm.

nucleosome (*nū′-klē-ō-sōm*)
The bead-like unit of DNA packing in a eukaryotic cell; consists of DNA wound around a protein core made up of eight histone molecules.

nucleotide (*nū′-klē-ō-tīd*)
An organic monomer consisting of a five-carbon sugar covalently bonded to a nitrogenous base and a phosphate group. Nucleotides are the building blocks of nucleic acids, including DNA and RNA.

nucleus
(plural, **nuclei**) (1) An atom's central core, containing protons and neutrons. (2) The genetic control center of a eukaryotic cell.

O

omnivore
An animal that eats both plants and animals. *See also* carnivore; herbivore.

oncogene (*on′-kō-jēn*)
A cancer-causing gene; usually contributes to malignancy by abnormally enhancing the amount or activity of a growth factor made by the cell.

operator
In prokaryotic DNA, a sequence of nucleotides near the start of an operon to which an active repressor can attach. The binding of repressor prevents RNA polymerase from attaching to the promoter and transcribing the genes of the operon.

operculum (*ō-per′-kyū-lum*)
(plural, **opercula**) A protective flap on each side of a bony fish's head that covers a chamber housing the gills.

operon (*op′-er-on*)
A unit of genetic regulation common in prokaryotes; a cluster of genes with related functions, along with the promoter and operator that control their transcription.

opportunistic life history
The pattern of reproducing when young and producing many offspring that receive little or no parental care; often seen in short-lived, small-bodied species.

order
In classification, the taxonomic category above family.

organelle (*ōr-guh-nel′*)
A membrane-enclosed structure with a specialized function within a eukaryotic cell.

organic compound
A chemical compound containing the element carbon.

organism
An individual living thing, such as a bacterium, fungus, protist, plant, or animal.

organismal ecology
The study of the evolutionary adaptations that enable individual organisms to meet the challenges posed by their abiotic environments.

osmoregulation
The control of the gain or loss of water and dissolved solutes in an organism.

osmosis (*oz-mō′-sis*)
The diffusion of water across a selectively permeable membrane.

ovary
(1) In animals, the female gonad, which produces egg cells and reproductive hormones. (2) In flowering plants, the base of a carpel in which the egg-containing ovules develop.

ovule (*ō′-vyūl*)
In a seed plant, a reproductive structure that contains the female gametophyte and the developing egg. An ovule develops into a seed.

P

P generation
The parent individuals from which offspring are derived in studies of inheritance. P stands for parental.

paleontologist
A scientist who studies fossils.

parasite
An organism that exploits another organism (the host) from which it obtains nourishment or other benefits; an organism that benefits at the expense of another organism, which is harmed in the process.

passive transport
The diffusion of a substance across a biological membrane without any input of energy.

pathogen
A disease-causing virus or organism.

pedigree
A family tree representing the occurrence of heritable traits in parents and offspring across a number of generations.

peer review
The evaluation of scientific work by impartial, often anonymous, experts in that same field. Peer review is considered a good means of recognizing valid scientific sources.

pelagic realm (*puh-laj'-ik*)
The open-water region of an ocean.

peptide bond
The covalent linkage between two amino acid units in a polypeptide, formed by a dehydration reaction between two amino acids.

periodic table of the elements
A table listing all of the chemical elements (both natural and human-made) ordered by atomic number (the number of protons in the nucleus of a single atom of that element).

permafrost
Continuously frozen subsoil found in the arctic tundra.

petal
A modified leaf of a flowering plant. Petals are the often colorful parts of a flower that advertise it to insects and other pollinators.

pH scale
A measure of the relative acidity of a solution, ranging in value from 0 (most acidic) to 14 (most basic).

phage (*fāj*)
See bacteriophage.

phagocytosis (*fag'-ō-sī-tō'-sis*)
Cellular "eating"; a type of endocytosis whereby a cell engulfs large molecules, other cells, or particles into its cytoplasm.

pharyngeal slit (*fuh-rin'-jē-ul*)
A gill structure in the pharynx, found in chordate embryos and some adult chordates.

phenotype (*fē'-nō-tīp*)
The expressed traits of an organism.

phospholipid (*fos'-fō-lip'-id*)
A molecule that is a part of the inner bilayer of biological membranes, having a hydrophilic head and a hydrophobic tail.

phospholipid bilayer
A double layer of phospholipid molecules (each molecule consisting of a phosphate group bonded to two fatty acids) that is the primary component of all cellular membranes.

photic zone (*fō'-tik*)
Shallow water near the shore or the upper layer of water away from the shore; region of an aquatic ecosystem where sufficient light is available for photosynthesis.

photon (*fō'-ton*)
A fixed quantity of light energy. The shorter the wavelength of light, the greater the energy of a photon.

photosynthesis (*fō'-tō-sin'-thuh-sis*)
The process by which plants, algae, and some bacteria transform light energy to chemical energy stored in the bonds of sugars. This process requires an input of carbon dioxide (CO_2) and water (H_2O) and produces oxygen gas (O_2) as a waste product.

photosystem
A light-harvesting unit of a chloroplast's thylakoid membrane; consists of several hundred molecules, a reaction-center chlorophyll, and a primary electron acceptor.

phylogenetic tree (*fī'-lō-juh-net'-ik*)
A branching diagram that represents a hypothesis about evolutionary relationships between organisms.

phylogeny (*fī-loj'-uh-nē*)
The evolutionary history of a species or group of related species.

phylum (*fī'-lum*)
(plural, **phyla**) In classification, the taxonomic category above class.

phytoplankton
Photosynthetic organisms, mostly microscopic, that drift near the surfaces of ponds, lakes, and oceans.

placebo
A harmless but ineffective procedure or treatment that is given purely for psychological reasons or to act as a control in a blind experiment.

placenta (*pluh-sen'-tuh*)
In most mammals, the organ that provides nutrients and oxygen to the embryo and helps dispose of its metabolic wastes. The placenta is formed from embryonic tissue and the mother's endometrial blood vessels.

placental mammal (*pluh-sen'-tul*)
Mammal whose young complete their embryonic development in the uterus, nourished via the mother's blood vessels in the placenta; also called a eutherian.

plant
A multicellular eukaryote that carries out photosynthesis and has a set of structural and reproductive terrestrial adaptations, including a multicellular, dependent embryo.

plasma membrane
The thin double layer of lipids and proteins that sets a cell off from its surroundings and acts as a selective barrier to the passage of ions and molecules into and out of the cell; consists of a phospholipid bilayer in which proteins are embedded.

plasmid
A small ring of self-replicating DNA separate from the larger chromosome(s). Plasmids are most frequently derived from bacteria.

plate tectonics (*tek-tahn'-iks*)
The theory that the continents are part of great plates of Earth's crust that float on the hot, underlying portion of the mantle. Movements in the mantle cause the continents to move slowly over time.

pleiotropy (*plī'-uh-trō-pē*)
The control of more than one phenotypic character by a single gene.

polar ice
A terrestrial biome that includes regions of extremely cold temperature and low precipitation located at high latitudes north of the arctic tundra and in Antarctica.

polar molecule
A molecule containing an uneven distribution of charge due to the presence of polar covalent bonds (bonds having opposite charges on opposite ends). A polar molecule will have a slightly positive pole (end) and a slightly negative pole.

pollen grain
In a seed plant, the male gametophyte that develops within the anther of a stamen. It houses cells that will develop into sperm.

pollination
In seed plants, the delivery, by wind or animals, of pollen from the male (pollen-producing) parts of a plant to the stigma of a carpel on the female part of a plant.

polygenic inheritance (*pol'-ē-jen'-ik*)
The additive effect of two or more genes on a single phenotypic character.

polymer (*pol'-uh-mer*)
A large molecule consisting of many identical or similar molecular units, called monomers, covalently joined together in a chain.

polymerase chain reaction (PCR) (*puh-lim'-uh-rās*)
A technique used to obtain many copies of a DNA molecule or many copies of part of a DNA molecule. A small amount of DNA mixed with the enzyme DNA polymerase, DNA nucleotides, and a few other ingredients replicates repeatedly in a test tube.

polynucleotide (*pol'-ē-nū'-klē-ō-tīd*)
A polymer made up of many nucleotides covalently bonded together.

polyp (*pol'-ip*)
One of two types of cnidarian body forms; a stationary (sedentary), columnar, hydra-like body.

polypeptide
A chain of amino acids linked by peptide bonds.

polyploid
An organism that has more than two complete sets of chromosomes as a result of an accident of cell division.

polysaccharide (*pol'-ē-sak'-uh-rīd*)
A carbohydrate polymer consisting of many monosaccharides (simple sugars) linked by covalent bonds.

population
A group of interacting individuals belonging to one species and living in the same geographic area at the same time.

population density
The number of individuals of a species per unit area or volume of the habitat.

population ecology
The study of how members of a population interact with their environment, focusing on factors that influence population density and growth.

population momentum

In a population in which fertility (the number of live births over a woman's lifetime) averages two children (replacement rate), the continuation of population growth as girls reach their reproductive years.

post-anal tail

A tail posterior to the anus, found in chordate embryos and most adult chordates.

postzygotic barrier (*pōst'-zī-got'-ik*)

A reproductive barrier that prevents development of fertile adults if hybridization occurs.

potential energy

Stored energy; the energy that an object has due to its location and/or arrangement. Water behind a dam and chemical bonds both possess potential energy.

predation

An interaction in which an individual of one species, the predator, kills and eats an individual of the other species, the prey.

prezygotic barrier (*prē'-zī-got'-ik*)

A reproductive barrier that prevents mating between species or hinders fertilization of eggs if members of different species attempt to mate.

primary consumer

An organism that eats only producers (autotrophs); an herbivore.

primary production

The amount of solar energy converted to the chemical energy stored in organic compounds by producers in an ecosystem during a given time period.

primary succession

A type of ecological succession in which a biological community begins in an area without soil. *See also* secondary succession.

primate

Member of the mammalian group that includes lorises, bush babies, lemurs, tarsiers, monkeys, apes, and humans.

primer

A short stretch of nucleic acid bound by complementary base pairing to a DNA sequence and elongated with DNA nucleotides. During PCR, primers flank the desired sequence to be copied.

prion (*prī'-on*)

An infectious form of protein that may multiply by converting related proteins to more prions. Prions

cause several related diseases in different animals, including scrapie in sheep, mad cow disease, and Creutzfeldt-Jakob disease in humans.

producer

An organism that makes organic food molecules from carbon dioxide, water, and other inorganic raw materials: a plant, alga, or autotrophic bacterium; the trophic level that supports all others in a food chain or food web.

product

An ending material in a chemical reaction.

prokaryote (*prō-kār'-ē-ōt*)

An organism characterized by prokaryotic cells. *See also* prokaryotic cell.

prokaryotic cell (*prō-kār'-ē-ot'-ik*)

A type of cell lacking a nucleus and other membrane-bound organelles. Prokaryotic cells are found only among organisms of the domains Bacteria and Archaea.

promoter

A specific nucleotide sequence in DNA, located at the start of a gene, that is the binding site for RNA polymerase and the place where transcription begins.

prophage (*prō'-fāj*)

Phage DNA that has inserted into the DNA of a prokaryotic chromosome.

prophase

The first stage of mitosis. During prophase, duplicated chromosomes condense to form structures visible with a light microscope, and the mitotic spindle forms and begins moving the chromosomes toward the center of the cell.

protein

A biological polymer constructed from hundreds to thousands of amino acid monomers. Proteins perform many functions within living cells, including providing structure, transport, and acting as enzymes.

proteomics

The systematic study of the full protein sets (proteomes) encoded by genomes.

protist (*prō'-tist*)

Any eukaryote that is not a plant, animal, or fungus.

proton

A subatomic particle with a single unit of positive electrical charge, found in the nucleus of an atom.

proto-oncogene (*prō'-tō-on'-kō-jēn*)

A normal gene that can be converted to a cancer-causing gene.

protozoan (*prō'-tō-zō'-un*)

A protist that lives primarily by ingesting food; a heterotrophic, animal-like protist.

provirus

Viral DNA that inserts into a host genome.

pseudopodium (*sū'-dō-pō'-dē-um*) (plural, **pseudopodia**) A temporary extension of an amoeboid cell. Pseudopodia function in moving cells and engulfing food.

pseudoscience

A field of study that is falsely presented or mistakenly regarded as having a scientific basis when it does not.

Punnett square

A diagram used in the study of inheritance to show the results of random fertilization.

Q

quaternary consumer (*kwot'-er-nār-ē*)

An organism that eats tertiary consumers.

R

radial symmetry

An arrangement of the body parts of an organism like pieces of a pie around an imaginary central axis. Any slice passing longitudinally through a radially symmetric organism's central axis divides the organism into mirror-image halves.

radiation therapy

Treatment for cancer in which parts of the body that have cancerous tumors are exposed to high-energy radiation to disrupt cell division of the cancer cells.

radioactive isotope

An isotope whose nucleus decays spontaneously, giving off particles and energy.

radiometric dating

A method for determining the ages of fossils and rocks from the ratio of a radioactive isotope to the nonradioactive isotope(s) of the same element in the sample.

radula (*rad'-yū-luh*)

A file-like organ found in many molluscs, typically used to scrape up or shred food.

ray-finned fish

A bony fish in which fins are webs of skin supported by thin, flexible skeletal rays. All but one living species of bony fishes are ray-fins. *See also* lobe-finned fish.

reactant

A starting material in a chemical reaction.

recessive allele

In heterozygotes, the allele that has no noticeable effect on the phenotype; the recessive version of a gene is usually represented with a lowercase italic letter (e.g., *f*).

recombinant DNA

A DNA molecule carrying genes derived from two or more sources, often from different species.

regeneration

The regrowth of body parts from pieces of an organism.

relative abundance

The proportional representation of a species in a biological community; one component of species diversity.

relative fitness

The contribution an individual makes to the gene pool of the next generation relative to the contribution of other individuals in the population.

repetitive DNA

Nucleotide sequences that are present in many copies in the DNA of a genome. The repeated sequences may be long or short and may be located next to each other or dispersed in the DNA.

repressor

A protein that blocks the transcription of a gene or operon.

reproductive barrier

A factor that prevents individuals of closely related species from interbreeding

reproductive cloning

Using a body cell from a multicellular organism to make one or more genetically identical individuals. *See also* therapeutic cloning.

reptile

Member of the clade of amniotes that includes snakes, lizards, turtles, crocodiles, alligators, birds, and a number of extinct groups (including most of the dinosaurs).

restoration ecology
A field of ecology that develops methods of returning degraded ecosystems to their natural state.

restriction enzyme
A bacterial enzyme that cuts up foreign DNA at one very specific nucleotide sequence. Restriction enzymes are used in DNA technology to cut DNA molecules in reproducible ways.

restriction fragment
A molecule of DNA produced from a longer DNA molecule cut up by a restriction enzyme.

restriction site
A specific sequence on a DNA strand that is recognized and cut by a restriction enzyme.

retrovirus
An RNA virus that reproduces by means of a DNA molecule. It reverse-transcribes its RNA into DNA, inserts the DNA into a cellular chromosome, and then transcribes more copies of the RNA from the viral DNA. HIV and a number of cancer-causing viruses are retroviruses.

reverse transcriptase (tran-skrip´-tās)
An enzyme that catalyzes the synthesis of DNA on an RNA template.

ribosomal RNA (rRNA)
(rī´-buh-sōm´-ul)
The type of ribonucleic acid that, together with proteins, makes up ribosomes.

ribosome (rī´-buh-sōm)
A cellular structure consisting of RNA and protein organized into two subunits and functioning as the site of protein synthesis in the cytoplasm. The ribosomal subunits are constructed in the nucleolus and then transported to the cytoplasm where they act.

ribozyme (rī´-bō-zīm)
An RNA molecule that functions as an enzyme.

RNA (ribonucleic acid)
(rī´-bō-nū-klā´-ik)
A type of nucleic acid consisting of nucleotide monomers, with a ribose sugar, a phosphate group, and the nitrogenous bases adenine (A), cytosine (C), guanine (G), and uracil (U); usually single-stranded; functions in protein synthesis and as the genome of some viruses.

RNA interference (RNAi)
A biotechnology technique used to silence the expression of specific genes. Synthetic RNA molecules with sequences that correspond to particular genes trigger the breakdown of the gene's mRNA.

RNA polymerase (puh-lim´-er-ās)
An enzyme that links together the growing chain of RNA nucleotides during transcription, using a DNA strand as a template.

RNA splicing
The removal of introns and joining of exons in eukaryotic RNA, forming an mRNA molecule with a continuous coding sequence; occurs before mRNA leaves the nucleus.

root
The underground organ of a plant. Roots anchor the plant in the soil, absorb and transport minerals and water, and store food.

rough ER (rough endoplasmic reticulum) (reh-tik´-yuh-lum)
A network of interconnected membranous sacs in a eukaryotic cell's cytoplasm. Rough ER membranes are studded with ribosomes that make membrane proteins and secretory proteins. The rough ER constructs membrane from phospholipids and proteins.

roundworm
An animal characterized by a cylindrical, wormlike body form and a complete digestive tract; also called a nematode.

rule of multiplication
A rule stating that the probability of a compound event is the product of the separate probabilities of the independent events.

S

saturated
Pertaining to fats and fatty acids whose hydrocarbon chains contain the maximum number of hydrogens and therefore have no double covalent bonds. Because of their straight, flat shape, saturated fats and fatty acids tend to be solid at room temperature.

savanna
A terrestrial biome dominated by grasses and scattered trees. The temperature is warm year-round. Frequent fires and seasonal drought are significant abiotic factors.

science
Any method of learning about the natural world that follows the scientific method.

seaweed
A large, multicellular marine alga.

secondary consumer
An organism that eats primary consumers.

secondary succession
A type of ecological succession that occurs where a disturbance has destroyed an existing biological community but left the soil intact. See also primary succession.

sedentarian
A member of a major annelid lineage that includes earthworms, leeches, and many tube-building marine worms.

seed
A plant embryo packaged with a food supply within a protective covering.

sepal (sē´-pul)
A modified leaf of a flowering plant. A whorl of sepals encloses and protects the flower bud before it opens.

sex chromosome
A chromosome that determines whether an individual is male or female; in mammals, for example, the X or Y chromosome.

sex-linked gene
A gene located on a sex chromosome.

sexual dimorphism
Marked differences between the secondary sexual traits of males and females.

sexual reproduction
The creation of genetically distinct offspring by the fusion of two haploid sex cells (gametes: sperm and egg), forming a diploid zygote.

sexual selection
A form of natural selection in which individuals with certain traits are more likely than other individuals to obtain mates.

shoot
The aerial organ of a plant, consisting of stem and leaves. Leaves are the main photosynthetic structures of most plants.

short tandem repeat (STR)
DNA consisting of tandem (in a row) repeats of a short sequence of nucleotides.

signal transduction pathway
A series of molecular changes that converts a signal received on a target cell's surface to a specific response inside the cell.

silencer
A eukaryotic DNA sequence that inhibits the start of gene transcription; may act analogously to an enhancer, binding a repressor.

single-blind experiment
A scientific experiment in which some information is withheld from the test subject.

sister chromatid (krō´-muh-tid)
One of the two identical parts of a duplicated chromosome. While joined, two sister chromatids make up one chromosome; chromatids are eventually separated during mitosis or meiosis II.

slime mold
A multicellular protist related to amoebas.

smooth ER (smooth endoplasmic reticulum) (reh-tik´-yuh-lum)
A network of interconnected membranous tubules in a eukaryotic cell's cytoplasm. Smooth ER lacks ribosomes. Enzymes embedded in the smooth ER membrane function in the synthesis of certain kinds of molecules, such as lipids.

solute (sol´-yūt)
A substance that is dissolved in a liquid (which is called the solvent) to form a solution.

solution
A liquid consisting of a homogeneous mixture of two or more substances: a dissolving agent, the solvent, and a substance that is dissolved, the solute.

solvent
The dissolving agent in a solution. Water is the most versatile known solvent.

somatic cell (sō-mat´-ik)
Any cell in a multicellular organism except a sperm or egg cell or a cell that develops into a sperm or egg; a body cell.

speciation (spē-sē-ā´-shun)
An evolutionary process in which one species splits into two or more species.

species
A group of populations the members of which have the potential to interbreed and produce viable, fertile offspring. See also biological species concept.

species diversity
The variety of species that make up a biological community; the number and relative abundance of species in a biological community.

species richness
The total number of different species in a community; one component of species diversity.

sponge
An aquatic stationary animal characterized by a highly porous body, choanocytes (specialized cells used for suspension feeding), and no tissues.

spore
(1) In plants and algae, a haploid cell that can develop into a multicellular haploid individual, the gametophyte, without fusing with another cell. (2) In fungi, a haploid cell that germinates to produce a mycelium.

sporophyte (spōr'-uh-fīt)
The multicellular diploid form in the life cycle of organisms undergoing alternation of generations; results from a union of gametes and meiotically produces haploid spores that grow into the gametophyte generation.

stabilizing selection
Natural selection that favors intermediate variants by acting against extreme phenotypes.

stamen (stā'-men)
A pollen-producing part of a flower, consisting of a stalk (filament) and an anther.

starch
A storage polysaccharide found in the roots of plants and certain other cells; a polymer of glucose.

start codon (kō'-don)
On mRNA, the specific three-nucleotide sequence (AUG) to which an initiator tRNA molecule binds, starting translation of genetic information.

steroid (stir'-oyd)
A type of lipid with a carbon skeleton in the form of four fused rings: three 6-sided rings and one 5-sided ring. Examples are cholesterol, testosterone, and estrogen.

stigma (stig'-muh)
(plural, **stigmata**) The sticky tip of a flower's carpel that traps pollen.

stoma (stō'-muh)
(plural, **stomata**) A pore surrounded by guard cells in the epidermis of a leaf. When stomata are open, CO_2 enters the leaf, and water and O_2 exit. A plant conserves water when its stomata are closed.

stop codon (kō'-don)
In mRNA, one of three triplets (UAG, UAA, UGA) that signal gene translation to stop.

STR analysis
A method of DNA profiling that compares the lengths of STR sequences at specific sites in the genome.

stroma (strō'-muh)
A thick fluid enclosed by the inner membrane of a chloroplast. Sugars are made in the stroma by the enzymes of the Calvin cycle.

substrate
(1) A specific substance (reactant) on which an enzyme acts. Each enzyme recognizes only the specific substrate of the reaction it catalyzes. (2) A surface in or on which an organism lives.

sugar-phosphate backbone
The alternating chain of sugar and phosphate to which DNA and RNA nitrogenous bases are attached.

survivorship curve
A plot of the number of individuals that are still alive at each age in the maximum life span; one way to represent the age-specific death rate.

sustainability
The goal of developing, managing, and conserving Earth's resources in ways that meet the needs of people today without compromising the ability of future generations to meet their needs.

sustainable development
Development that meets the needs of people today without limiting the ability of future generations to meet their needs.

sustainable resource management
Management practices that allow use of a natural resource without damaging it.

swim bladder
A gas-filled internal sac that helps bony fishes maintain buoyancy.

symbiosis (sim'-bē-ō'-sis)
An interaction between organisms of different species in which one species, the symbiont, lives in or on another species, the host.

sympatric speciation
The formation of a new species in populations that live in the same geographic area. *See also* allopatric speciation.

systematics
A discipline of biology that focuses on classifying organisms and determining their evolutionary relationships.

systems biology
An approach to studying biology that aims to model the dynamic behavior of whole biological systems based on a study of the interactions among the system's parts.

T

taiga (tī'-guh)
The northern coniferous forest, characterized by long, snowy winters and short, wet summers. Taiga extends across North America and Eurasia, to the southern border of the arctic tundra; it is also found just below alpine tundra on mountainsides in temperate zones.

taxonomy
The branch of biology concerned with naming and classifying the diverse forms of life.

telophase
The fourth and final stage of mitosis, during which daughter nuclei form at the two poles of a cell. Telophase usually occurs together with cytokinesis.

temperate broadleaf forest
A terrestrial biome located throughout midlatitude regions where there is sufficient moisture to support the growth of large, broadleaf deciduous trees.

temperate grassland
A terrestrial biome located in a temperate zone and characterized by low rainfall and nonwoody vegetation. Tree growth is hindered by occasional fires and periodic severe drought.

temperate rain forest
A coniferous forest of coastal North America (from Alaska to Oregon) supported by warm, moist air from the Pacific Ocean.

temperate zones
Latitudes between the tropics and the Arctic Circle in the north and the Antarctic Circle in the south; regions with milder climates than the tropics or polar regions.

terminator
A special sequence of nucleotides in DNA that marks the end of a gene. It signals RNA polymerase to release the newly made RNA molecule, which then departs from the gene.

tertiary consumer (ter'-shē-ār-ē)
An organism that eats secondary consumers.

testcross
The mating between an individual of unknown genotype for a particular character and an individual that is homozygous recessive for that same character.

tetrapod
A vertebrate with four limbs. Tetrapods include mammals, amphibians, and reptiles (including birds).

theory
A widely accepted explanatory idea that is broader in scope than a hypothesis, generates new hypotheses, and is supported by a large body of evidence.

therapeutic cloning
The cloning of human cells by nuclear transplantation for therapeutic purposes, such as the replacement of body cells that have been irreversibly damaged by disease or injury. *See also* nuclear transplantation; reproductive cloning.

threatened species
As defined in the U.S. Endangered Species Act, a species that is likely to become endangered in the near future throughout all or a significant portion of its range.

three-domain system
A system of taxonomic classification based on three basic groups: Bacteria, Archaea, and Eukarya.

thylakoid (thī'-luh-koyd)
One of a number of disk-shaped membranous sacs inside a chloroplast. Thylakoid membranes contain chlorophyll and the enzymes of the light reactions of photosynthesis. A stack of thylakoids is called a granum.

thymine (**T**) (thī'-mēn)
A single-ring nitrogenous base found in DNA.

trace element
An element that is essential for the survival of an organism but is needed in only minute quantities. Examples of trace elements needed by people include iron and zinc.

trait
A variant of a character found within a population, such as purple flowers in pea plants or blue eyes in people.

trans fat
An unsaturated fatty acid produced by the partial hydrogenation of vegetable oils

and present in hardened vegetable oils, most margarines, many commercial baked foods, and many fried foods.

transcription
The synthesis of RNA on a DNA template.

transcription factor
In the eukaryotic cell, a protein that functions in initiating or regulating transcription. Transcription factors bind to DNA or to other proteins that bind to DNA

transfer RNA (tRNA)
A type of ribonucleic acid that functions as an interpreter in translation. Each tRNA molecule has a specific anticodon, picks up a specific amino acid, and conveys the amino acid to the appropriate codon on mRNA.

transgenic organism
An organism that contains genes from another organism, typically of another species.

translation
The synthesis of a polypeptide using the genetic information encoded in an mRNA molecule. There is a change of "language" from nucleotides to amino acids. *See also* genetic code.

triglyceride (*trī-glis′-uh-rīd*)
A dietary fat that consists of a molecule of glycerol linked to three molecules of fatty acids.

trisomy 21
See Down syndrome.

trophic structure (*trō′-fik*)
The feeding relationships among the various species in a community.

tropical forest
A terrestrial biome characterized by warm temperatures year-round.

tropics
The region between the Tropic of Cancer and the Tropic of Capricorn; latitudes between 23.5° north and south.

tumor
An abnormal mass of cells that forms within otherwise normal tissue.

tumor-suppressor gene
A gene whose product inhibits cell division, thereby preventing uncontrolled cell growth.

tundra
A terrestrial biome characterized by bitterly cold temperatures. Plant life is limited to dwarf woody shrubs, grasses, mosses, and lichens. Arctic tundra has permanently frozen subsoil (permafrost); alpine tundra, found at high elevations, lacks permafrost.

tunicate
One of a group of stationary invertebrate chordates.

U

unsaturated
Pertaining to fats and fatty acids whose hydrocarbon chains lack the maximum number of hydrogen atoms and therefore have one or more double covalent bonds. Because of their bent shape, unsaturated fats and fatty acids tend to stay liquid at room temperature.

uracil (U) (*yū′-ruh-sil*)
A single-ring nitrogenous base found in RNA.

V

vacuole (*vak′-ū-ōl*)
A membrane-enclosed sac, part of the endomembrane system of a eukaryotic cell, having diverse functions.

variable
A factor or condition of an experiment that is changed, often while keeping all other factors or conditions constant.

vascular tissue
Plant tissue consisting of cells joined into tubes that transport water and nutrients throughout the plant body. Xylem and phloem make up vascular tissue.

vector
A piece of DNA, usually a plasmid or a viral genome, that is used to move genes from one cell to another.

vertebrate (*ver′-tuh-brāt*)
A chordate animal with a backbone. Vertebrates include lampreys, cartilaginous fishes, bony fishes, amphibians, reptiles (including birds), and mammals.

vesicle
A membranous sac in the cytoplasm of a eukaryotic cell.

vestigial structure (*ve-sti′-gē-al*)
A structure of marginal, if any, importance to an organism. Vestigial structures are historical remnants of structures that had important functions in the organism's ancestors.

virus
A microscopic particle capable of infecting cells of living organisms and inserting its genetic material. Viruses have a very simple structure and are generally not considered to be alive because they do not display all of the characteristics associated with life.

W

warning coloration
The bright color pattern, often yellow, red, or orange in combination with black, of animals that have effective chemical defenses.

water vascular system
In echinoderms, a radially arranged system of water-filled canals that branch into extensions called tube feet. The system provides movement and circulates water, facilitating gas exchange and waste disposal.

wavelength
The distance between crests of adjacent waves, such as those of the electromagnetic spectrum including light.

wetland
An ecosystem intermediate between an aquatic ecosystem and a terrestrial ecosystem. Wetland soil is saturated with water permanently or periodically.

whole-genome shotgun method
A method for determining the DNA sequence of an entire genome by cutting it into small fragments, sequencing each fragment, and then placing the fragments in the proper order.

wild type
The trait most commonly found in nature.

Z

zooplankton
In aquatic environments, free-floating animals, including many microscopic ones.

zygote (*zī′-gōt*)
The fertilized egg, which is diploid, that results from the union of haploid gametes (sperm and egg) during fertilization.

Index

Page numbers followed by *f* indicate figures; *t* indicate tables; those in bold indicate defined key term.

A

Abalones, eyes of, 285–286, 285*f*
Abiotic factors, **375**
   of biosphere, 376–377, 376*f*, 377*f*
   in ecosystem ecology, 440–442
Abiotic reservoirs, **440**–442
ABO blood groups, **159**, 159*f*
Abortion, spontaneous, 139
Absorption, **329**
Acclimation, **378**, 378*f*
Acetic acid, 97, 97*f*
Acetyl CoA, 97, 97*f*, 100*f*
Achondroplasia, 154*t*, 155, 155*f*, 221–222
Acidification, ocean,
   carbon dioxide and, 32, 32*f*
   coral bleaching, 32
Acidity, in food digestion, 31
Acid precipitation, 372, 391
Acids
   acetic, 97, 97*f*
   amino (*see* Amino acids)
   bases, pH, and, 31–32, 32*f*
   fatty, 43–44, 43*f*
   lactic, 91, 101–102, 101*f*
   nucleic (*see* Nucleic acids)
   pyruvic (*see* Pyruvic acid)
Acquired immunodeficiency syndrome (AIDS),
     190–191, 191*f*
Actinomycetes, 300*f*
Activation energy, enzymes and, **80**, 80*f*
Activators, DNA, **201**
Active site, enzyme, **82**, 82*f*
Active transport, **86**, 86*f*
Adaptations, evolutionary. *See* Evolutionary
     adaptations
Adaptive coloration, predation and, 430–431
Adelie penguins, 373, 399, 399*f*
Adenine (A), 49, 49*f*, **173**
Adenosine, 79
Adenosine diphosphate (ADP), **79**, 79*f*
Adenosine triphosphate. *See* ATP
Adenoviruses, 186*f*
Adipose tissue, 43
ADP (adenosine diphosphate), **79**, 79*f*
Adult stem cells, therapeutic cloning using, **208**
Aerobic capacity, athletic conditioning and, 91
Aerobic processes, **94**
   anaerobic processes vs., 91, 101, 103
AFP (alpha-fetoprotein), 157
African violets, 123*f*
Age structure, population, **405**, 405*f*
   human, 418–419, 418*f*, 419*f*
Agriculture
   angiosperms in, 326
   biological control of pests in, 414–415, 414*f*, 415*f*

deforestation for, 327, 327*f*, 392, 392*f*
   genetically modified organisms in, 222–223,
     222*f*, 223*f*
   integrated pest management in, 416, 416*f*
   in temperate grassland, 388
AIDS (acquired immunodeficiency syndrome),
     190–191, 191*f*
Albinism, 154*t*, 155
Alcohol, ethyl, 103, 103*f*
Alcoholic fermentation, 102–103, 103*f*
Algae, **307**
   biofuels from, 116, 116*f*, 310
   photosynthesis experiment with, 111, 111*f*
   plant origin from, 318, 318*f*
   seaweeds, 307, 307*f*, 310–311, 311*f*
   unicellular and colonial, 307, 310, 310*f*
Alleles, **148**
   ABO blood groups as example of codominant,
     159, 159*f*
   biological diversity and, 268
   dominant vs. recessive, 148
   gene pools of, 256–258, 257*f*
   homologous chromosomes and, 149, 149*f*
   human genetic disorders and, 154–157
   law of segregation and, 148–149, 152*f*, 162*f*
   multiple, and codominance, 159
   sickle-cell disease as example of pleiotropy by, 160
Allopatric speciation, **274**–275, 274*f*, 275*f*
Alpha-fetoprotein (AFP), 157
Alpine bumblebee, 399, 399*f*
Alpine tundra, 390
Alternation of generations, **320**, 320*f*, 322–323,
     322*f*, 323*f*
Alternative RNA splicing, **202**, 202*f*
Altitude, effects of, on terrestrial biome distribution,
     384, 384*f*
Alzheimer's disease, 154*t*
American pika, 377, 377*f*
Amino acids, **46**–48, 46*f*
   joining, 47*f*
   as monomers of proteins, 46
   origin of, 296–297, 297*f*
   substitution of, 48, 48*f*
Amino groups, 46, 46*f*
Ammonites, 245*f*
Amniocentesis, 157, 157*f*
Amniotes, **358**
Amniotic eggs, **358**, 358*f*
Amoebas, 69, 69*f*, 123*f*, **308**, 308*f*
Amoebocytes, 341, 341*f*
Amoxicillin, 61
Amphibians, **357**, 357*f*
Ampicillin, 55, 61
Anabolic steroids, athletic abuse of, 45, 45*f*
Anaerobic processes, **101**
   aerobic processes vs., 91, 101, 103
Analogy, **287**

Anaphase, mitosis, **127**, 127*f*
Anaphase I, meiosis I, 132*f*
Anaphase II, meiosis II, 133*f*
Anatomical responses to environmental conditions,
     379, 379*f*
Anecdotal evidence, 10, 10*f*
Angiosperms, **319**, 319*f*
   agriculture and, 326
   fruits of, 326, 326*f*
   life cycle of, 324–326, 324*f*, 325*f*, 326*f*
   reproduction of, 324–326, 324*f*, 325*f*, 326*f*
Animals, **338**
   appearance on Earth, 294–295, 294*f*–295*f*
   artificial selection of, 252–253
   asexual reproduction by, 121–122, 137
   body plans of, 339–340, 340*f*, 341*f*
   breeding of, by humans, 145, 156, 156*f*
   Cambrian explosion and, 339, 339*f*
   cell cycle of (*see* Cell cycle)
   cells of, 59*f*
   chordates, 354–355, 354*f*, 355*f*
   classification of, 288–289, 288*f*
   communities of (*see* Communities)
   digestive systems of (*see* Digestive systems)
   diversity of, 336, 338–340, 338*f*, 339*f*,
     340*f*, 341*f*
   early, 339, 339*f*
   endangered and threatened (*see* Endangered
     species)
   evolution of, 338–340, 338*f*, 339*f*, 340*f*, 341*f*
     (*see also* Evolution)
   genetically modified pharmaceutical, 218, 222,
     222*f*
   homeotic genes of, 204, 204*f*
   humans and primates (*see* Humans; Primates)
   invertebrates (*see* Invertebrates)
   life cycle of, 338, 338*f*
   nutrition for, 338, 338*f*
   osmoregulation in cells of, 85
   phylogeny of, 340, 340*f*, 341*f*
   populations of (*see* Populations)
   prion infections of, 48, 48*f*, 192
   regeneration of body parts by, 205
   reproductive cloning of, 206–207, 206*f*, 207*f*
   sexual reproduction by, 121–122, 140 (*see also*
     Meiosis)
   urban-adapted, 289, 289*f*
   vertebrates (*see* Vertebrates)
   viruses of, 188–189, 188*f*, 189*f*
Annelids (Annelida), **345**–346, 345*f*
Antarctica, polar ice biome of, 390, 390*f*
Anthers, flower, **324**, 324*f*
Anthrax, 233–234, 233*f*, 305
Anthropocene, 269, 277, 289, 289*f*
Anthropoids, **361**–362, 361*f*, 362*f*
Antibiotic-resistant bacteria, 17, 17*f*
   evolution of, 265, 265*f*

Antibiotics, 55, 61, 61*f*
  bacterial resistance to (*see* Antibiotic-resistant bacteria)
  from fungi, 332, 332*f*
Anticodon, **181**, 181*f*
Apes, 361–362, 361*f*, 362*f*
Aphids, 411, 411*f*, 416, 416*f*
Aphotic zone, **380**, 380*f*, 382, 382*f*
Apicomplexans, **308**, 309*f*
Aquatic biomes. *See also* Freshwater biomes; Marine biomes
  abiotic factors in, 376–377, 376*f*, 377*f*
  freshwater, 380–381, 380*f*, 381*f*, 393, 393*f*
  human impacts on, 380–381, 381*f*, 393, 393*f*
  marine, 382–383, 382*f*, 383*f*
Aqueous solutions, **31**, 31*f*
Arachnids, **348**, 348*f*. *See also* Spiders
Archaea, 292, **304**–306, 304*f*, 305*f*. *See also* Prokaryotes
  cells of (*see* Prokaryotic cells)
  domain of, 288, 288*f*, 304
Archaeopteryx, 284, 284*f*
Arctic, effect of global climate change on, 373, 373*f*
Arctic foxes, 379*f*
Arctic tundra, 390, 390*f*
Ardipithecus ramidus, 363*f*, 364
Aristotle, idea of fixed species and, 245
Arthropods (Arthropoda), **347**
  arachnids, 348, 348*f*
  Burgess Shale fossils of, 339, 339*f*
  crustaceans, 347, 347*f*, 349, 349*f*
  general characteristics of, 347, 347*f*
  insects, 350–352, 350*f*, 351*f*, 352*f*
  millipedes and centipedes, 350, 350*f*
Artificial corridor, 445f
Artificial selection, **252**
  dog breeding as, 145, 156, 156*f*
  in enzyme engineering, 81
  evolution and, 242, 252–253
Asexual reproduction, 121–**122**, 130, 134
  artificial (*see* Cloning)
  of bdelloid rotifers, 137
  evolutionary advantages of, 140
  sexual reproduction vs., 137, 140
Asian lady beetles, 253*f*
Asian palm civet, 326*f*
Aspartame, 257
Atherosclerosis, 44
Athletes, aerobic capacity and, 91
Atmosphere
  as abiotic reservoir for carbon, 427, 440
  as abiotic reservoir for nitrogen, 440
  on early Earth, 294–296
  energy flow through, 438
  greenhouse gas accumulation in, 395, 395*f*
Atomic mass, **25**
Atomic model, DNA, 174*f*

Atomic number, **25**
Atoms, **25**
  isotopes and, 25–26
  as level of life, 15*f*
  structure of, 25, 25*f*
ATP (adenosine triphosphate), **79**
  active transport and, 86, 86*f*
  in Calvin cycle, 110, 115, 115*f*
  in cellular respiration, 79–80, 79*f*, 80*f*, 95, 100, 100*f*
  in chemical cycling between photosynthesis and cellular respiration, 93
  in citric acid cycle, 97, 97*f*
  cycle of, 80, 80*f*
  in electron transport, 98–99, 98*f*, 99*f*
  energy-consuming activities and, 100
  in fermentation, 101
  in glycolysis, 96, 96*f*
  in ion transport, 79
  mitochondrial production of, 68
  nanotechnology and production of, 75
  in photosynthesis, 109, 109*f*, 113–114, 113*f*, 114*f*
  structure of, 79, 79*f*
ATP cycle, 80, 80*f*
ATP synthases, **99**, 99*f*
Atropine, 327*t*
Australia, marsupials on, 282, 283*f*, 360, 360*f*
Australopithecus afarensis, 363*f*, 364, 364*f*
Autosomal disorders, 154*t*
Autosomes, **131**
Autotrophs, **92**, 108
Avatar cancer treatment, 210, 210*f*
AZT, anti-AIDS drug, 191, 191*f*

B

Baby boomers, 418–419, 419*f*
Bacilli, **300**, 300f
Bacillus anthracis. See Anthrax
Bacitracin, 55
Bacteria, **304**–306, 305*f*
  antibiotic-resistant (*see* Antibiotic-resistant bacteria)
  cells of (*see* Prokaryotic cells)
  domain of, 288, 288*f*, 304
  giant, 300*f*
  in human body, 293, 299, 306, 306*f*, 311, 311*f*
  lactose and, 37
  pathogenic, 304–306, 305*f*
  photosynthesis experiment with, 111, 111*f*
  ribosomes, 55
Bacterial plasmids, 218*f*
Bacteriophages, **186**–187, 186*f*, 187*f*
Balanus, 429, 429*f*
Bald eagles, 359*f*
Ball-and-stick models of molecules, 27*f*, 28, 38*f*

Bananas, 426, 426*f*
Bangladesh study, bacterial resistance, 71
Bark beetles, global climate change effects on, 396, 396*f*
Barnacles, 349*f*, 429, 429*f*
Barriers, reproductive. *See* Reproductive barriers
Base pairing, DNA, 50, 50*f*, 174–175, 174*f*, 175*f*
Bases, **31**
  acids and, 31–32, 32*f*
Base substitution, 184*f*
Bdelloid rotifers, asexual reproduction of, 137, 137*f*
Beagle voyage, Charles Darwin's, 246–247, 246*f*, 247*f*
Behavioral isolation, 272, 273*f*
Behavioral responses to environmental conditions, 379, 379*f*
Behcet's disease, 232, 232*f*
Benign tumors, **129**
Benthic realm, **380**, 380*f*, 382, 382*f*
Bilateral symmetry, **340**, 340*f*
Binary fission, **301**
Binomial, **244**
Biocapacity, **419**
Biodiesel, 107
Biodiversity, **426**
  animal, 336, 338–340, 338*f*, 339*f*, 340*f*, 341*f*
  causes of loss of, 427–428
  classification of, 244, 244*f*, 286–289, 286*f*, 287*f*, 288*f*
  Darwin's observations of, 247, 247*f*, 270, 270*f*, 278, 278*f*
  early explanations of, 245, 245*f*
  ecosystem diversity and, 445*f*, 446, 449
  evolutionary history and, 268
  evolution of biological novelty and, 284–286, 284*f*, 285*f*
  fish, 356, 356*f*
  forest fragmentation and, 446
  of fungi, 328, 328*f*
  hot spots of, 444, 444*f*
  insect, 351, 351*f*
  loss of, 269, 327, 327*f*, 421, 421*f*, 425, 426–428
  mammalian, 360, 360*f*
  mass extinctions and, 283, 283*f*
  naming of, 244, 244*f*
  plant, 318–319, 318*f*, 319*f*, 327, 327*f*
  pollution effects on, 269
  primate, 362*f*
  reptile, 358, 358*f*–359*f*
  value of, 425
  vertebrate, 354–355, 354*f*, 355*f*
Biodiversity hot spots, **444**, 444*f*
Bioethanol, 107
Biofilms, **301**, 301*f*, 311
Biofuels
  algae and, 116, 116*f*, 310
  biodiesel, 107
  bioethanol, 107

factory improvement for, 115–116
photosynthesis and, 107
Biogeochemical cycles, **440**, 440*f*
Biogeography, **282**
plate tectonics and, 281–282, 281*f*, 282*f*, 283*f*
Bioinformatics, **229**
DNA sequencing, 229–230, 229*f*
ethical issues regarding, 236–237
Biological communities. *See* Communities
Biological control, of pests, **414**–415, 414*f*, 415*f*
Biological diversity. *See* Biodiversity
Biological magnification, **433**, 433*f*
Biological species concept, **271**, 271*f*
Biological systems, interactions within, as major
theme in biology. *See* Major themes in
biology, Interactions within biological
systems
Biology, **4**
conservation, 412, 412*f*
conservation biology, 444–449
ecology (*see* Ecology)
everyday life and, 3, 25
evo-devo (evolutionary developmental),
283–284, 284*f*
evolution as unifying theme of, 16–17, 16*f*, 17*f*
genetics in (*see* Genetics)
major themes in, 11–18, 17*f* (*see also* Major themes
in biology)
restoration biology, 444–449
scientific study of life, 4
taxonomy as branch of, 244
Biology, Why It Matters, features
animal diversity, 336
biological diversity, 268
cells, 54
cellular function, 74
cellular reproduction, 120
cellular respiration, 90
chemistry, 22
communities and ecosystems, 424
DNA technology, 216
ecology, 372
gene regulation, 196
genetics, 144
introduction to, 2
macromolecules, 36
microorganisms, 292
molecular biology, 170
photosynthesis, 106
plants and fungi, 314
population ecology, 402
populations, evolution of, 242
Biology and Society essays
asexual and sexual reproduction, 121
biodiversity, 425
biofuels, 107
breast cancer, 197

climate change, 373, 373*f*
DNA profiling, 217, 217*f*
dog breeding, 145
evolution, 243
exercise science, 91
human evolution, 337
humanity's footprint on evolution, 269
human microbiota, 293
humans vs. bacteria, 55
lactose intolerance, 37
lionfish invasion, 403
nanotechnology, 75
nuclear medicine, 23, 23*f*
plant-fungi interactions, 315
Zika virus, 171
Biomass, **438**
Biomes, **380**
abiotic factors in, 376–377, 376*f*, 377*f*
freshwater, 380–381, 380*f*, 381*f*, 393, 393*f*
global climate change effects on, 396, 396*f*
human impacts on, 380–381, 381*f*, 383, 392–393,
392*f*, 393*f*
marine, 382–383, 382*f*, 383*f*
terrestrial (*see* Terrestrial biomes)
water cycle and, 391, 391*f*
Bioprospecting, 426
Bioremediation, **303**, **447**, 447*f*
using prokaryotes, 302*f*, 303–304, 303*f*
Biosphere, **14**, **375**
abiotic factors of, 376–377, 376*f*, 377*f*
adaptations of organisms to variability of, 378–
379, 378*f*, 379*f*
biomes in (*see* Biomes)
climate change and, 373, 373*f*
distribution of life in, 376, 376*f*, 397, 397*f*
interactions within, 14
levels of life in, 15*f*
study of, 372 (*see also* Ecology)
Biotechnology, **218**
DNA technology (*see* DNA technology)
genetic testing and, 157
human gene therapy and, 224, 224*f*
pharmaceutical applications, 221–222, 221*f*, 222*f*
safety and ethical issues in, 235–237
Bioterrorism, anthrax killer investigation,
233–234, 233*f*
Biotic factors, **375**
Bipedalism, 364, 364*f*
Birds, **359**, 359*f*
brown tree snakes and, 449, 449*f*
cladistics of, 288, 288*f*
evolution of, 284–285, 284*f*, 285, 358
finches (*see* Finches, Galápagos Islands)
urban-adapted, 289, 289*f*
wetlands and, 381
Birth defects, trisomy 21 and Down syndrome,
139, 139*f*

Birth rate, human population growth and, 417, 417*f*,
417*t*
Bivalves, **343**, 343*f*
Black spider monkey, 362*f*
Black widow spiders, 348*f*
Blastula, **338**, 338*f*
Blood
ABO blood groups of, 159, 159*f*
glucose homeostasis of, 100
hemoglobin in (*see* Hemoglobin)
transfusion of, 81
Blood types, 81, 81*f*
Blowflies, 352
Blue dasher dragonflies, 351*f*
Blue-footed boobies, 16*f*, 261, 261*f*
Blue whales, 360
Body cavity, **340**, 341*f*
Body plans, animal, 339–340, 340*f*, 341*f*
Body segmentation, **345**, 345*f*
Boll weevil, 416, 416*f*
Bonds, chemical, 13, **27**
Bony fishes, **356**, 356*f*
Booms, population growth, 411
Borneo rain forest, 386*f*
Bottleneck effect, **259**–260, 259*f*
Botulism, 302, 305–306
Bracket fungi, 328*f*
Brain, human, 337, 364
cellular respiration and, 90, 100
Brazilian rain forest, 292*f*
BRCA1 gene, mutations in, 211
BRCA2 gene, mutations in, 211
Breast cancer, 129*f*, 197
Breathing, cellular respiration and, 94, 94*f*
Breeding. *See* Artificial selection
Bristlecone pine trees, 322
Brown algae, 310–311, 311*f*
Brown tree snakes, 449, 449*f*
Bryophytes, **318**–320, 319*f*, 320*f*
Buckeye butterfly, 351*f*
Buffers, pH and, **32**
Bugs, 350–352, 350*f*, 351*f*, 352*f*
Burgess Shale, 339, 339*f*
Burmese pythons (*Python bivittatus*), as invasive
species, 414, 414*f*
Busts, population growth, 411
Butterflies, 351*f*, 352*f*, 397, 397*f*
Bwindi Impenetrable National Park, 327

C

Cactus finch, 278, 278*f*, 405, 405*f*
Caffeine, 54, 67
Calcium (Ca), in human body composition, 24
Calorie (cal), **78**, 78*f*
Calvin cycle, **110**, 115, 115*f*

Cambrian explosion, 295, 339, 339*f*
Cambrian period, 280*t*, 339, 339*f*
Camouflage, predation and, 430–431
Cancer, **129**
    avatar treatment for, 210, 210*f*
    breast (*see* Breast cancer)
    as cell cycle disease, 128–129
    cells, human, 197*f*
    colon, 211, 211*f*
    development of, 211, 211*f*
    evolution of, 213
    genetic basis of, 209–212
    inherited, 211
    leukemia, 208
    lifestyle choices and risk of, 196, 212
    mutagens and, 185
    prevention and survival of, 129
    prostate, 26, 26*f*
    radiation therapy for, 23, 26, 26*f*, 129
    risk and prevention of, 212, 212*f*
    smoking and, 185, 212
    treatments for, 129
    in United States, 212*f*
Cane toads, 415, 415*f*
Canines. *See* Dogs
Cap, RNA, 180, 180*f*
Capsules, prokaryotic cell, 58
Carbohydrates, **40**
    as macromolecules, 39–40
Carbo-loads, 42
Carbon (C)
    in ecosystem chemical cycling, 440–441, 441*f*
    radiometric dating and, 33, 281
Carbon-12, 25–26, 33
Carbon-13, 25–26
Carbon-14, 25–26, 33, 281
Carbon-based molecules. *See* Organic compounds
Carbon cycle, 440–441, 441*f*
Carbon dioxide (CO_2)
    in cellular respiration, 93–94
    in chemical cycling between photosynthesis and
        cellular respiration, 92–93, 93*f*
    fossil fuels and, 321
    as greenhouse gas, 395, 395*f*
    in photosynthesis, 108–109, 109*f*
    in redox reactions, 95
    ocean acidification and, 32
    sugar production from, in Calvin cycle, 110, 115,
        115*f*
Carbon fixation, **110**
Carbon footprints, **398**, 398*f*
Carboniferous period, 280*t*, 321–322, 321*f*
Carbon skeletons, 38, 38*f*
Carboxyl groups, 38, 46, 46*f*
Carcinogens, **212**
Cardiovascular disease
    cholesterol and, 45
    saturated fats and, 44

Carnivores, **433**
Carolina anole lizards, 377*f*
Carotenoids, 112
Carpels, flower, 146, 146*f*, **324**, 324*f*
Carriers, recessive allele, **155**
Carrying capacity, **409**
    of human population, 417
    logistic population growth and, 409, 409*f*
    sustainable resource management and, 412–413
Cartilaginous fishes, **356**, 356*f*
Cas9, 220–221, 221*f*
Cattle, lactose tolerance and domestication of, 51
Caulerpa, 307*f*
Cavalier King Charles Spaniels, 145
Cavendish bananas, 426, 426*f*
cDNA (complementary DNA), **204**
Cell cycle, **125**
    cancer as disease of, 128–129
    mitosis and cytokinesis in, 126–128, 126*f*–127*f*,
        128*f*
    phases of, 125, 125*f*
Cell cycle control system, **128**
Cell division, **122**, 124*f*
    functions of, 122
    mitosis vs. meiosis in, 132, 134–135, 134*f*
    in sexual and asexual reproduction, 121
Cell junctions, 61
Cell plates, **128**
Cells
    animal, 59*f*
    cellular work of (*see* Cellular respiration)
    chloroplasts, mitochondria, and energy conversion
        in, 68, 68*f*, 108
    cytoskeletons of, 54, 69, 69*f*
    endomembrane systems of, 64–67, 64*f*–67*f*
    genetic potential of, 205
    as level of life, 11*f*, 15*f*
    membrane structure and function in, 60
    microscope study of, 4, 56–57, 56*f*
    nuclei and ribosomes of, 62–63, 62*f*–63*f*
    origin of, 297
    plant, 59*f*
    prokaryotic vs. eukaryotic, 57–58, 57*t*
    reproduction of (*see* Cellular reproduction)
    size range of, 56*f*
    structure of, 75, 75*f*
    surfaces of, 61
Cell signaling, 203, 203*f*
Cell theory, **57**
Cellular reproduction. *See also* Cell cycle; Meiosis;
    Mitosis
    animal asexual reproduction by bdelloid rotifers
        and, 137
    functions of cell division in, 122*f*–123*f*
    sexual vs. asexual reproduction and, 122
Cellular respiration, **94**
    as aerobic harvest of food energy, 94
    aerobic vs. anaerobic lifestyles and, 91, 101, 103

    ATP in, 79–80, 79*f*, 80*f*, 95, 100, 100*f*
    breathing and, 94, 94*f*
    in chemical cycling, 92–93
    chemical energy and, 78
    citric acid cycle in, 95, 97, 97*f*
    electron transport in, 95, 98–99, 98*f*
    equation for, 94–95
    fermentation vs., 101–103
    glycolysis in, 95
    human muscles and, 91, 102
    map of, 95*f*
    by mitochondria, 68
    overview of, 94–95, 95*f*
    oxygen role in, 94–95
    photosynthesis and, 109
    three stages of, 95*f*, 96–99, 100*f*
Cellular slime molds, 309, 309*f*
Cellular work. *See* Cellular respiration
Cellulose, **42**, 42*f*, 61
Cell walls
    of plant cells, 59
    of prokaryotic cells, 58
Cenozoic era, 279, 280*t*
Centipedes, **350**, 350*f*
Central vacuoles, **67**, 67*f*
Centromeres, **125**
Centrosome, 69
Cephalopods, **343**, 343*f*
CF (cystic fibrosis), 154*t*, 155
Chanterelle mushrooms, 332*f*
Chaparral, **388**, 388*f*
Characters, heritable, **146**
    law of independent assortment and, 150–151
    law of segregation and, 147–149
    traits vs., 146
Charophytes, **318**, 318*f*
Cheetahs, 421
    bottleneck effect in, 259–260, 259*f*
    population ecology, 404*f*
Chemical bonds, 13, **27**
Chemical composition of human body, 24, 24*f*
Chemical cycling, **437**
    biogeochemical cycles and abiotic reservoirs in,
        440–442
    carbon cycle in, 440–441, 441*f*
    energy flow and, in ecosystems, 437, 440–443
    general scheme of, 440, 440*f*
    nitrogen cycle in, 442, 442*f*
    nutrient pollution and, 443, 443*f*
    phosphorus cycle in, 441, 441*f*
    by prokaryotes, 303
Chemical energy, **77–78**
    as abiotic factor in biosphere, 376, 376*f*
    in cellular respiration, 98
    in ecosystem energy flow, 437–439, 438*f*
    photosynthesis and, 108
    solar energy conversion to, by light reactions, 110
Chemical reactions, **28**, 80, 80*f*

Chemistry
  basic, 24–28, 24f–28f
  of water and life, 29–33, 29f–33f
Chemotherapy, cancer, **129**, 197
Chernobyl nuclear reactor explosion, 26
Chicken of the woods, 332f
Chimpanzees (Pan troglodytes), 361–363, 361f, 362f
  genome of, 232
  Goodall's observations of, 4f
  skulls of, 283–284, 284f
China, ecological footprint of, 420, 420f
Chinese hibiscus, 275f
Chlorophyll, **108**, 111–112, 111f
Chlorophyll a pigment, 111–112
Chlorophyll b pigment, 112
Chloroplasts, 59, **68**, 68f, 108
  light energy and, 92
  pigments of, 111–112, 112f
Choanocytes, 341, 341f
Cholera, 71, 216
Cholesterol, cardiovascular disease and, 45
Chondrodysplasia, 156, 156f
Chordates (Chordata), **354**–355, 354f, 355f
Chorionic villus sampling, 157
Chromatin, **62, 123**
Chromosomal basis of inheritance, 162f
Chromosomes, **57, 122**
  altered numbers of, 138, 138f
  alternative arrangements of, 135f
  autosomes, 131
  in cells of selected mammals, 123f
  crossing over between, 132, 136–137, 163
  duplication and distribution of, 125, 125f
  eukaryotic, 123–124
  homologous, in sexual reproduction, 130–131, 130f
  independent assortment of, 135–136
  law of segregation and homologous, 147–149, 162f
  mutations of, 255
  number of, in cells, 62, 120, 216, 231
  in prokaryotic and eukaryotic cells, 57t
  sex (see Sex chromosomes)
  triploid, 139
Chromosome theory of inheritance, **162**
Chthamalus, 429, 429f
Cigarettes. See Smoking
Cilia, **70**, 70f
Ciliates, **308**, 309f
Ciprofloxacin, 55, 61
Citric acid cycle, **95**, 97, 97f, 100
Clades, **287**
Cladistics, 287–288, 287f, 288f
Clams, 343
Classes (taxonomy), **244**
Classification, **286**
Cleavage, **128**, 128f
Cleavage furrows, 128
Climate, population growth and, 411, 411f
Climates

effects of, on terrestrial biome distribution, 384–385, 384f, 385f
  global changes of (see Global climate change)
Clones, **122, 219**
Cloning
  DNA, 218
  of genes, 218–219
  of giant pandas, 196, 206
  of humans, 207
  by nuclear transplantation, 206, 206f
  plant, 205, 205f
  reproductive, of animals, 206–207, 206f, 207f
  test-tube, 205, 205f
  therapeutic, 208, 208f
Clostridium botulinum, 302, 305–306
Clothing, origin of, 366, 366f
Clouded leopards, 426f
Cnidarians (Cnidaria), **342**, 342f
CO₂. See Carbon dioxide
CoA (coenzyme A), 97
Coal, 314, 321
Coal forests, 321, 321f
Cocci, **300**, 300f
Coccidioidomycosis, 331
Cocoa butter, 36, 44
CODIS (Combined DNA Index System) database, 226
Codominant alleles, **159**, 159f
Codons, **177**–178, 178f, 182
Coelacanth, 356
Coenzyme A (CoA), 97
Cohesion, 29–30, 29f
Cold-blooded animals, 358
Collagen, 61
Colonial algae, 307, 310, 310f
Coloration, predation and adaptive, 430–431
Colorblindness, 163–164, 163f, 164f
Colors, of light and photosynthesis, 111, 111f
Columbia River basin, 381f
Combined DNA Index System (CODIS) database, 226
Commensalism, **432**
Communication, scientific, 5–6, 6f
Communities, **375, 428**
  as level of life, 15f
Community ecology, **375**, 375f
  community disturbances in, 436
  ecological succession in, 436–437, 436f, 437f
  interspecific interactions in, 428–432
  species diversity in, 435–436
  trophic structure in, 432–434
Comparative embryology, 251, 251f
Community
Competition
  interspecific, 429, 429f
  intraspecific, 410, 410f
  sexual selection and, 263–264, 264f
Competitive exclusion principle, **429**, 429f
Complementary DNA (cDNA), **204**
Complete digestive tracts, **345**
Compost, decomposers in, 433f

Compounds, **25**
Computer model, DNA, 174f
Concentration gradients, passive transport and, **84**
Conception. See Fertilization
Cones, gymnosperm, 323, 323f
Coniferous forests, 322, 322f, **389**, 389f
Conifers, **319**, 322, 322f
Connell, Joseph, 429
Conservation biology, 412, 412f, **444**
  biodiversity hot spots in, 444, 444f
  endangered and threatened species and (see Endangered species)
  forest fragmentation, 446
  on Guam, 449, 449f
  restoration ecology in, 447, 447f
  sustainable development and, 448, 448f
Conservation of energy, **76**
Consumers, 92, 92f
  animals as, 13
  in chemical cycling, 92
  producers vs., 13, 92
Continental drift, 281–282, 281f, 282f
Contractile proteins, 46f
Contractile vacuoles, 67, 67f
Control group, **8**
Controlled experiments, **8**–9, 8f, 9f
Convergent evolution, 286–**287**
  in seaweeds, 310
Copper (Cu), 22, 24
Coral bleaching, 32, 427
Coral reefs, **382**, 382f, 383f
  biodiversity of, 427, 427f
  global climate change and, 32
  lionfish as threats to, 403
Corals, 342, 342f
Corn, phylogeny of, 286
Corn smut, 328f
Covalent bonds, **27**
Crabs, 349f
Crayfish, 349f
Crested anole lizards, 289, 289f
Cretaceous period, 280t, 283
Creutzfeldt-Jakob disease, 192
Crick, Francis, 173f
  discovery of structure of DNA by, 173–175
CRISPR-Cas9 system, **220**–221, 221f
Cristae, mitochondrial, 68
Crocodiles, 358, 359f
Crops. See also Agriculture; Food
  angiosperms as, 326
  genetically modified GM (see Genetically modified organisms)
  integrated pest management and, 416, 416f
Crosses, genetic, **147**
  dihybrid, 150–151, 150f
  monohybrid, 150–151
  technique for, 146f, 147
  testcrosses, 152

Crossing over, 132, **136**–137, 136*f*, 163
Crustaceans, 347, 347*f*, **349**, 349*f*
Cryptic coloration, **430**, 430*f*
Cu (copper), **22**, 24
Cuticle, **316**
Cyanide, 14
    electron transport and, 99
Cyanobacteria, 300*f*, 302–303, 302*f*
Cystic fibrosis (CF), 154*t*, 155
Cytokinesis, **125**
    animal cell, 128*f*
    in meiosis II, 133*f*
    plant cell, 128*f*
Cytoplasm, **59**
    cellular respiration in, 95, 95*f*
Cytosine (C), 49, 49*f*, **173**
Cytoskeletons, 54, **69**, 69*f*
Cytosol, 59, 63–64

D

Dams, 381, 381*f*
Dandelions, 407*f*
Darwin, Charles, 268
    artificial selection observations of, 252–253
    on *Beagle* voyage, 246–247, 246*f*, 247*f*
    descent with modification theory of, 16, 247
    as early ecologist, 378
    evolutionary trees and, 251, 252*f*
    Mendel and, 146
    observations of, on biodiversity of species, 247,
        247*f*, 270, 270*f*, 278, 278*f*
    *On the Origin of Species by Means of Natural
        Selection*, 16, 16*f*, 245–248, 246*f*, 247*f*, 254,
        270, 289
    theory of evolution by natural selection, 16–17,
        16*f*, 17*f*, 33, 247–248, 261, 261*f*, 263–264,
        263*f*, 264*f*
Darwin's dogs, 145
Darwin's finches. *See* Finches, Galápagos Islands
Data, **4**
Dating methods, fossil. *See* Radiometric dating
Daughter cells, 122
DDT, 243, 255, 256
Dead zone, Gulf of Mexico, 443, 443*f*
Death rate, human population growth and, 417,
    417*f*, 417*t*
Decomposers, **433**
    fungi as, 328, 331
    prokaryotes as, 303
    roundworms as, 346
    waste products and, 14
Deep-sea hydrothermal vents, 376*f*, 382–383
    prokaryotes living in, 299, 299*f*, 302
Deepwater Horizon oil spill, 304, 304*f*, 374*f*, 383
Deforestation, 110, 327, 327*f*, 392, 392*f*
Dehydration reactions, **39**

formation of polymers from monomers by,
    39, 43, 43*f*
formation of polynucleotides by, 50, 50*f*
Density-dependent factors, population growth,
    410, 410*f*
Density-independent factors, population growth,
    411, 411*f*
Dental plaque, 301, 301*f*, 311, 311*f*
Dependent assortment, 150*f*
Dependent variable, **9**, 9*f*
Descent with modification theory, 16, 247
Deserts, **387**, 387*f*
Detoxification. *See also* Bioremediation
    smooth ER and, 65
Detritus, **433**, 433*f*
Development, macroevolution by small changes in,
    283–284, 284*f*
Devonian period, 280*t*
Diabetes (diabetes mellitus)
    genetic engineering of insulin for, 216, 221, 221*f*
    glucose homeostasis and, 100
Diana fritillary butterflies, 397, 397*f*
Diatoms, **310**, 310*f*
Dictionary of genetic code, 178*f*
Dietary fats, 44
Dietary fiber, 42, 212
Diets. *See also* Food
    cardiovascular disease and, 44–45
    goiter and, 24, 25*f*
    lactose intolerance and human, 37, 51, 51*f*
Diffusion, **83**
Digestive systems, complete tracts, 345
Digitalin, 327*t*
Dihybrid crosses, **150**–151, 150*f*
Dinoflagellates, **310**, 310*f*
Dinosaurs, 358–359, 359*f*
    bird evolution from, 285
    footprints, 249*f*
    mass extinction of, 283, 283*f*
Diploid organisms, **131**
Directional selection, **264**, 264*f*
Disaccharides, **41**, 41*f*
Discovery science, 374*f*
Diseases and disorders, human
    abnormal numbers of sex chromosomes and,
        139–140, 140*t*
    achondroplasia, 154*t*, 155, 155*f*, 221–222
    Alzheimer's disease, 154*t*
    antibiotic-resistant bacteria and, 17, 61
    atherosclerosis, 44
    Behcet's disease, 232, 232*f*
    botulism, 302, 305–306
    cancer (*see* Cancer)
    cardiovascular disease (*see* Cardiovascular disease)
    cholera, 71
    cystic fibrosis, 154*t*, 155
    diabetes (*see* Diabetes)
    dwarfism, 154*t*, 155, 155*f*, 221–222

goiter, 24, 25*f*
hemophilia, 164, 164*f*
Huntington's disease, 154*t*, 155
hypercholesterolemia, 154*t*, 158, 158*f*
influenza viruses, 188, 188*f*
lactose intolerance, 37, 51, 51*f*
lysosomal storage diseases, 66
meningitis, 305, 305*f*
misfolded proteins and, 48
parasitic roundworms, 346, 346*f*
phenylketonuria, 154*t*, 257, 257*f*
prions and Creutzfeldt-Jakob disease,
    48, 48*f*, 192
red-green colorblindness, 163–164, 163*f*
severe combined immunodeficiency disease, 224
sickle-cell disease, 48, 48*f*, 154*t*, 160, 160*f*, 184,
    184*f*, 255
single gene, 154–157
Tay-Sachs disease, 66, 154*t*
trisomy 21 and Down syndrome, 139, 139*f*
Zika virus, 171
Dispersal, seed, 326, 326*f*
Disruptive selection, **264**, 264*f*
Disturbances, **436**–437, 436*f*
Diversity, biological. *See* Biodiversity
DNA (deoxyribonucleic acid), **49**, **173**
    in cellular reproduction, 120, 123
    chromatin and, 62*f*
    chromosomes and, 62*f*
    in eukaryotic chromosomes, 123–124
    as evidence of evolution, 250
    flow of genetic information from, to RNA and
        proteins, 177, 177*f*
    human genome and, 231–232
    humans vs. other animal species, 36
    microarray, 196
    molecular building blocks, 12–13, 13*f*
    mutations of, 184–185, 184*f*
    Neanderthal, 365
    noncoding, 231–232
    as nucleic acid, 49, 49*f*, 172–173
    operator of, 199, 199*f*
    in protein production, 63, 63*f*
    replication of, 175, 175*f*
    representations of, 174*f*
    restriction enzyme cutting and pasting of,
        219–220, 220*f*
    sequencing of, 229–230, 229*f*
    STR analysis and, 226, 227*f*, 228*f*
    structure of, 50, 50*f*, 172–175, 172*f*, 173*f*, 174*f*
    transcription of, to RNA, 176, 176*f*, 179, 179*f*,
        201, 201*f*
    vaccine, 190, 190*f*
    Zika virus and, 171
DNA chip, 196
DNA cloning, **218**
DNA ligase, **220**, 220*f*
DNA microarrays, 204, 204*f*

DNA packing
  in cell division, 124, 124*f*
  gene regulation of, 200, 201*f*
DNA polymerases, **175**
  in DNA replication, 175
  in polymerase chain reaction, 225
DNA profiling, **225**
  ancient samples and, 229
  endangered species and, 229
  in forensics (*see* Forensics)
  gel electrophoresis, 227–228, 227*f*, 228*f*
  paternity and, 229
  polymerase chain reaction in, 225–226, 225*f*
  short tandem repeat analysis in, 226,
      227*f*, 228*f*
  techniques of, 225–228
  victims of World Trade Center attack and, 228
  of XIAP protein mutation, 233, 233*f*
DNA → RNA → Protein, 63*f*, 183, 183*f*, 198. *See also*
      Transcription; Translation
DNA sequencing, **229**–230, 229*f*
DNA synthesis, 125, 125*f*
DNA technology, 216. *See also* Bioinformatics;
      Biotechnology; Genetically modified
      organisms
  CRISPR-Cas9 system, 220–221, 221*f*
  DNA profiling (*see* DNA profiling)
  ethical questions and, 236–237
  genetic code and, 170, 178
  genomics, 230, 230*f*
  human gene therapy, 224, 224*f*
  profiling with (*see* DNA profiling)
  proteomics, 234, 234*f*
  recombinant DNA technology, 218–220, 219*f*, 220*f*
  safety and ethical issues of, 235–237
  vaccines and, 222
Dogs
  artificial selection of, 145, 156
  evolution of, 165, 165*f*
  genetic basis of leg variation in, 156, 156*f*
Dolly (sheep), 206–207, 207*f*
Domains (taxonomy), **57**, **244**, 288–289, 288*f*, 304
Dominant alleles, **148**
  human genetic disorders and, 154*t*, 155, 155*f*
  recessive alleles vs., 148
Dominant disorders, 154*t*, 155, 155*f*
Dominant traits, 153, 153*f*, 154*f*
Dorsal, hollow nerve cord, **354**, 354*f*
Double-blind experiment, **9**, 9*f*
Double helix, DNA, **50**, 50*f*, 173–175, 174*f*
Down, John Langdon, 139
Down syndrome, **139**, 139*f*
Drosophila melanogaster. See Fruit flies
Duck-billed platypus, 360, 360*f*
Dust mites, 348*f*
Dutch elm disease, 331, 331*f*
Dwarfism (achondroplasia), 154*t*, 155, 155*f*,
      221–222

E

Earth. *See also* Biosphere
  early conditions on, 294–296, 296*f*
  early life on, 29
  history of, 279–283, 279*f*, 280*t*, 281*f*, 282*f*, 283*f*,
      292, 294–295, 294*f*–295*f*
Earthworms, 345–346
Eating, as nutritional mode, 338, 338*f*
Eat Lionfish campaign, 403
Ebola virus, 192*f*
Echinoderms (Echinodermata), **353**, 353*f*
E. coli bacteria. *See Escherichia coli*
Ecological footprint, **419**–420, 420*f*
Ecological niches, **429**, 429*f*
Ecological pyramids, 438–439
Ecological succession, **436**–437, 436*f*, 437*f*
Ecology, **374**
  abiotic factors in, 375–377, 376*f*, 377*f*
  biomes (*see* Biomes)
  climate change and (*see* Global climate change)
  environmentalism and, 374, 374*f*
  environmental variability and, 378–379, 378*f*,
      379*f*
  hierarchy of interactions in, 375, 375*f*
  importance of, 372
  overview of, 374–375, 374*f*, 375*f*
  population (*see* Population ecology)
Ecosystem diversity, 427, 427*f*
Ecosystem ecology, **375**, 375*f*, 437
  chemical cycling in, 440–443
  energy flow in, 438–439
Ecosystems, **375**, **437**
  biodiversity of, 446
  changes and industrial processes, 32
  conservation biology, restoration ecology, and,
      444–445
  energy flow and chemical cycling in, 93*f*,
      437–439, 438*f*
  global climate change effects on, 396, 396*f*
  human resource use, 439, 439*f*
  as level of life, 15*f*
  nutrient and energy flow, 14*f*
  restoration of, 447, 447*f*
  solar power and, 13, 106, 108
Ecosystem services, **427**
Ectotherms, **358**, 378, 378*f*
Eggs, amniotic, 358, 358*f*
Electromagnetic spectrum, **110**–111, 110*f*
Electron configuration of molecules, 27*f*
Electron microscopes, 56*f*
Electrons, **25**
  in cellular respiration, 98
  in fermentation, 101
  in ionic bonds, 27, 27*f*
  in photosynthesis, 113–115, 113*f*, 114*f*, 115*f*
Electron transfer

  ionic bonding and, 27, 27*f*
  photosynthesis and, 109
Electron transport, **95**, 98–99, 99*f*
Electron transport chains
  in cellular respiration, 98–99, 99*f*
  in light reactions, 113–114, 113*f*, 114*f*
Elements, **24**–25
Elephants, 407*f*
Elongation phase
  transcription, 179, 179*f*
  translation, 182, 182*f*
Embryonic development
  evidence of evolution in comparative, 251, 251*f*
  lysosome functions in, 66
Embryonic stem cells (ES cells), therapeutic cloning
      using, **208**, 208*f*
Embryos, plant, 317, 317*f*
Emergent properties, 14, 15*f*
  life as, 296
Emerging viruses, **192**, 192*f*
Enceladus, 11
Endangered species, **412**
  bottleneck effect in, 259–260, 259*f*
  conservation of, 412, 412*f*
  DNA profiling and, 229
  population ecology and, 404, 404*f*
  repopulating by cloning, 207
Endemic species, **444**
Endocytosis, **86**, 86*f*
Endomembrane systems, **64**–67, 64*f*–67*f*
Endoplasmic reticulum (ER), 63, **64**–65, 64*f*
Endoskeleton, **353**
  vertebrate, 354, 354*f*
Endosperm, **325**
Endospores, **301**–302
Endotherms, **359**, 378
Endotoxins, **305**
Energy, **76**
  as abiotic factor in biosphere, 376, 376*f*
  carbon footprints and consumption of, 398
  cellular, and active transport, 86, 86*f*
  chemical, 77–78
  chemical bond storage, 13
  concepts of, 76–80
  conservation of, 76
  conversions of, 76, 76*f*, 77*f*
  entropy and, 77
  flow of, in biosphere, 92–93, 93*f*
  from food, 100, 100*f*
  food calories and, 78
  input and transformation of, 13–14, 14*f*
  in muscles, 91
  from sun, 106, 108
  transformation of, 68, 94
  trophic structures and, 432–433
Energy and matter transformations, as major theme
      in biology. *See* Major themes in biology,
      Pathways that transform energy and matter

Energy flow, **437**
    chemical cycling and, in ecosystems, 92–93, 437
    ecological pyramids of production in, 438–439,
        438*f*
    human resource use and, 372
Energy processing. *See also* Cellular respiration;
        Chemical energy; Photosynthesis
    cellular (*see* Cellular respiration)
    chemical cycling between photosynthesis and
        cellular respiration in, 92–93
    by chloroplasts, 111–112
    producers vs. consumers in, 92
    as property of life, 10, 11*f*
Energy pyramid, **439**, 439*f*
Engelmann, Theodor, 111
Engineering, genetic. *See* Genetic engineering
Enhancers, DNA, **201**
Entropy, **77**
Environmentalism, ecology and, 374, 374*f*
Environments. *See also* Biomes; Ecosystems
    abiotic factors in, 376–377, 376*f*, 377*f*
    carrying capacities of, 409, 409*f*, 412–413, 417
    epigenetics and factors of, 161
    evolutionary adaptations to, 378–379, 378*f*, 379*f*
    exponential population growth in unlimited, 408,
        408*f*
    influences of, 161
    inheritance role in, 144, 161
    organism interactions with (*see* Ecology)
    traits of people and, 161
    variability, 278
Enzyme inhibitors, **82**, 82*f*
Enzymes, **80**
    activation energy and, 80, 80*f*
    in cellular work, 80–82
    chemical reactions and, 80, 80*f*
    digestive, in lysosomes, 66, 66*f*
    in DNA replication, 175
    engineering of, 81, 81*f*
    inhibitors of, 82, 82*f*
    lactase, 37
    in metabolism, 80
    nanotechnology and, 75
    as proteins, 46*f*
    restriction, 219–220, 220*f*
    in smooth ER, 65
    substrate binding and, 82, 82*f*
Eocene epoch, 280*t*
Epigenetic inheritance, **161**
Epigenetics
    environment and role of, 161
    genotypes and phenotypes, 161
EPO. *See* Erythropoietin
Epulopiscium fishelsoni, 300*f*
Equilibrial life histories, **407**, 407*f*, 407*t*
ER (endoplasmic reticulum), 63, **64**–65, 64*f*
Errantians, **345**–346, 345*f*
Erythrocytes. *See* Red blood cells

Erythromycin, 55
Erythropoietin (EPO), genetic engineering of, 222
ES cells (embryonic stem cells), therapeutic cloning
        using, **208**, 208*f*
Escherichia coli, 218
    bacteriophages and, 186, 187*f*
    lac operon and, 199, 199*f*
Estuaries, **383**, 383*f*
Ethyl alcohol, 103, 103*f*
Euglena, 307, 307*f*
Eukarya, domain of, 288, 288*f*, 304
Eukaryotes, **295**
    appearance on Earth, 294–295, 294*f*–295*f*
    chromosomes of, 123–124
    evolution of, 307
    protists (*see* Protists)
Eukaryotic cells, **57**–59
    gene expression in, 200, 200*f*
    gene regulation in, 200–203, 200*f*–202*f*
    gene transcription of, 201, 201*f*
    prokaryotic cells vs., 57*t*
European blackbirds, 289*f*
European grass snakes, 358, 358*f*
Eutherian (placental) mammals, 282, 283*f*, **360**, 360*f*
Evaporative cooling, **30**, 30*f*
Evidence, anecdotal, **10**, 10*f*
Evo-devo biology. *See* Evolutionary developmental
        biology
Evolution, **16**, **248**. *See also* Macroevolution;
        Microevolution
    advantages of asexual and sexual reproduction
        for, 137
    of animals, 338–340, 338*f*, 339*f*, 340*f*, 341*f*
    of antibiotic-resistant bacteria, 17, 265, 265*f*
    artificial selection and, 242, 252–253
    bacterial resistance in humans and, 71, 71*f*
    of bdelloid rotifers without sexual
        reproduction, 137
    biodiversity and (*see* Biodiversity)
    of birds, 358
    from chemical to Darwinian, 298
    of chloroplasts and mitochondria, 68
    convergent, 286–287, 310
    Darwin's theory of, 16–17, 16*f*, 17*f*, 33, 247–248,
        261, 261*f*, 263–264, 263*f*, 264*f*
    directed, 81, 81*f*, 115–116
    DNA profiling in researching, 217
    of dogs, 145, 156, 156*f*
    of emerging viruses, 192
    of eukaryotes, 307
    evidence for, 248–251, 249*f*, 250*f*, 251*f*, 252*f*
    evolutionary trees representing, 251, 252*f*
    of eyes, 285–286, 285*f*
    of feathers, 284–285, 284*f*
    gene flow as mechanism of, 260–261, 260*f*
    genetic drift as mechanism of, 258–260, 258*f*,
        259*f*, 260*f*
    human (*see* Human evolution)

    in human-dominated world, 269, 289, 289*f*
    of human lactose intolerance, 51
    mass extinctions and, 283, 283*f*
    of membranes, 87
    of microbial life, 292–298, 294*f*–295*f*, 296*f*,
        297*f*, 298*f*
    natural selection as mechanism for, 252–254, 253*f*,
        254*f*, 258, 261, 261*f*, 263–264, 263*f*, 264*f*
    origin of life, 296–298, 296*f*, 297*f*, 298*f*
    pesticide resistance and, 243, 253–256, 254*f*
    of plants, 318–319, 318*f*, 319*f*
    of populations, 254–258, 255*f*, 256*f*, 257*f*
    populations as units of, 255–256, 256*f*
    of primates, 361–362, 361*f*, 362*f*
    as property of life, 10, 11*f*
    of reproductive barriers, 274, 275*f*
    of tetrapods, 357, 357*f*
    as unifying theme of biology, 17*f*
    of vertebrates, 354–355, 354*f*, 355*f*
Evolution, as major theme in biology. *See* Major
        themes in biology, Evolution
Evolutionary adaptations, **247**
    of birds, 284–285, 284*f*, 359
    to desert biomes, 387
    to environmental variability, 378–379, 378*f*, 379*f*
    to global climate change, 399, 399*f*
    homologies and, 250–251, 250*f*, 251*f*
    of humans, 337
    Lamarck's theory on, 245
    life history traits as, 406–407, 407*f*, 407*t*
    natural selection and, 247, 261, 261*f*, 263–264,
        263*f*, 264*f*
    ongoing human, 367, 367*f*
    of plants, 316–317, 316*f*, 317*f*, 322–323,
        322*f*, 323*f*
Evolutionary clock, 33
Evolutionary developmental (evo-devo) biology,
        283–284, 284*f*
Evolutionary fitness, 261
Evolutionary history, 268
    of species, 286–289, 286*f*, 287*f*, 288*f*
Evolutionary trees, **251**
    of dog breeds, 165, 165*f*
    of green sea turtles, 18, 18*f*
Evolution Connection essays
    antibiotic resistance, 265, 265*f*
    bacterial resistance in humans, 71, 71*f*
    biodiversity, 449
    biofuels, 115–116
    cancer, 213
    climate change as agent for natural selection,
        399, 399*f*
    DNA profiling, Y chromosome, 237
    dog breeding, 165, 165*f*
    emerging viruses, 192
    evolution in Anthropocene epoch, 289, 289*f*
    human microbiota, 311, 311*f*
    humans as invasive species, 421, 421*f*

lactose intolerance in humans, 51
membrane origin, 87, 87f
ongoing human evolution, 367, 367f
oxygen importance, 103, 103f
plant-fungi interactions, 332–333, 333f
radioactivity, 33
sexual reproduction, evolutionary advantage of, 140
Exaptations, **285**
Exercise, science of, 91, 102
Exocytosis, **86**, 86f
Exons, **180**, 180f
Exoskeleton, **347**, 347f
Exotoxins, **305**
Experimental group, **8**
Experiments, **5**, 5f
controlled, 8–9, 8f, 9f
double-blind, 9, 9f
single-blind, 9, 9f
Exponential population growth model, **408**, 408f
Extinctions
humanity's footprint and, 269
humans as invasive species and, 421, 421f
mass (see Mass extinctions)
Extracellular matrix, **61**
Extraterrestrial life, search for,, 11
Eyes, evolution of, 285–286, 285f

F

F₁ generation, **147**
F₂ generation, **147**
Facilitated diffusion, transport proteins and, **84**
Facts, **7**
FADH₂, 97, 97f, 99–100, 99f
Fairy rings, 328f
Families (taxonomy), **244**
Family pedigrees (family trees), **153**–154, 153f
Farmers' markets, 398f
Farming. See Agriculture
Fats, 38, **43**, 43f, 44f
Fatty acids, as lipids, 43–44, 43f
Feathers, 359, 359f
evolution of, 284–285, 284f
Females, human, abnormal numbers of sex
chromosomes in, 139–140, 140t
Fermentation, **101**
alcoholic, 102–103, 103f
as anaerobic process, 101
evolution of glycolysis for cellular respiration and,
103
glucose in, 96, 101
in human muscle cells, 101–102
lactic acid and, 101f, 102–103
in microorganisms, 102–103
muscle fatigue and, 102
Ferns, **319**, 319f, 321, 321f
Fertilization, **131**

Mendel's experiments with plants, 146–147
process of, 136f
in sexual reproduction, 131
Fibers, cytoskeleton, 69, 69f
Finches, Galápagos Islands, 262–263, 262f–263f, 270
age structure of, 405, 405f
speciation of, 278, 278f
Fishes, **356**, 356f
contaminates and, 424, 433, 433f
overharvesting of, 402, 413, 413f
sustainable catch rates of, 413, 413f
Fitness
evolutionary, 261
relative, 261
Flagella, 58, **70**, 70f, 301
Flagellates, **308**, 308f
Flatworms (Platyhelminthes), **344**, 344f
Flowering plants. See Angiosperms
Flowers, **319**
structure and diversity of, 324, 324f
Fluid mosaics, plasma membranes as, **60**, 60f
Flukes, 344, 344f
Fluorescence, 112f
Flu viruses. See Influenza viruses
Flying squirrels (Glaucomys sp.), 289
Food. See also Diets; Digestive systems
carbon footprints and, 398, 398f
crops (see Agriculture; Crops)
energy from, 100, 100f
fungi as, 314–315, 328, 331–332, 332f
genetically modified, 216, 222–223, 222f, 223f,
235–236
ingestion of, 338, 338f
plants as, 314, 326
resources required to produce, 424, 439, 439f
Food calories, 78
Food chains, **432**, 432f, 434, 434f
Food labels, 41, 41f
Food webs, **434**, 434f
Forams, **308**, 309f
Forensics, **225**
DNA profiling in, 217, 217f, 225–229
Forests. See also Deforestation; Trees
clear-cutting of, 392, 392f
coniferous, 389, 389f
effects of fragmentation on biodiversity of, 446
global climate change effects on, 396, 396f
human impacts on, 392, 392f
temperate broadleaf, 389, 389f
temperate rain, 389
tropical, 386, 386f
Formaldehyde, 27f
Form and function. See Structure and function
Fossil fuels, **321**
biofuels and, 107
from ferns, 321
greenhouse gas accumulation and, 395, 395f
oil spills and, 428

Fossil record, **248**–249
biological diversity and, 268
evidence for evolution in, 33, 248–249, 249f
evidence for speciation in, 270, 279
as macroevolution archive, 279–281, 279f, 280t
Fossils, **245**
of early animals, 339, 339f
fixed species idea and, 245, 245f
of fungi, 328
methods of dating, 33
Founder effect, **260**, 260f
Franklin, Rosalind, 173, 173f
Freckles, 153, 153f
Freshwater biomes, 380–381, 380f, 381f
pollution of, 393
water shortages and, 393, 393f
Frogs, 357, 357f
polyploid speciation in, 275f
Frozen Zoo, 206–207
Fructose, 40, 40f
Fruit flies (Drosophila melanogaster), homeotic genes
of, 204, 204f
Fruits, **326**, 326f
Fuel, burning efficiency, 90, 94
Fuel molecules, cellular respiration and, 93
Fukushima Nuclear Power Plant, 26
Functional groups, **38**
Fungi, **328**
appearance on Earth, 294–295, 294f–295f
cells of (see Eukaryotic cells)
classification of, 288–289, 288f
commercial uses of, 331–332, 332f
as decomposers, 328, 331
diversity of, 328, 328f
ecological impact of, 331, 331f
as food, 314–315, 328, 331–332, 332f
nutrition of, 329
parasitic, 328, 328f, 331, 331f
plant interactions with, 315–317, 317f, 330, 330f,
332–333, 333f
reproduction of, 329, 329f
spores, 329
structure of, 329, 329f
truffles, 315
Fur, 379, 379f

G

G3P (glyceraldehyde 3-phosphate), 115, 115f
Galápagos Islands
Darwin's study of biodiversity of, 247, 247f, 270,
270f, 278, 278f
finches of, 262–263, 262f–263f, 270, 278, 278f,
405, 405f
iguanas of, 270, 270f
speciation on, 278, 278f
tortoises of, 247, 247f, 270

Gametes, **122**, 131, 138
Gametic isolation, 273, 273*f*
Gametophytes, **320**
  angiosperm, 325, 325*f*
  bryophyte, 320, 320*f*
  gymnosperm, 322–323, 322*f*, 323*f*
Gannets, population growth of, 410, 410*f*
Garbage, 372, 402
  carbon footprints and, 398, 398*f*
  in ocean, 383
Gastropods, **343**, 343*f*
Gastrovascular cavity, **342**, 342*f*
Gastrula, **338**, 338*f*
Gause, G. F., 429
Gel electrophoresis, **227**–228, 227*f*, 228*f*
GenBank, 231
Gene cloning, **219**
Gene expression, **198**, 198*f*
  in eukaryotic cells, 200, 200*f*, 203
  visualization of, 204, 204*f*
Gene flow, **260**
  as mechanism of evolution, 260–261, 260*f*
Gene pools, **256**
  analyzing, 256–257, 257*f*
  changes in, 257–258
  in study of evolution of populations,
    256–257, 257*f*
Gene regulation, **198**. *See also* Genes
  in bacteria, 198–199, 199*f*
  cell signaling, 203, 203*f*
  in eukaryotic cells, 200–203, 200*f*–202*f*
  homeotic genes in, 204, 204*f*
Genes, **12**–13, **49**, 148. *See also* Genomes; Humans,
    genome of
  alleles of (*see* Alleles)
  cancer and, 209
  cloning of, 218–219
  editing of, 220–221, 221*f*
  homeotic, 204, 204*f*
  human number of, 123
  linked, 162–163
  locus, 149, 149*f*
  mutations of (*see* Mutations)
  origin of, 297–298, 298*f*
  sex-linked, 163–164
  tumor-suppressor, **209**, 209*f*, 211
Genetically modified (GM) organisms, **218**, 218*f*
  in agriculture, 222–223, 222*f*, 223*f*
  as foods, 216, 222, 222*f*, 235–236
  protests over, 235–236, 235*f*
Genetic code, 170, **178**, 178*f*
  evidence of evolution in, 250
Genetic cross, **147**
Genetic diversity. *See also* Genetic variation
  in biodiversity, 426
  lack of, 259–260, 260*f*
  populations, lack of in, 242

Genetic drift, **259**
  bottleneck effect in, 259–260, 259*f*
  founder effect in, 260, 260*f*
  as mechanism of evolution, 258–260, 258*f*,
    259*f*, 260*f*
Genetic engineering, **218**
  in agriculture, 222–223, 223*f*
  controversy over, 236
  gene editing, 220–221, 221*f*
  genetic code and, 178
  human gene therapy, 224, 224*f*
  of insulin, 221, 221*f*
  pharmaceutical applications of, 221–222, 221*f*, 222*f*
  recombinant DNA techniques, 218–220, 219*f*, 220*f*
Genetic information, flow of, 176, 176*f*
Genetic Information Nondiscrimination Act of 2008,
    237
Genetic profiling, 236–237, 236*f*. *See also* Forensics
Genetics, **146**
  DNA technology (*see* DNA technology)
  dog breeding as longest-running genetic
    experiment, 145, 156
  human genetic disorders, 154–155, 157
  population, in health science, 257, 257*f*
Genetic testing. *See also* Forensics
  for human genetic disorders, 157, 157*f*
  personal kits, 237, 237*f*
Genetic variation
  crossing over and, 136–137, 136*f*
  independent assortment of chromosomes and,
    135–136
  sources of, 254–255, 255*f*
Genghis Khan, 237
Genomes. *See also* Genes
  Neanderthal, 365
  sequencing of, 230, 230*f*
Genomics, **230**
  applied, 233–234
  Human Genome Project, 231–232
  sequencing of whole genomes of organisms for,
    230, 230*f*
  systems biology and, 234
  techniques for, 231, 231*f*
  XIAP protein and, 233, 233*f*
Genotypes, **149**
  determination of unknown, by testcrosses, 152
  environment and, 161
  family pedigrees and, 153–154
  gene expression as determination of phenotypes
    from, 176
  phenotypes vs., 176
  population frequencies of, 256–257, 257*f*
  in transcription and translation, 183
Genus, **244**
Geographic distribution of life, 376, 376*f*, 397, 397*f*
Geologic time scale, **279**–281, 279*f*, 280*t*
Germination, **323**

Ghost crabs, 349*f*
Giant panda, cloning and, 196, 206
Giant puffball, 332*f*
Giant sequoias, 322
Giant tube worms, 302, 302*f*, 376*f*
Giardia, 308, 308*f*
Gibbons, 361–362, 361*f*, 362*f*
Gila monsters, 358*f*
GLO (enzyme), 251
Global biogeochemical cycles, 440
Global climate change
  as agent of natural selection, 399, 399*f*
  ecosystem effects of, 396, 396*f*
  evolution spurred by, 289
  greenhouse effect and, 394, 394*f*
  greenhouse gas accumulation in, 395, 395*f*
  human carbon footprints and, 398, 398*f*
  human ecological footprint and, 419–420
  krill habitat and, 390
  polar bear and penguin effects of, 373, 373*f*,
    399, 399*f*
  reduction of, 106, 110
  species distribution effects of, 397, 397*f*
  water resources and, 393, 393*f*
Global warming. *See* Global climate change
Global water cycle, 391, 391*f*
Glucose
  cellular energy conversion of, 40
  in cellular respiration, 92, 94–95, 100
  in chemical cycling between photosynthesis and
    cellular respiration, 92–93, 93*f*
  cyanide and, 14
  in fermentation, 96, 101
  in glycolysis, 96, 96*f*
  insulin and, 12
  linear and ring structure of, 40, 40*f*
  in muscle cells, 91
Glyceraldehyde 3-phosphate (G3P), 115, 115*f*
Glycogen, **42**, 42*f*
Glycolysis, **95**
  in cellular respiration, 96, 96*f*, 100
  in fermentation, 101
  oxygen and, 103
Glycoside hydrolase, 81, 81*f*
GM (genetically modified) foods. *See* Genetically
    modified organisms
GMO (genetically modified organisms). *See*
    Genetically modified organisms
Goatsbeard (*Tragopogon* sp.), 276–277, 277*f*
Goiter, iodine deficiencies and, 24, 25*f*
Golden rice, 223, 223*f*
Golgi apparatus, 64–**65**, 65*f*
Goodall, Jane, 4*f*
Gorillas, 361–362, 361*f*, 362*f*
Grana, 68, **109**
Grand Canyon, 268, 274*f*
  fossil record in, 279, 279*f*

Grant, Peter and Rosemary, 262–263, 262*f*–263*f*
Grasshoppers, 350*f*
Grasslands, 388, 388*f*
Graves' disease, 23
Gray tree frogs, 275*f*
Greater arid-land katydid, 351*f*
Great Lakes, 380, 380*f*
Green algae, **310**–311, 310*f*, 311*f*
    plant origin from, 318, 318*f*
Green energy. *See* Renewable energy
Greenhouse effect, **394**, 394*f*
Greenhouse gases, **394**
    accumulation in, 395, 395*f*
Green sea turtles, 3
    evolutionary tree of, 18, 18*f*
    major themes of biology applied to, 17*f*
    properties of life and, 10, 11*f*
    tracking of, 8–9, 8*f*, 9*f*
Growth, as property of life, 10, 11*f*
Growth factors, **209**
Gram bridled white-eye, 449*f*
Guam kingfisher, 449*f*
Guam rail, 449*f*
Guam, conservation biology on, 449, 449*f*
Guanine (G), 49, 49*f*, **173**
Gulf of Mexico
    dead zone in, 443, 443*f*
    oil spill in, 428*f*
Gymnosperms, **319**, 319*f*, 322–323, 322*f*, 323*f*

H

Habitats, **375**
    of Archaea, 304, 304*f*, 305*f*
    destruction of, and declining biodiversity, 427, 427*f*
    fragmentation, 446, 446*f*
    isolation of, 272, 272*f*
    of krill, global climate change and, 390
    of prokaryotes, 299, 299*f*
    protist, 307
    of red-cockaded woodpecker, 412, 412*f*
Hagfishes, 356, 356*f*
Halophiles, 304
Haploid cells, **131**
Hardy-Weinberg equation, 257
Hardy-Weinberg equilibrium, **257**
Harris's antelope squirrel (*Ammospermophilus harrisii*), 274*f*
He (helium), 25, 25*f*
Health, human. *See* Diseases and disorders, human
Heat, **77**
    energy conversions and generation of, 13–14
Hector's dolphin, 426*f*
Helicobacter pylori, 57*f*
Hemoglobin, 48, 48*f*

Hemophilia, 164, 164*f*
Herbivores, **432**
Herbivory, **431**, 431*f*
Heredity, **146**. *See also* Genetics
Heritable traits, 16, 146
Herpesvirus, 189
Herrerasaurus, 359*f*
Heterotrophs, **92**
Heterozygotes, 158
Heterozygous gene, **148**
HGH (human growth hormone), 222, 236
High-fructose corn syrup, 41
Hill, A. V., 102, 102*f*
Hirudo medicinalis, 346
Histones, **124**
History, of Earth, 279–283, 279*f*, 280*t*, 281*f*, 282*f*, 283*f*, 292, 294–295, 294*f*–295*f*
HIV (human immunodeficiency virus), **190**–191
    nucleic acid in an infected cell, 191*f*
    as retrovirus causing AIDS, 190
    as RNA virus, 190
Holocene epoch, 280*t*
Homeostasis, of glucose, 100
Homeotic genes, **204**
    in gene regulation, 204
    in macroevolution, 284
Hominins, **363**, 363*f*. *See also* Human Evolution
Homo erectus, 363*f*, 364–365
Homo habilis, 363*f*, 364
Homologies, **250**, 286
    as evidence of evolution, 250–251, 250*f*, 251*f*
    inferring phylogeny from, 287–288, 287*f*, 288*f*
Homologous chromosomes, **130**, 130*f*
    alleles and, 149, 149*f*
    in asexual reproduction, 137
    independent assortment of, 135–136
    Mendel's law of segregation and, 149
    in sexual reproduction, 130–131
Homo sapiens. *See* Humans
Homozygous dominant
    genotypes, 155
    individuals, 158
Homozygous gene, **148**
Honey, 85
Hookworms, 346*f*
Hosts, parasite, **432**
Hot spots, biodiversity, 444, 444*f*, 449, 449f
HPV (human papillomavirus), 209
Huitlacoche, 328*f*
Human DNA technologies, 236–237
Human evolution, 337
    Australopithecus in, 363*f*, 364, 364*f*
    bipedalism in, 364, 364*f*
    emergence of humankind, 363–365, 363*f*, 364*f*, 365*f*
    first clothing, 366f
    Homo erectus in, 363*f*, 364–365

    Homo habilis in, 363*f*, 364
    Homo neanderthalensis in, 363*f*, 365
    lice and, 365, 365*f*
    ongoing, 367, 367*f*
    primate evolution and, 361–362, 361*f*, 362*f*
Human gene therapy, **224**, 224*f*
Human Genome Project, **231**–232
Human growth hormone (HGH), 222, 236
Human microbiota, 293
Human papillomavirus (HPV), 209
Human population, 402
    age structures of, 418–419, 418*f*, 419*f*
    ecological footprint of, 419–420, 420*f*
    growth of, 417–420, 417*f*, 417*t*, 418*f*, 419*f*, 420*f*
    history of, 417, 417*f*, 417*t*
    population ecology and, 404
Humans (*Homo sapiens*)
    ABO blood groups of, as example of codominant alleles, 159, 159*f*
    appearance on Earth, 294–295, 294*f*–295*f*
    behavioral responses of, to environmental variability, 379, 379*f*
    biological species concept applied to, 271, 271*f*
    brains of (*see* Brain, human)
    breeding of plants and animals by, 242, 252
    carbon footprints of, 398, 398*f*
    cloning of, 207
    disorders of, single gene, 154–157
    evolution caused by, 269, 289, 289*f*
    evolution of (*see* Human evolution)
    genome of, 231–232
    global climate change effects on, 396, 396*f*
    impacts of
        on accumulation of greenhouse gases, 395, 395*f*
        on biomes, 380–381, 381*f*, 383, 392–393, 392*f*, 393*f*
        on global climate change, 398, 398*f*
        on global water cycle, 391
        on loss of biodiversity, 446
    importance of fungi for, 328
    as invasive species, 421, 421*f*
    microbiota of, 293, 299, 306, 306*f*, 311, 311*f*
    origin and dispersal of, 365, 365*f*
    sex chromosome abnormalities in, 139–140, 140*t*
    skulls of, 283–284, 284*f*
    speciation facilitated by, 276–277, 277*f*
    survivorship curves for, 406, 406*f*
Humulin, genetic engineering of, 221, 221*f*
Huntington's disease, 154*t*, 155
Hybrid breakdown, 273, 273*f*
Hybrids, **147**
    cross-fertilization and, 146*f*, 147
    human activities facilitating, 276–277, 277*f*
    polyploid speciation and, 275–276, 276*f*
    postzygotic reproductive barriers and, 273, 273*f*
Hydras, 342, 342*f*

Hydrocarbons, as fuel, 38*f*
Hydrocodone, 222
Hydrogenation, **45**
Hydrogen bonds, **28**, 28*f*
Hydrogen ions (H⁺), 99
Hydrolysis, **39**, 42
Hydrophilic molecules, **43**, 43*f*
Hydrophobic molecules, **43**, 43*f*
Hydrothermal vents, 376*f*, 382–383
    prokaryotes living in, 299, 299*f*, 302
Hydroxyl group, 38
Hypercholesterolemia, 154*t*, 158, 158*f*
Hypertonic solutions, **85**, 85*f*
Hyphae, **329**, 329*f*
Hypotheses, **5**–7
Hypothesis-driven science, 4–7
Hypotonic solutions, **85**, 85*f*

Ice, biological significance of floating of, 30–31, 30*f*
Ichthyosaur fossils, 245, 245*f*
Identical twins, environmental influences and, 161, 161*f*
Iguanas, on Galápagos Islands, 270, 270*f*
Immunotherapy, 129
Incomplete dominance, **158**
    in human hypercholesterolemia, 158*f*
    in snapdragons, 158*f*
Independent assortment, 150*f*
    in Labrador retrievers, 151, 151*f*
    law of, 151, 162*f*, 163
Independent variable, **9**, 9*f*
India, ecological footprint of, 420, 420*f*
Indian mongooses, 414, 414*f*
Induced fit, enzyme, **82**, 82*f*
Influenza viruses, 188, 188*f*
Information flow, as major theme in biology. *See* Major themes in biology, Information Flow
Inheritance. *See also* DNA; Heredity
    chromosomal basis of, 162–164
    dog breeding and, 145, 156, 156*f*
    Mendelian (*see* Mendelian inheritance)
Inorganic nutrients. *See* Nutrients, inorganic
Insecticides, enzyme inhibition of, 74, 82
Insects, **350**–352, 350*f*, 351*f*, 352*f*
Insulin
    from bacteria, 216, 221, 221*f*
    genetic engineering of, 221, 221*f*
    glucose regulation, 12
Insulin-dependent diabetes. *See* Diabetes; Type 1 insulin-dependent diabetes
Integrated pest management (IPM), 416, 416*f*
Interactions within biological systems, as major theme in biology. *See* Major themes in biology, Interactions within biological systems

Intermediate filaments, cytoskeleton, 69
Interphase, **125**, 125*f*
    meiosis, 131, 131*f*, 132*f*
    mitosis, 126*f*
Interspecific competition, **429**, 429*f*
Interspecific interactions, **428**
    classification of, 428–429
    commensalism, 432
    herbivory, 431, 431*f*
    interspecific competition, 429, 429*f*
    mutualism, 430, 430*f*
    parasites and pathogens, 432
    predation, 430–431
Intertidal zone, **383**, 383*f*
Intraspecific competition, **410**, 410*f*
Introns, **180**, 180*f*
Invasive species, 413–414, 413*f*, 414*f*
    declining biodiversity and, 427
    humans as, 421, 421*f*
    lionfish as, 403, 413
Invertebrates, **341**
    annelids, 345–346, 345*f*
    arthropods (*see* Arthropods)
    chordates, 355, 355*f*
    cnidarians, 342, 342*f*
    echinoderms, 353, 353*f*
    flatworms, 344, 344*f*
    molluscs, 285–286, 285*f*, 343, 343*f*
    roundworms, 346, 346*f*
    sponges, 341, 341*f*
Iodine (I), as trace element in human diets, 24
Iodine deficiency, 24, 25*f*
Ionic bonds, **27**
Ionic compounds, 27, 27*f*
Ions, **27**, 79
Iron (Fe), as micronutrient, 24–25
Isomers, **40**, 40*f*
Isotonic solutions, **85**, 85*f*
Isotopes, **25**–26

Jaguars (*Panthera onca*), 244
Jellies, 342, 342*f*
Jolie, Angelina, 211*f*
J-shaped curve, population growth, 408, 408*f*
Jurassic period, 280*t*

Kangaroos, 360, 360*f*
Karyotypes, **130**, 130*f*
Kelp, 310–311, 311*f*
Kelp perch, population growth of, 410, 410*f*
Keystone species, **435**

Kilocalories (kcal), food, 78, 78*f*
Kinetic energy, **76**–77
Kingdoms (taxonomy), **244**, 288, 288*f*
Kissimmee River Restoration Project, 447, 447*f*
Klinefelter syndrome, 140, 140*t*
Koalas, 283*f*
Komodo dragons, 121
Kopi luwak, 326*f*
Krebs cycle. *See* Citric acid cycle
Krill, 390
Kuru, 192
Kuruman River Reserve, 404*f*

Labels, food, 41, 41*f*
Labrador retrievers, 151, 151*f*, 152*f*
Lacks, Henrietta, 129
lac operon, 199, 199*f*
Lactase enzyme, 37, 47, 51, 51*f*, 82, 82*f*
Lactic acid, muscle fatigue as accumulation of, 91, 101–102, 101*f*
Lactose
    lactase and, 47
    tolerance, genetic mutation that allows, 51
Lactose intolerance, 37, 51, 51*f*
Ladybird beetles, 416, 416*f*
Lake Erie, 443*f*
Lake Mead, 393, 393*f*
Lakes, 380, 380*f*
Lamarck, Jean-Baptiste, 245
Lampreys, 356, 356*f*
Lancelets, **355**, 355*f*
Land
    biomes (*see* Terrestrial biomes)
    plant colonization of, 316–318, 316*f*, 317*f*, 318*f*
Landscape ecology, **445**
Landscapes, **445**
Larvae, **338**, 338*f*, 352, 352*f*
Las Vegas, population of and water consumption in, 393, 393*f*
Lateral line system, **356**
Law of independent assortment, 150–**151**, 151*f*, 162*f*, 163
Law of segregation, 147–149, **148**, 149*f*, 162*f*
    monohybrid crosses and, 150–151
    seven characters of pea plants and, 147*f*
Leaves
    chloroplast pigments and colors of autumn, 111–112, 112*f*
    photosynthesis in chloroplasts of, 68, 108
Leeches, 346
Lemba people, 237
Lemmings, 411
Lemurs, 282, 361–362, 361*f*, 362*f*
Leopards (*Panthera pardus*), taxonomy of, 244, 244*f*

Leukemia, 208
Lice, ancient humans and, 366, 366*f*
Lichens, 302
Life
    classification of, 244, 244*f*, 286–289, 286*f*,
        287*f*, 288*f*
    Darwin's theory of evolution of, by natural
        selection, 16–17, 16*f*, 17*f*, 33, 247–248, 261,
        261*f*, 263–264, 263*f*, 264*f*
    diversity of, 245, 245*f* (*see also* Biodiversity)
    domains of, **57, 243**, 287–288, 287*f*, 304
    evolution of (*see* Evolution)
    geographic distribution of, 376, 376*f*, 397, 397*f*
    hierarchy of, 15*f*
    microbial (*see* Microbial life)
    naming of, 244, 244*f*
    origin of, 296–298, 296*f*, 297*f*, 298*f*
    passion for, 3
    properties of, 10–11, 11*f*
    scientific study of life, 4
    synthesis of organic compounds and, 296–297, 297*f*
Life cycles, **131**
    angiosperm, 324–326, 324*f*, 325*f*, 326*f*
    animal, 338, 338*f*
    human, 131*f*
    plant, 320, 320*f*, 322–323, 322*f*, 323*f*
Life history, **406**–407, 407*f*, 407*t*
Life tables, **406**, 406*t*
Light. *See also* Sunlight
    electromagnetic spectrum of, 110–111, 110*f*
    energy of (*see* Solar energy)
Light micrograph (LM), 56*f*
Light reactions, **109**–110
    chloroplast pigments in, 112
    generation of ATP and NADPH by, 113–114,
        113*f*, 114*f*
    harvest of light energy by photosystems in,
        112–113, 113*f*
    wavelengths of light and, 111, 111*f*
Lignin, **317**
Limiting factors, **409**
    logistic population growth and, 409, 409*f*
Limpets, eyes of, 285–286, 285*f*
Linked genes, 162–**163**
Linnaeus, Carolus, 244, 288
Lionfish, 403, 413
Lions (*Panthera leo*), 244
Lipids, 40, **43**, 43*f*
Liposomes, nanotechnology and creation of, 87
Liverworts, 332–333, 333*f*
Lizards, 358, 358*f*
    environmental adaptations of, 378, 378*f*
    urban-adapted, 289, 289*f*
LM (light micrograph), 56*f*
Lobe-finned fishes, **356**
Lobsters, 347, 347*f*
Locus (loci), gene, **149**, 149*f*

Logistic population growth model, **409**, 409*f*
Loofah, 336
Lucy (*Australopithecus afarensis*), 363*f*, 364, 364*f*
Lungfishes, 251, 252*f*, 356
Lungs, structure of, 12, 12*f*
Lycopene, 112
Lyell, Charles, 247
Lyme disease, 305, 305*f*, 424, 446, 446*f*
Lynxes, population cycles of, 411, 411*f*
Lysogenic cycle, bacteriophage, **186**, 187*f*
Lysosomal storage diseases, 66
Lysosomes, 64, **66**, 66*f*
Lysozyme, 222, 222*f*
Lytic cycle, bacteriophage, **186**, 187*f*

M

Macroevolution, **279**
    biological novelty and, 284–286, 284*f*, 285*f*
    fossil record of, 279–281, 279*f*, 280*t*
    mass extinctions and, 283, 283*f*
    mechanisms of, 283–286, 284*f*, 285*f*
    plate tectonics and, 281–282, 281*f*, 282*f*, 283*f*
    by small changes in developmental processes,
        283–284, 284*f*
Macromolecules, 36, **39**
Mad cow disease, 48*f*, 170, 192
Maize, phylogeny of, 286
Major themes in biology
    Evolution
        adaptations to environmental variability, 378–
            379, 378*f*, 379*f*
        Darwinian evolution, 16–17, 16*f*, 17*f*, 33, 213
        Darwin's theory on, 247–248
        humans compared with related species, 232
        life history traits as adaptations, 406–407,
            407*f*, 407*t*
        mechanisms of speciation, 274–279, 274*f*, 275*f*,
            276*f*, 278*f*
        terrestrial adaptations of seed plants, 322–323,
            322*f*, 323*f*
        tetrapods, 357, 357*f*
    Information Flow
        animal body plans, 339
        cell signaling, 203, 203*f*
        DNA → RNA → Protein, 63, 176, 176*f*
        duplicating chromosomes, 125
        for orderly process of life functions, 12–13,
            13*f*, 17*f*
    Interactions within biological systems, 14–16, 15*f*,
        17*f*
        cells, 59, 59*f*
        hierarchy of interactions, 375, 375*f*
        insects, 352
        photosynthesis, carbon fixation, and global
            climate, 110

    polygenic inheritance, 160, 160*f*
    population cycles, 411, 411*f*
    Pathways that transform energy and matter
        abiotic factors of biosphere, 376, 376*f*
        ATP and cellular work, 79–80
        cellular respiration, 94–95
        chloroplasts and mitochondria, 68
        energy flow in ecosystems, 438–439
        pathways that transform energy and matter,
            13–14, 17*f*
        photosynthesis, 108
    Relationship of structure to function
        adaptation of old structures for new functions,
            284–285, 284*f*
        adaptations of plant body, 316–317, 316*f*, 317*f*
        characteristics of fungi, 329, 329*f*
        chloroplast, 109
        chromosomes, 124, 124*f*
        DNA structure, 175
        enzyme activity, 82, 82*f*
        invasive species, 413–414, 413*f*
        mitochondrial membranes, 99
        plasma membrane, 60
        pleiotropy and sickle-cell disease, 160
        prokaryotes, 300–304, 300*f*, 301*f*, 302*f*,
            303*f*, 304*f*
        protein shape, 46–47
        relationship of structure to function, 12,
            12*f*, 17*f*
        scales, 377, 377*f*
        tapeworms, 344, 344*f*
        water, unique properties of, 29
Malaria, 243
Males, human
    abnormal numbers of sex chromosomes in,
        139–140, 140*t*
    infertility of, 70
Malignant tumors, **129**, 211
Malthus, Thomas, 253
Maltose, 41
Mammals, **360**, 360*f*
    cladistics of, 287–288, 287*f*
    diversification of, 283, 283*f*
    primates, 361–362, 361*f*, 362*f* (*see also* Humans)
Mammary glands, 360
Mammoth, 249*f*
Mantle, **343**, 343*f*
Marine biomes, 382–383, 382*f*, 383*f*
Marine fisheries, 428
Marine iguanas, 270*f*
Mars rover Curiosity, 2, 11, 11*f*
Marsupials, 282, 283*f*, **360**, 360*f*
Mass, **24**
Mass extinctions. *See also* Extinctions
    explosive diversifications after, 283, 283*f*
    geologic time scale and, 279
Mass number, **25**

Master control genes. *See* Homeotic genes
Mating. *See also* Sexual reproduction; Sexual selection
    natural selection and, 263–264, 263*f*–264*f*
Matter, **24**
    composition of, 13, 24–25
    recycling of, 14
    three physical states of, 24
Meadowlarks, 271*f*
Mechanical isolation, 273, 273*f*
Medicines
    from fungi, 332, 332*f*
    from plants, 327, 327*t*
Medusas, **342**, 342*f*
Meiosis, 130, **132**
    crossing over in, 132, 136–137, 136*f*
    errors in, 138–140, 138*f*, 139*f*, 140*t*
    exchange of genetic material in, 132
    functions of, 122
    interphase before, 131
    mitosis vs., 132, 134–135, 134*f*
    origins of genetic variation in, 255
    stages of, 131*f*, 132*f*–133*f*
Meiosis I, 131–132, 132*f*–133*f*
    nondisjunction in, 138, 138*f*
Meiosis II, 131–132, 133*f*
    nondisjunction in, 138, 138*f*
Membrane proteins, 83, 83*f*
Membranes. *See* Plasma membranes
Men. *See* Males, human
Mendel, Gregor, 146*f*, 254
    cross tracking, 148*f*
    Darwin, Charles, and, 146
    experiments of, 146–151, 172
    law of independent assortment by, 150–151,
        162*f*, 163
    law of segregation by, 147–149, 162*f*
    variations on laws of, 158–159
Mendelian inheritance
    chromosomal basis of (*see* Chromosomal basis of
        inheritance)
    dog breeding and, 145, 156, 156*f*
    family pedigrees in, 153–154
    Gregor Mendel's experiments and, 172
    human ABO blood groups and codominance in,
        159, 159*f*
    incomplete dominance in plants and people in, 158
    law of independent assortment in, 135–136,
        162*f*, 163
    law of segregation in, 147–149, 162*f*
    Mendel's experiments and, 146–151
    polygenic inheritance in, 160–161, 160*f*
    recessive human genetic disorders in, 154–155
    rules of probability, 152–153, 152*f*
    sickle-cell disease and pleiotropy in, 160
    testcrosses to determine unknown genotypes in,
        152
Meningitis, 305, 305*f*
Menthol, 327*t*

Mescaline, 431
Mesozoic era, 279, 280*t*, 282, 295
Messenger RNA (mRNA), 63, **180**, 180*f*, 182*f*, 202
Metabolic processes, cellular respiration, 90, 102
Metabolism, 14, **80**
    enzymes and, 80
    exercise and, 91
Metagenomics, 234
Metamorphosis, **338**, 338*f*, 352, 352*f*
Metaphase, mitosis, **127**, 127*f*
Metaphase I, meiosis I, 132*f*
Metaphase II, meiosis II, 133*f*
Metastasis, cancer, **129**
Methane (CH_4), 38, 38*f*
Methanogens, 304, 305*f*
Methicillin-resistant *Staphylococcus aureus* (MRSA),
    61, 265
Mexico, population growth in, 418, 418*f*
Mice, fluorescent proteins and, 178, 178*f*
Microbial life
    appearance on Earth, 294–295, 294*f*–295*f*
    Archaea (*see* Archaea)
    bacteria (*see* Bacteria)
    evolution of, 292–298, 294*f*–295*f*, 296*f*, 297*f*, 298*f*
    in history of life on Earth, 292, 294–295,
        294*f*–295*f*
    in human body, 293, 299, 306, 306*f*, 311, 311*f*
    origin of, 296–298, 296*f*, 297*f*, 298*f*
    prokaryotes (*see* Prokaryotes)
    protists (*see* Protists)
Microbiome, **299**
Microbiota, 293, **299**
    obesity and, 306, 306*f*
Microevolution, **258**
    as changes in gene pools, 257–258
Microfilaments, cytoskeleton, 69
Micrographs, 56*f*
MicroRNAs (miRNAs), **202**, 202*f*
Microscopes, 56, 56*f*
Microtubules, **69**, 69*f*
Milk, lactose digestion and, 37
Milkweed, 326*f*
Miller, Stanley, 296–297, 297*f*
Miller-Urey experiments, 296-97, 297f
Millipedes, **350**, 350*f*
Mimicry, predation and, 430–431, 431*f*
Miocene epoch, 280*t*
miRNAs (microRNAs), **202**, 202*f*
Miscarriage, 139
Misfolded proteins, diseases and, 48
Mites, 348, 348*f*
Mitochondria, **68**, 68*f*
    cellular respiration in, 95, 95*f*
Mitochondrial matrix, 68, 68*f*
Mitosis, **125**
    cancer and, 128–129, 129*f*
    cytokinesis and, 126–128, 128*f*
    functions of, 122, 122*f*

    meiosis vs., 132, 134–135, 134*f*
    phases of, 126*f*–127*f*, 127
Mitotic (M) phase, cell cycle, **125**, 125*f*
Mitotic spindles, **127**
Molds, 328, 328*f*, 332, 332*f*
Molecular biology, 170, **250**
    evidence for speciation in, 270
    homologies and, 250–251, 250*f*, 251*f*
Molecular clock, 366, 366*f*
Molecular formula, 27*f*
Molecules, 15*f*, **27**
Molecules, alternative ways to represent
    ball-and-stick model, 27*f*
    electron configuration, 27*f*
    molecular formula, 27*f*
    space-filling model, 27*f*
    structural formula, 27*f*
Molluscs (Mollusca), **343**, 343*f*
    eye complexity among, 285–286, 285*f*
Momentum, population, 418, 418*f*
Monarch butterflies, 352*f*
Mongooses, 414, 414*f*
Monkeys, 361–362, 361*f*, 362*f*
Monoculture agriculture, 416
Monohybrid crosses, **150–151**
Monomers, **39**–40
Monosaccharides, **40**, 40*f*
Monotremes, **360**, 360*f*
Monozygotic twins. *See* Identical twins
Morphine, 327*t*, 431
Mosquitoes, evolution of, 243, 253–256, 254*f*
Mosses, **318**–320, 319*f*, 320*f*
Mountain ranges, climate and, 385, 385*f*
Movement corridors, **445**
M (mitotic) phase, cell cycle, **125**, 125*f*
mRNA (messenger RNA), 63, **180**, 180*f*, 182*f*, 202
MRSA (methicillin-resistant *Staphylococcus aureus*),
    61, 265
Multicellular organisms, appearance on Earth,
    294–295, 294*f*–295*f*
Mumps virus, 189, 189*f*
Muscle burn, 102
Muscles
    ATP requirement, 101
    cellular respiration in, 91
    fermentation in, 101–102
Mushrooms, 54, 57, 328–329, 328*f*, 329*f*, 331–332,
    332*f*. *See also* Fungi
Mutagens, **185**
Mutations, **184**
    cancers caused by, 211, 213
    cholera resistance and, 71
    chromosomal, 255
    in directed evolution, 81, 81*f*
    frameshift, 185
    nonsense, 185
    as source of genetic variation, 255
    types of, 184–185, 184*f*

Mutualism, **430**, 430*f*
Mycelium, **329**, 329*f*
Mycorrhizae, **316**, 317*f*
    liverworts, 332–333, 333*f*
    in pine forests, 330, 330*f*

N

NAD⁺ (nicotinamide adenine dinucleotide), 95
    in citric acid cycle, 97, 97*f*
    in electron transport, 99, 99*f*
    in glycolysis, 96, 96*f*
NADH, **95**
    in citric acid cycle, 97, 97*f*
    in electron transport, 99–100, 99*f*
    in glycolysis, 96, 96*f*
NADPH, 109*f*, **110**, 113–115, 113*f*,
        114*f*, 115*f*
Nanotechnology. *See also* DNA technology
    enzyme engineering, 81, 81*f*
    evolution of membranes and, 87
    harnessing cellular structures with, 75
Natural selection, **16**, **247**
    action of, 253, 254*f*
    antibiotic-resistant bacteria and, 17, 17*f*
    bacterial resistance in humans and, 71
    climate change as agent of, 399, 399*f*
    Darwin's theory of evolution by, 16–17, 16*f*, 17*f*,
        33, 247–248, 261, 261*f*, 263–264, 263*f*,
        264*f*, 270
    descent with modification theory as, 16
    in evolution of cancer, 213
    evolution of resistance to pesticides and, 243,
        253–256, 254*f*
    in Galápagos finches, 262–263, 262*f*–263*f*
    general outcomes of, 264, 264*f*
    key points about, 254, 254*f*
    life history traits and, 406–407, 407*f*, 407*t*
    as mechanism for evolution, 252–254, 253*f*, 254*f*,
        258, 261, 261*f*, 263–264, 263*f*, 264*f*
    in origin of life, 298
    sexual selection and, 263–264, 263*f*–264*f*
Nautiluses, 343*f*
    eyes of, 285–286, 285*f*
Neanderthals (*Homo neanderthalensis*), 232, 268, 270,
        287*f*, 336, 363*f*, 365
Neisseria meningitidis, 305*f*
Nematodes (Nematoda), **346**, 346*f*
Nerve gases, enzyme inhibition of, 74, 82
Neutrons, **25**
New Zealand wilderness, 276*f*
Niall of the Nine Hostages, 237
Nicotinamide adenine dinucleotide. *See* NAD⁺
Nicotine, 431
Nile crocodile, 359*f*
Nitrogen (N)
    in ecosystem chemical cycling, 442, 442*f*

    in freshwater pollution, 393
        pollution, source of, 443, 443*f*
Nitrogen cycle, 442, 442*f*
Nitrogen fixation, 303, 303*f*, **442**
Nitrogen gas (N₂), in nitrogen cycle, 442, 442*f*
Nitrogenous bases, DNA and RNA, 49, 49*f*, 173, 177
Nondisjunction, **138**, 138*f*
Nonrenewable resources, plants as, 327, 327*f*
Northern quoll, 283*f*
Notochord, **354**, 354*f*
Nuclear disasters
    Chernobyl nuclear reactor explosion, 26
    Fukushima Nuclear Power Plant, 26
Nuclear envelope, **62**, 62*f*, 64
Nuclear medicine, 23
Nuclear transplantation, **206**, 206*f*
Nucleic acids, **49**–50
    as macromolecules, 39–40
Nucleoids, **58**
Nucleolus, **62**, 62*f*
Nucleosomes, DNA, **124**
Nucleotides, **49**, **172**, 184*f*
Nucleus (nuclei), **25**, **58**, 62, 62*f*
Nutrients, as abiotic factor in biosphere, 377
Nutrients, inorganic
    chemical cycling of (*see* Chemical cycling)
    pollution of, 443, 443*f*
Nutrition
    animal, 338, 338*f*
    fungal, 329
    prokaryotic, 302–303, 302*f*
Nutritional modes. *See* Trophic structures

O

Obesity, intestinal microbiota role in, 306, 306*f*
Ocean acidification
    carbon dioxide and, 32, 32*f*
    coral bleaching, 32
Oceans, 382–383, 382*f*, 383*f*
    climate effects of, 384
Octopuses, 343, 343*f*
Oil sands, 269
Oligocene epoch, 280*t*
Omega-3 fatty acids, 45
Omnivores, **434**
Oncogenes, **209**, 209*f*
On the Origin of Species by Means of Natural Selection
        (Darwin), 16, 16*f*, 146, 245–248, 246*f*, 247*f*,
        254, 270, 289
Operators, DNA, **199**, 199*f*
Operculum, **356**
Operons, DNA, **199**, 199*f*
Opinions, 7
Opportunistic life history, **407**, 407*f*, 407*t*
Orangutans, 361–362, 361*f*, 362*f*
Order, as property of life, 10, 11*f*

Orders (taxonomy), **244**
Ordovician period, 280*t*
Organelles, **58**–59
    chloroplasts and mitochondria, 68, 68*f*
    cytoskeleton, 69–70, 69*f*–70*f*
    of endomembrane system, 64–67, 64*f*–67*f*
    as level of life, 15*f*
    nucleus and ribosomes, 62–63, 62*f*, 63*f*
Organic compounds, **38**
    in origin of life, 296–297, 297*f*
Organismal ecology, **375**, 375*f*
Organisms, **375**
    communities of (*see* Communities)
    environmental adaptations of, 378–379, 378*f*, 379*f*
    environmental interactions of (*see* Ecology)
    as level of life, 15*f*
    size range, 56, 56*f*
Organs, as level of life, 15*f*
Organ systems, as level of life, 15*f*
Oscillatoria, 302*f*
Osmoregulation, animal, **85**
Osmosis, **84**, 84*f*, 85*f*
    in food preservation, 74, 85
    water balance and, 84–85
Outcomes, scientific, 5–6, 6*f*
Ovaries, flower, **324**, 324*f*
Overconsumption, ecological footprints and, 419–
        420, 420*f*
Overexploitation
    declining biodiversity through, 428
    overharvesting of fish, 402, 413, 413*f*
    overharvesting of plants, 428
Overnourishment. *See* Obesity
Overweight. *See* Obesity
Ovules, **323**, 323*f*
Oxygen gas (O₂)
    in aerobic lifestyles, 91
    in cellular respiration, 93–94
    in chemical cycling between photosynthesis and
        cellular respiration, 92–93, 93*f*
    evolution of glycolysis in absence of, 103
    in exercise science, 102
    in harvesting food energy, 98*f*
    in photosynthesis, 108–109, 109*f*
    time line on Earth, 103*f*
Oysters, 343

P

Paclitaxel, 327*t*
Paine, Robert, 435
Paleocene epoch, 280*t*
Paleontologists, **248**
Paleozoic era, 279, 280*t*, 282, 295
Palm oil plantation, 392*f*
Pangaea, 282, 282*f*
Panthera genus, 244, 244*f*

Papaya ringspot potyvirus (PRSV), 188
Paramecium, 67*f*, 308, 309*f*, 429, 429*f*
Paramecium aurelia, 429, 429*f*
Paramecium caudatum, 429, 429*f*
Paranthropus boisei, 363*f*, 364
Paranthropus robustus, 363*f*, 364
Parasites, **307**, **432**
    in community ecology, 432
    flatworm, 344, 344*f*
    fungi as, 328, 328*f*, 331, 331*f*
    protists as, 307
    roundworm, 346, 346*f*
Parental care, life histories and, 407,
        407*f*, 407*t*
Parthenogenesis, 121
Passive transport, 83–**84**, 84*f*
Patas monkey, 362*f*
Paternity, 229
Pathogens, **304**, **432**
    bacterial, 304–306, 305*f*
    in community ecology, 432
PCBs, 433, 433*f*
PCR (polymerase chain reaction), **225**–226, 225*f*
Pea flower, structure of, 146*f*
    Mendel's study of, 146–147, 147*f*
    technique for cross-fertilizing, 146*f*
Peat moss, 319*f*
Pediculus humanus, 366, 366*f*
Pedigrees, **153**–154
Peer review, **6**–7, 10
Pelagic realm, **382**, 382*f*
Penguins, 373, 375*f*, 399, 399*f*
Penicillin, 55, 61, 82, 332, 332*f*
Penicillium, 332, 332*f*
Peptide bonds, amino acid, **47**, 47*f*
Periodic table of the elements, **24**, 24*f*
Permafrost, **390**, 396
Permian period, 280*t*, 283
Pesticides
    evolution of resistance to, 243, 253–256, 254*f*
    integrated pest management and, 416
Pests
    biological control of, 414–415, 414*f*, 415*f*
    integrated management of, 416, 416*f*
Petals, flower, **324**, 324*f*
PET scan, 23, 23*f*
P generation, **147**
Phages, **186**–187, 186*f*, 187*f*
Phagocytosis, **86**, 86*f*
Pharmaceutical applications, of biotechnology,
        221–222, 221*f*, 222*f*
Pharyngeal slits, **354**, 354*f*
Phenotypes, **149**
    family pedigrees of, 153–154
    gene expression as determination of, from
        genotypes, 176
    genetic variation and, 254, 255*f*

genotypes vs., 176
    in transcription and translation, 183
Phenylalanine, 257
Phenylketonuria (PKU), 154*t*, 257, 257*f*
Phosphate pollution, 443, 443*f*
Phosphates, 443
Phosphate transfer, ATP cycle and, 79, 79*f*
Phospholipid bilayers, **60**, 60*f*
Phospholipids, **60**, 87
Phosphorus (P)
    biogeochemical cycling of, 441, 441*f*
    in freshwater pollution, 393
Phosphorus cycle, 441, 441*f*
Photic zone, **380**, 380*f*, 382, 382*f*
Photoautotrophs, 108, 108*f*
Photons, **112**–113
Photosynthesis, **92**, **108**
    basics of, 108–109
    biofuels from, 107
    Calvin cycle of, 115, 115*f*
    cellular respiration and, 109
    in chemical cycling, 92–93, 93*f*
    chloroplasts as sites of, 68
    green energy and, 115–116
    light reactions of, 109–114, 109*f*–114*f*
    map of, 109*f*
    overview of stages of, 109–110
    process of, 109*f*
    by prokaryotes, 302, 302*f*
    protist, 307
    wavelengths of light driving, 110
Photosynthetic organisms, 13–14
Photosynthetic pigments, 112, 112*f*
Photosystems
    generation of ATP and NADPH by, 113–114,
        113*f*, 114*f*
    harvest of light energy by, 112–113, 113*f*
Photovoltaic (PV) solar panels, 107, 107*f*
pH scale, **31**–32, 32*f*
Phyla (taxonomy), **244**
Phylogenetic trees, **286**–288, 286*f*, 287*f*. See also
        Evolutionary trees
    changes in, 288–289, 288*f*
Phylogeny, **286**–288, 286*f*, 287*f*, 288*f*
    animal, 340, 340*f*, 341*f*
    as work in progress, 288–289, 288*f*
Physiological responses to environmental conditions,
        378, 378*f*
Phytoplankton, **310**, **380**, 382, 382*f*
    in freshwater biomes, 380–381
Pigments, photosynthetic, 111–112, 112*f*
Pill bugs, 349*f*
Pine forests, mycorrhizae in, 330, 330*f*
Pisaster sea stars, 435, 435*f*
PKU (phenylketonuria), 154*t*, 257, 257*f*
Placebo, **9**
Placenta, **360**

Placental mammals, 282, 283*f*, **360**, 360*f*
Planarians, 344, 344*f*
Plants, **316**
    alternation of generations in, 320, 320*f*, 322–323,
        322*f*, 323*f*
    angiosperms (*see* Angiosperms)
    appearance on Earth, 294–295, 294*f*–295*f*
    artificial selection of, 242, 252
    bryophytes, 318–320, 319*f*, 320*f*
    cells of, 59*f*
    central vacuoles in cells of, 67
    classification of, 288–289, 288*f*
    cloning of, 205, 205*f*
    cohesion and water transport in, 29, 29*f*
    colonization of land by, 316–318, 316*f*, 317*f*, 318*f*
    diversity of, 318–319, 318*f*, 319*f*, 327, 327*f*
    evolution of, 318–319, 318*f*, 319*f*
    ferns, 319, 319*f*, 321, 321*f*
    as food, 314, 326
    fungus interactions with, 315–317, 317*f*, 330,
        330*f*, 332–333, 333*f*
    gymnosperms, 319, 319*f*, 322–323, 322*f*, 323*f*
    medicines from, 327, 327*t*
    Mendelian inheritance and (*see* Mendelian
        inheritance)
    as nonrenewable resource, 327, 327*f*
    origin of, 318, 318*f*
    overharvesting of, 428
    parasitic fungi and, 328, 328*f*
    as photoautotrophs, 108
    photosynthesis in (*see* Photosynthesis)
    polyploid speciation in, 275–276, 275*f*, 276*f*
    populations of (*see* Populations)
    reproductive adaptations of, 317, 317*f*
    terrestrial adaptations of, 316–317, 316*f*, 317*f*,
        322–323, 322*f*, 323*f*
    in terrestrial biomes, 385–386, 385*f*, 386*f*
    viruses that infect, 188, 188*f*
Plasma membranes, **57**
    active transport across, 86, 86*f*
    endomembrane systems and, 64–67, 64*f*–67*f*
    exocytosis, endocytosis, and transport of large
        molecules across, 86, 86*f*
    functions of, 60, 83–86
    nanotechnology and evolution of, 87
    origin of, 297
    passive transport as diffusion across, 83–84, 84*f*
    spontaneous formation of, 87*f*
    structure of, 60, 60*f*
    transport functions of, in cellular work, 83–86
    water balance and osmosis across, 84–85
Plasmids, **218**, 218*f*
Plasmodial slime molds, 309, 309*f*
Plasmodium, 308
Pleiotropy, **160**
Pleistocene epoch, 280*t*, 421
Pliocene epoch, 280*t*

Polar bears, 373, 373*f*
Polar ice, **390**, 390*f*
Polarity of water, 28, 28*f*
Polar molecules, **28**
Pollen grains, **323**
Pollination, **323**
Pollution, 372
  bioremediation of, 302*f*, 303–304, 303*f*
  chemical cycling and nutrient, 443, 443*f*
  declining biodiversity and, 269, 428, 428*f*
  of freshwater biomes, 380–381, 393
  of marine biomes, 383
Polygenic inheritance, **160**–161, 160*f*
Polymerase chain reaction (PCR), **225**–226, 225*f*
Polymers, **39**, 39*f*
  abiotic synthesis of, 297
Polynucleotides, **172**
Polypeptides, **47**–48, 47*f*, 48*f*
Polyploid speciation, **275**–276, 275*f*, 276*f*
  human activities facilitating, 276–277, 277*f*
Polyps, **342**, 342*f*
Polysaccharides, **42**, 42*f*
Polyunsaturated fats, 43, 44*f*
Ponderosa pine trees, 323*f*
Ponds, 380, 380*f*
Population cycles, 411, 411*f*
Population density, **405**, 405*f*
Population ecology, **375**, 375*f*, **404**
  applications of, 412–416, 412*f*, 413*f*, 414*f*, 415*f*, 416*f*
  biological control of pests in, 414–415, 414*f*, 415*f*
  human population growth in, 417–420, 417*f*, 417*t*, 418*f*, 419*f*, 420*f*
  humans as invasive species in, 421, 421*f*
  integrated pest management in, 416, 416*f*
  invasive species management in, 413–414, 413*f*, 414*f*
  life history traits as evolutionary adaptations in, 406–407, 407*f*, 407*t*
  life tables and survivorship curves, 406, 406*f*, 406*t*
  overview of, 404–407, 404*f*, 405*f*, 406*f*, 406*t*, 407*f*, 407*t*
  population age structure, 405, 405*f*
  population growth models in, 408–411, 408*f*, 409*f*, 410*f*, 411*f*
  species conservation in, 412, 412*f*
  sustainable resource management in, 412–413, 413*f*
Population growth
  exponential model of, 408, 408*f*
  human, 417–420, 417*f*, 417*t*, 418*f*, 419*f*, 420*f*
  logistic model of, 409, 409*f*
  regulation of, 410–411, 410*f*, 411*f*
Population momentum, **418**, 418*f*
Populations, **255**, **375**, **404**
  analyzing gene pools of, 256–257, 257*f*
  evolution of, 254–258, 255*f*, 256*f*, 257*f*

genetics, in health science, 257, 257*f*
  as level of life, 15*f*
  as units of evolution, 255–256, 256*f*
Post-anal tail, **354**, 354*f*
Postzygotic barriers, 272*f*, **273**, 273*f*
Potatoes, 216, 222
Potential energy, **76**
Prairie, 388, 388*f*
Prairie dogs, 405*f*
Praying mantises, 351*f*
Precambrian, 279, 280*t*
Pre-cells, 297–298
Precipitation, effects of, on terrestrial biomes, 386, 386*f*
Predation, **430**
  in community ecology, 430–431
  by lionfish, 403
Prezygotic barriers, **272**–273, 272*f*–273*f*
Primary consumers, **432**
Primary production, ecosystem energy flow and, **438**, 438*f*
Primary succession, **436**–437, 436*f*
Primates, **361**–362, 361*f*, 362*f. See also* Humans
Primers, **226**
Principles of Geology (Lyell), 247
Prions, 48, 48*f*, **192**
The Process of Science, essays
  avatar cancer treatment, 210, 210*f*
  biodiversity, 446
  biological control of pests, 415, 415*f*
  DNA and RNA vaccines, 190, 190*f*
  DNA profiling, genomics, 233, 233*f*
  dog breeding, 156, 156*f*
  enzyme engineering, 81, 81*f*
  exercise science, 102
  global climate change effects on species distribution, 397, 397*f*
  human activities facilitating speciation, 276–277, 277*f*
  human microbiota and obesity, 306, 306*f*
  humans vs. bacteria, 61, 61*f*
  lactose intolerance, 51
  lice and ancient humans, 365, 365*f*
  natural selection, 262–263, 262*f*–263*f*
  plant-fungi interactions, 330, 330*f*
  radiation therapy, 26, 26*f*
  sexual and asexual reproduction, 137
  solar energy, 111
Producers, **92**, 92*f*, **432**
  in chemical cycling, 92
  consumers vs., 13, 92
  photosynthetic autotrophs as, 92
  plants as, 13
  in trophic structure food chains, 432
Products, chemical reaction, **28**, 28*f*
Profiling, DNA. *See* DNA profiling
Prokaryotes, **294**

appearance on Earth, 294–295, 294*f*–295*f*
  Archaea branch of, 304–306, 304*f*, 305*f (see also* Archaea)
  bacteria branch of, 304–306, 305*f (see also* Bacteria)
  bioremediation using, 302*f*, 303–304, 303*f*
  chemical cycling by, 303
  classification of, 288, 288*f*
  ecological impact of, 303–304, 303*f*
  evolution of, 292–298, 294*f*–295*f*, 296*f*, 297*f*, 298*f*
  evolution of glycolysis in, 103, 103*f*
  as first life on Earth, 57
  forms of, 300–301, 300*f*, 301*f*
  habitats of, 299, 299*f*
  nutrition of, 302–303, 302*f*
  reproduction of, 301–302, 301*f*
  structure and function of, 300–304, 300*f*, 301*f*, 302*f*, 303*f*, 304*f*
Prokaryotic cells, **57**–58
  eukaryotic cells vs., 57*t*
Promoters
  DNA, **179**, **199**, 199*f*
  transcription, **179**
Pronghorn antelope (*Antilocapra americana*), 421, 421*f*
Properties, of life, 10–11, 11*f*
Prophages, **187**
Prophase, mitosis, 126*f*, **127**
Prophase I, meiosis I, 132*f*
Prophase II, meiosis II, 133*f*
Prostate cancer, 26, 26*f*
Proteins, **46**
  activation and breakdown, 203
  amino acids as monomers of, 46–48
  ATP and cellular work of, 79–80
  building of, 49*f*
  from DNA, flow of genetic information, 177
  enzymes as, 80
  exocytosis, endocytosis, and transport of, across membranes, 86*f*
  as macromolecules, 39–40
  of membrane, 83, 83*f*
  in phospholipid bilayers, 60, 62
  prions and misfolded, 48, 48*f*
  production, 63, 63*f*
  proteomics and study of whole sets of, 234, 234*f*
  roles played by, 46*f*
  shape and structure of, 47, 47*f*
Proteomes, 234
Proteomics, **234**, 234*f*
Protists, **307**
  appearance on Earth, 294–295, 294*f*–295*f*
  cells of (*see* Eukaryotic cells)
  cilia and flagella in movement of, 70
  classification of, 288–289, 288*f*
  contractile vacuoles of, 67
  cytoskeletons and movement of, 69, 69*f*

Protists (*continued*)
  diversity and characteristics of, 307, 307*f*
  protozoans, 308, 308*f*–309*f*
  seaweeds, 307, 307*f*, 310–311, 311*f*
  slime molds, 309, 309*f*
  unicellular and colonial algae, 307, 310, 310*f*
Protons, **25**
Proto-oncogenes, **209**, 209*f*
Protozoans, **308**, 308*f*–309*f*
Proviruses, **191**
PRSV (papaya ringspot potyvirus), 188
Pseudopodia, **308**, 308*f*
Pseudoscience, **9**, 10*f*
Puerto Rico, pine forests in, 330, 330*f*
Punnett squares, **148**
  law of independent assortment and,
    150–151, 150*f*
  law of segregation and, 148–149, 149*f*
Pupa, 352, 352*f*
PV (photovoltaic) solar panels, 107, 107*f*
Pygmy seahorse, 430*f*
Pyramid of production, ecosystem energy flow and,
    438–439, 439*f*
Pyruvic acid
  in cellular respiration, 96–97, 96*f*, 97*f*
  in fermentation, 101
Pythons
  food ingestion by, 338, 338*f*
  as invasive species, 414, 414*f*
Pytilia, green-winged, 263*f*

Q

Quad screen test, 157
Quaternary consumers, **433**
Quaternary period, 280*t*
Quechuans, 367*f*
Quinine, 327*t*

R

Radial symmetry, **340**, 340*f*
Radiation
  sunlight as, 110
  ultraviolet (*see* Ultraviolet light)
Radiation therapy, cancer, 23, 26, 26*f*, **129**
Radioactive decay, 33
Radioactive isotopes, **26**
  radiometric dating using, 33, 33*f*, 281
Radioactive seed implantation, 26, 26*f*
Radioactivity, 33
Radiometric dating, 33, 33*f*, **281**
Radula, **343**, 343*f*
Ragworms, 345, 345*f*
Rainbow shield bugs, 351*f*

Rain forests. *See* Temperate rain forests; Tropical
    rain forests
Ratios, genotypes to phenotypes, 151
Ray-finned fishes, **356**
Reactants, chemical reaction, 28, 28*f*
Recessive alleles, **148**
  carriers of, 155
  dominant alleles vs., 148
  human genetic disorders and, 154–155, 154*t*, 155*f*
Recessive disorders, 154–155, 154*t*, 155*f*
Recessive traits, 153, 153*f*, 154*f*
Recombinant chromosomes, 136
Recombinant DNA, **218**
Recombinant DNA technology
  recombinant plasmids and, 219, 219*f*
  restriction enzyme cutting and pasting DNA in,
    219–220, 220*f*
  techniques of, 218, 218*f*
Recycling, carbon footprints and, 398, 398*f*
Red algae, 310–311, 311*f*
Red blood cells
  blood types, 81, 81*f*
  structure and function of, 12, 12*f*
Red-cockaded woodpeckers, population ecology and,
    412, 412*f*
Red-crowned cranes, 359*f*
Red-green colorblindness, 163–164, 163*f*
Red lionfish, 403
Red ruffed lemur, 362*f*
Red tide, 310
Reduced hybrid fertility, 273, 273*f*
Reduced hybrid viability, 273, 273*f*
Redwoods, 322
Regeneration, 123*f*, **205**
Regulation
  of population growth, 410–411, 410*f*, 411*f*
  as property of life, 10, 11*f*
Relationship of structure to function, as major theme
    in biology. *See* Major themes in biology,
    Relationship of structure to function
Relative abundance, species, **435**, 435*f*
Relative fitness, **261**
Reliable source, 10, 10*f*
Renewable energy, 107. *See also* Biofuels
  photosynthesis and, 115–116
Repetitive DNA, **226**
Replication
  of DNA, 175, 175*f*
  origin of, 297–298, 298*f*
Repressor proteins, 199
Reproduction
  angiosperm, 324–326, 324*f*, 325*f*, 326*f*
  cellular (*see* Cellular reproduction)
  fungal, 329, 329*f*
  life history traits and, 406–407, 407*f*, 407*t*
  plant, 317, 317*f*
  prokaryotic, 301–302, 301*f*

  as property of life, 10, 11*f*
Reproductive barriers, **272**–273, 272*f*, 273*f*
  evolution of, 274, 275*f*
Reproductive cloning, **206**, 206*f*. *See also* Cloning
Reproductive compatibility, biological species concept
    and, 271, 271*f*
Reptiles, **358**–359, 358*f*, 359*f*. *See also* Birds
Resources
  ecological footprints and, 419–420, 420*f*
  sustainable management of, 412–413, 413*f*
Respiration, cellular. *See* Cellular respiration
Response, of organisms to environments as property
    of life, 10, 11*f*
Restoration ecology, **444**
Restriction enzymes, 219–**220**, 220*f*
Restriction fragments, **220**, 220*f*
Restriction sites, **220**, 220*f*
Retroviruses, **190**
Reverse transcriptase, **190**–191
Review, peer, **6**–7, 10
Rhinoceros beetle, 351*f*
Ribbon model, DNA, 174*f*
Ribonucleic acid. *See* RNA
Ribosomal RNA (rRNA), **181**
Ribosomes, 55, **57**, 63, 63*f*, 181, 181*f*
Ribozymes, **298**
Ringworm, 331
Rivers, 381, 381*f*
RNA (ribonucleic acid), **49**
  as evidence of evolution, 250
  as nucleic acid, 49–50, 49*f*, 50*f*
  in origin of life, 297–298, 298*f*
  polynucleotide, 173, 173*f*
  processing and breakdown of, 202, 202*f*
  self-replicating, 297–298, 298*f*
  transcription of DNA to, 176–177, 179, 179*f*
  vaccine, 190, 190*f*
RNA interference (RNAi), 202
RNA monomers, 298*f*
RNA polymerase, **179**
RNA polymers, 298*f*
RNA splicing, **180**
RNA world, 298
Robber fly, 351*f*
Roots, **316**–317, 317*f*
Rough ER, **64**
Roundworms (Nematoda), **346**, 346*f*
rRNA (ribosomal RNA), **181**
Rule of multiplication, **152**
Rules of probability, 152–153, 152*f*, 158

S

Salamanders, 357, 357*f*
Salmon, 223
Salmonella, 302*f*, 305

Salts. *See* Solutes; Table salt
San Bernadino County wildfire, 396*f*
Sand dollars, 353, 353*f*
Saturated fatty acids, **43**–44, 44*f*
Savannas, **387**, 387*f*
Scales, 377, 377*f*
Scallops, 343, 343*f*
Scanning electron micrograph (SEM), 56*f*
Schistosomes, 344, 344*f*
Schistosomiasis, 344
SCID (severe combined immunodeficiency
   disease), 224
Science, **4**
   process of, 4–7, 5*f*, 7*f*
Scientific claims, 9–10, 10*f*
Scientific communication, 5–6, 6*f*
Scientific method, 2, 5–7, 5*f*, 7*f*
Scientific outcomes, 5–6, 6*f*
Scorpions, 348*f*
Sea anemones, 342, 342*f*
Sea cucumbers, 353, 353*f*
Sea slugs, 253*f*
Sea sponges, 336
Sea stars, 353, 353*f*
   Pisaster, 435, 435*f*
Sea turtles. *See* Green sea turtles
Sea urchins, 353, 353*f*
Seaweeds, 307, 307*f*, 310–311, 311*f*
Secondary consumers, **433**
Secondary succession, **437**, 437*f*
Sedimentary fossil, 249*f*
Sedimentary rock, 249*f*
Sedentarians, 345*f*, **346**
Seedless vascular plants. *See* Ferns
Seed plants, 322–323, 322*f*, 323*f*. *See also*
   Angiosperms; Gymnosperms
Seeds, **319**
   dispersal of, 326, 326*f*
   evolution of, 319
   germination of, 323
Selection
   natural (*see* Natural selection)
   sexual (*see* Sexual selection)
Self-fertilization, Gregor Mendel's experiments
   with, 146, 146*f*
Self-replicating RNA, 297–298, 298*f*
SEM (scanning electron micrograph), 56*f*
Sepals, flower, **324**, 324*f*
Sequencing, of genome, 231–232
Serengeti Plain, Tanzania, 387*f*
Severe combined immunodeficiency disease (SCID),
   224
Sewage treatment, prokaryotes in, 303, 303*f*
Sex chromosomes, **130**
   abnormal numbers of, 139–140, 140*t*
Sex determination in humans, 144, 163, 163*f*
Sex-linked genes, **163**–164, 163*f*

Sexual dimorphism, **263**–264, 263*f*–264*f*
Sexual reproduction, **122**
   animal, 121–122, 140 (*see also* Meiosis)
   asexual reproduction vs., 137, 140
   evolutionary advantages of, 140
   evolution of bdelloid rotifers without, 137
   gametes in, 122, 131
   meiosis in (*see* Meiosis)
   as source of genetic variation, 255
Sexual selection, **263**–264, 263*f*–264*f*
   in sympatric speciation, 276
Sharks, 121, 356, 356*f*
Shoots, **316**–317
Short tandem repeat (STR), **226**, 227*f*, 228*f*
Shrimps, 349*f*
Sickle-cell disease, 48, 154*t*, 160, 160*f*
   molecular basis of, 184, 184*f*
   mutation and, 255
Side chains, amino acid, 46, 46*f*
Signal transduction pathways, 203, 203*f*
Silencers, DNA, **201**
Silk Road, 232, 232*f*
Silurian period, 280*t*
Single-blind experiment, **9**, 9*f*
Sinosauropteryx, 285
siRNAs (small interfering RNAs), 202
Sister chromatids, **125**
Skeletal systems, vertebrate, 354, 354*f*
Skin, color of, 367, 367*f*
Skulls
   human compared with chimpanzee, 283–284, 284*f*
   vertebrate, 354, 354*f*
Slime molds, **309**, 309*f*
Slugs, 343, 343*f*
Small interfering RNAs (siRNAs), 202
Small intestine, bacteria living in, 293, 299, 306, 306*f*
Smoking
   cancer and, 212
   as mutagen, 185
Smooth ER, 64–**65**, 64*f*
Snails, 343, 343*f*
   eyes of, 285–286, 285*f*
   variations in populations of, 255*f*
Snakes, 358, 358*f*
   brown tree snakes, 449, 449*f*
   mimicry in, 430, 431*f*
Snakestone, 245
Snapdragons, 158, 158*f*
Snowshoe hares, population cycles of, 411, 411*f*
Sodium chloride (NaCl)
   bonding of, 27, 27*f*
   as table salt, 22, 25, 31*f*
Solar energy. *See also* Light; Sunlight
   as abiotic factor in biosphere, 376, 376*f*
   effects of, on distribution of terrestrial biomes,
      384, 384*f*
   photosynthesis and, 111

Solar panels, 107, 107*f*
Solutes, **31**, 84
Solutions, **31**
Solvent, water as, **31**
Somatic cells, **130**
Sonoran desert, 387*f*, 434*f*
Space-filling model of molecules, 27*f*, 38*f*
Spanish ibex, 264*f*
Speciation, **270**
   on Galápagos Islands, 270, 270*f*
   human activities facilitating, 276–277, 277*f*
   islands and, 277–278, 278*f*
   mechanisms of, 274–279, 274*f*, 275*f*, 276*f*, 278*f*
   observing in progress, 278–279
   reproductive barriers between species and, 272–
      274, 272*f*, 273*f*, 275*f*
   species concepts and, 271, 271*f*
Species, **271**
   biodiversity hot spots and endemic, 444, 444*f*
   classification of, 244, 244*f*, 286–289, 286*f*, 287*f*,
      288*f*
   concepts of, 271, 271*f*
   conservation of, 412, 412*f*
   distribution, 397, 397*f*
   ecological succession of, after disturbances,
      436–437
   endangered (*see* Endangered species)
   evolutionary history of, 286–288, 286*f*, 287*f*, 288*f*
   fixed species idea about, 245, 245*f*
   geographic distribution of, as evidence of
      evolution, 246–247
   global climate change effects on, 397, 397*f*
   invasive (*see* Invasive species)
   naming of, 244, 244*f*
   origin of (*see* Speciation)
   plate tectonics and geographic distribution of, 283,
      283*f*
   sequencing of whole genomes of, 230, 230*f*
Species diversity, **435**
   in biodiversity, 426, 426*f*
   in communities, 435–436
   forest fragmentation and, 446
Species richness, **435**
Sperm cells, human
   energy-producing capability, 75
   flagella of, 70, 70*f*
Sphagnum mosses, 319*f*
S phase, **125**, 125*f*
Spiders, 348, 348*f*
   raft spiders, 29*f*
Sponges (*Porifera*), **341**, 341*f*
Spontaneous abortion, 139
Spores, **320**, 329
Sporophytes, **320**
   angiosperm, 325, 325*f*
   bryophyte, 320, 320*f*
   gymnosperm, 322–323, 322*f*, 323*f*

Spotted owl, 404*f*
Squid, 343
 eyes of, 285–286, 285*f*
Squirrels
 allopatric speciation of, 274*f*
 climate-related adaptations of, 289
S-shaped curve, population growth, 409, 409*f*
Stabilizing selection, **264**, 264*f*
Stamens, flower, 146, 146*f*, **324**, 324*f*
Staphylococci, 300
Staphylococcus aureus, 55*f*, 61, 305
 methicillin-resistant, 265
Starches, **42**, 42*f*
Start codons, **182**
Stem cells
 embryonic, 208, 208*f*
 therapeutic cloning using, 208, 208*f*
Steroid hormones, **45**, 45*f*. *See also* Anabolic steroids
 smooth ER and, 65
Stigma, flower, **324**, 324*f*
St. John's wort, 414
Stomata, **108**, 316–317
 in photosynthesis, 108
Stop codons, **182**
Storage proteins, 46*f*
STR (short tandem repeat), **226**, 227*f*, 228*f*
STR analysis, **226**
Streams, 381, 381*f*
Streptococci, 300
Streptococcus mutans, 311, 311*f*
Streptomycin, 55, 57
Stroma, 68, 68*f*, **108**
Structural formulas of molecules, 27*f*, 38*f*
Structural proteins, 46*f*
Structure and function, relationship of, as major theme in biology. *See* Major themes in biology, Relationship of structure to function
Strychnine, 431
Strychnos toxifera, 431
Substrate imposters, 82
Substrates, enzyme, **82**, 82*f*
Sucrose, as disaccharide, 41
Sugar gliders, 283*f*
Sugar-phosphate backbones of DNA and RNA, **50**, 50*f*, **172**
Sugars. *See also* Glucose
 carbohydrates as, 40–41
 from carbon dioxide, 110, 115, 115*f*
Sunlight. *See also* Light; Light reactions; Solar energy; Ultraviolet light
 ecosystem energy and, 13–14, 14*f*, 106, 108
 photosynthesis and, 111
Sunscreens, 106
Superbugs, 265, 265*f*
Surface tension, water, 29–30, 29*f*
Survivorship curves, **406**, 406*f*
Sustainability, **392**

ecological footprints and, 419–420, 420*f*
 human impacts on biomes and, 392–393, 392*f*, 393*f*
 resource management and, 412–413, 413*f*
Sustainable development, **448**, 448*f*
Sustainable resource management, **412**–413, 413*f*
Swim bladder, **356**
Symbiosis, **302**
 in mitochondria and chloroplasts, 68
 plant-fungi interactions, 315–317, 317*f*, 330, 330*f*, 332–333, 333*f*
 prokaryotes and, 302–303, 302*f*
Symmetry, animal body, 340, 340*f*
Sympatric speciation, **274**–276, 275*f*, 276*f*
Synthesis (S) phase, **125**, 125*f*
Systematics, **286**
Systems biology, **234**

T

Table salt, 22, 25, 31*f*, 85. *See also* Sodium chloride
Table sugar, 41
Taiga, **389**, 389*f*
Tail, RNA, 180, 180*f*
Tapeworms, 336, 344, 344*f*
Tarantulas, 348*f*
Tar sands, 269
Tarsier (*Tarsius syrichta*), 361–362, 361*f*, 362*f*
Tasmanian devils, 425, 425*f*
Taxonomy, **244**
 of leopard, 244, 244*f*
 phylogeny compared with, 286
 as work in progress, 288–289, 288*f*
Tay-Sachs disease, 66, 154*t*
Technology. *See* Biotechnology; DNA technology; Forensics; Genetic engineering; Nanotechnology
Teixobactin, 61
Telophase, mitosis, **127**, 127*f*
Telophase II, meiosis II, 133*f*
TEM (transmission electron micrograph), 56*f*
Temperate broadleaf forests, **389**, 389*f*
Temperate grasslands, **388**, 388*f*
Temperate rain forests, **389**
Temperate zones, **384**
Temperature
 as abiotic factor in biosphere, 377, 377*f*
 animal thermoregulation of body (*see* Thermoregulation)
 effects of, on terrestrial biomes, 386, 386*f*
 global climate change and, 394, 394*f*
 water moderation of, 30, 30*f*
Temporal isolation, 272, 272*f*
Termination phase
 transcription, 179, 179*f*
 translation, 182

Terminators, transcription, **179**
Termites, protists and, 308
Terrestrial biomes
 abiotic factors in, 376–377, 376*f*, 377*f*
 chaparral, 388, 388*f*
 climate effects on distribution of, 384–385, 384*f*, 385*f*
 coniferous forests, 389, 389*f*
 deserts, 387, 387*f*
 global climate change effects on, 396, 396*f*
 human impacts on, 392–393, 392*f*, 393*f*
 polar ice, 390, 390*f*
 savannas, 387, 387*f*
 temperate broadleaf forests, 389, 389*f*
 temperate grasslands, 388, 388*f*
 tropical forests, 386, 386*f*
 tundra, 390, 390*f*
 vegetation types in, 385–386, 385*f*, 386*f*
Tertiary consumers, **433**
Tertiary period, 280*t*
Testcrosses, **152**, 152*f*
Testing
 genetic (*see* Genetic testing)
 of hypotheses, 5
Testosterone, 45
Tetrapods, **357**
 evolutionary tree of, 251, 252*f*
 evolution of, 357, 357*f*
Themes in biology. *See* Major themes in biology
Theories, **7**, **248**
Therapeutic cloning, **208**, 208*f*
Thermophiles, 304, 304*f*
Thermoregulation, 378–379, 378*f*, 379*f*
Threatened species, **412**
Three-domain system, **288**–289, 288*f*
Thylakoid membranes
 chloroplast pigments in, 111–112
 light reactions in, 114, 114*f*
Thylakoids, **109**
Thymine (T), 49, 49*f*, **173**
Thyroid gland, 23–24
Ticks, 348, 348*f*
 Lyme disease and, 446, 446*f*
Tigers (*Panthera tigris*), 244
Tissues, in animals, as level of life, 15*f*
Tissues, in plants, as level of life, 15*f*
T nucleotides, 191, 191*f*
Tobacco. *See* Smoking
Tobacco mosaic virus, 188, 188*f*
Tomcods, 289
Tooth decay, 311, 311*f*
Tortoises, on Galápagos Islands, 247, 247*f*, 270
Tosanoides obama, 244
Toxins. *See* Pollution
Toxoplasma, 308, 309*f*
Trace elements, 24–25
Tracking of sea turtles, 8–9

Traits, **146**
  characters vs., 146
  family pedigrees of, 153–154
  heritable, 16, 146
  inherited human, 153, 153*f*
  life history, 406–407, 407*f*, 407*t*
  sex-linked, 164, 164*f*
  wild-type, 154, 164
Transcription, **176**
  in eukaryotic cells, 201, 201*f*
  in gene expression, 176, 176*f*, 177*f*
  initiation of, 201, 201*f*
  phases of, 179, 179*f*
  summary of, 183*f*
Transcription factors, DNA, **201**
Transduction. *See* Signal transduction pathways
Trans fats, **45**
Transfer RNA (tRNA), **180**–181, 181*f*
Transformation of energy, 68, 94
Transgenic animals, 222–223, 222*f*
Transgenic organisms, **218**
Translation, **176**, 176*f*
  in gene expression, 177, 177*f*
  initiation of, 182, 182*f*, 203
  messenger RNA in, 180
  phases of, 182, 182*f*
  ribosomes in, 181, 181*f*
  summary of, 183*f*
  transfer RNA in, 180, 181*f*
Transmission electron micrograph (TEM), 56*f*
Transport proteins, 46*f*
Transport regulation, plasma membranes and, 83–86
Transport vesicles, 64–65, 64*f*
Trash. *See* Garbage
Trees. *See also* Deforestation; Forests
  wind responses of, 379, 379*f*
Triassic period, 280*t*
Trichinosis, 346*f*
Trichomonas, 308, 308*f*
Triglycerides, **43**, 43*f*
Trilobites, 339*f*
Triphosphate tail, 79, 79*f*
Trisomy 21, **139**, 139*f*
tRNA (transfer RNA), **180**–181, 181*f*
Trophic structures, 432–434
Tropical forests, **386**, 386*f*
Tropical rain forests
  deforestation and loss of plant diversity in, 327, 327*f*
  medicines from plants of, 327, 327*t*
Tropics, **384**
Truffles, 315
Trypanosomes, 307, 307*f*
Tube-dwellers, 345*f*, 346
Tubocurarine, 327*t*
Tumors, **129**
  benign, 129
  development of colon cancer, 211, 211*f*

  evolution of cancer cells in, 213
  malfunction in cell division, 120, 129
  malignant, 129, 129*f*
Tumor-suppressor genes, **209**, 209*f*, 211
Tundra, **390**, 390*f*
Tunicates, **355**, 355*f*
Turgor, plant, 85, 85*f*
Turner syndrome, 140, 140*t*
Turtles. *See* Green sea turtles
Type 1 insulin-dependent diabetes, 13
Tyrosinemia, 221

U

Ultraviolet (UV) light. *See also* Sunlight
  as carcinogen, 185, 212
  human evolution and, 367, 367*f*
Umbilical cord blood banking, 208, 208*f*
Unicellular algae, 307, 310, 310*f*
United States
  ecological footprint of, 420, 420*f*
  population growth in, 418–419, 419*f*
Unsaturated fatty acids, **43**–45, 44*f*
Uracil (U), 50, 50*f*, **173**
Urban-adapted animals, 289, 289*f*
Urey, Harold, 296–297, 297*f*
UV light. *See* Ultraviolet light

V

Vaccines
  DNA and RNA, 190, 190*f*
  DNA technology and, 222
Vacuoles, 64, 66–**67**, 67*f*
Valley fever, 331
Vampire devices, energy use by, 372, 398
Variables, **8**
  dependent, 9, 9*f*
  independent, 9, 9*f*
Variation, genetic. *See* Genetic variation
Vascular tissue, **317**, 317*f*
Vectors, gene, **219**
Vertebrates, **355**
  amphibians, 357, 357*f*
  evolution and diversity of, 354–355, 354*f*, 355*f*
  fishes (*see* Fishes)
  genealogy of, 355*f*
  mammals (*see* Mammals)
  reptiles, 358–359, 358*f*, 359*f* (*see also* Birds)
Vesicles, **64**
Vestigial structures, **251**
Vibrio cholerae, 71*f*
Vinblastine, 327*t*
Viruses, **186**
  animal, 188–189, 188*f*, 189*f*

  bacteriophages, 186–187, 186*f*, 187*f*
  emerging, 192, 192*f*
  enveloped, reproductive cycle, 189, 189*f*
  HIV and AIDS, 190–191, 190*f*
  influenza, 188, 188*f*
  mumps, 189, 189*f*
  plant, 188
  tobacco mosaic virus, 188, 188*f*
  West Nile, 190, 190*f*
Vitamin A, 223
Volvox, 310, 310*f*

W

Wallace,, Alfred Russel, 248
Warbler finch, 278, 278*f*
Warning coloration, **430**, 430*f*
Water
  as abiotic factor in biosphere, 377, 377*f*
  acids, bases, and pH of solutions of, 31–32, 32*f*
  biological significance of floating of ice as solid, 30–31, 30*f*
  biomes (*see* Aquatic biomes)
  bioremediation of, 303–304, 303*f*, 304*f*
  in cellular respiration, 95
  cohesion of, 29–30, 29*f*
  hydrogen bonding and polarity of, 28, 28*f*
  hydrophilic molecules vs. hydrophobic molecules and, 43
  in photosynthesis, 108–109, 109*f*
  shortage of, 393, 393*f*
  as solvent of life, 31, 31*f*
  temperature moderation by, 30
  three forms of, 29*f*
  unique life-supporting properties, 29
Water balance, osmosis and, 84–85
Water cycle, 391, 391*f*
Water fern (*Azolla*), 303*f*
Water vascular system, **353**
Watson, James, 173*f*
  discovery of structure of DNA by, 173–175
Wavelength of light, **110**–111, 110*f*, 111*f*
Weather, population growth and, 411, 411*f*
Weevils, 351*f*
West Nile virus, 190, 190*f*
Wetlands, **381**, 381*f*
Whales, 360
  evolution of, 250–251, 250*f*
White-tailed antelope squirrel (*Ammospermophilus leucurus*), 274*f*
Whole-genome shotgun method, **231**, 231*f*
Why It Matters chapter photo collage topics
  animal diversity, 336
  biological diversity, 267
  cells, 54
  cellular function, 74

Why It Matters chapter photo collage topics (*continued*)
    cellular reproduction, 120
    cellular respiration, 90
    chemistry, 22
    communities and ecosystems, 424
    DNA technology, 216
    ecology, 372
    gene regulation, 196
    genetics, 144
    introduction to, 2
    macromolecules, 36
    microorganisms, 292
    molecular biology, 170
    photosynthesis, 106
    plants and fungi, 314
    population ecology, 402
    populations, evolution of, 241
Widow's peak, 154, 154*f*

Wildfires, global climate change and, 396, 396*f*
Wild-type traits, **154**, 164
Wilkins, Maurice, 173–175
Wind
    as abiotic factor in biosphere, 377
    anatomical responses to, 379, 379*f*
Women. *See* Females, human
Woodpecker finch, 278, 278*f*
Wood ticks, 348*f*
Worms. *See* Flatworms

X

X chromosomes, 130
    inactivation of, 200, 201*f*
    nondisjunction errors and abnormal numbers of,
        139–140, 140*t*

Y

Y chromosomes, 130
    DNA profiling of male ancestry using, 237
    nondisjunction errors and abnormal numbers of,
        139–140, 140*t*
Yeasts, 328, 328*f*, 332
    fermentation and, 102–103
Yellowstone Park, Wyoming, 304*f*

Z

Zebra mussels, 413–414, 413*f*
Zika virus, 171
Zooplankton, **382**
Zygotes, **131**